Technologies for Solar Thermal Energy

Technologies for Solar Thermal Energy

Theory, Design, and Optimization

Edited by

Md Hasanuzzaman

UM Power Energy Dedicated Advanced Centre (UMPEDAC),
Higher Institution Centre of Excellence (HICoE), University of Malaya,
Kuala Lumpur, Malaysia

ELSEVIER

ACADEMIC PRESS

An imprint of Elsevier

Academic Press is an imprint of Elsevier
125 London Wall, London EC2Y 5AS, United Kingdom
525 B Street, Suite 1650, San Diego, CA 92101, United States
50 Hampshire Street, 5th Floor, Cambridge, MA 02139, United States
The Boulevard, Langford Lane, Kidlington, Oxford OX5 1GB, United Kingdom

Notices

Knowledge and best practice in this field are constantly changing. As new research and experience broaden our understanding, changes in research methods, professional practices, or medical treatment may become necessary.

Practitioners and researchers must always rely on their own experience and knowledge in evaluating and using any information, methods, compounds, or experiments described herein. In using such information or methods they should be mindful of their own safety and the safety of others, including parties for whom they have a professional responsibility.

To the fullest extent of the law, neither the Publisher nor the authors, contributors, or editors, assume any liability for any injury and/or damage to persons or property as a matter of products liability, negligence or otherwise, or from any use or operation of any methods, products, instructions, or ideas contained in the material herein.

Library of Congress Cataloging-in-Publication Data

A catalog record for this book is available from the Library of Congress

British Library Cataloguing-in-Publication Data

A catalogue record for this book is available from the British Library

ISBN: 978-0-12-823959-9

For information on all Academic Press publications visit our website at https://www.elsevier.com/books-and-journals

Publisher: Charlotte Cockle
Acquisitions Editor: Lisa Reading
Editorial Project Manager: Alice Grant
Production Project Manager: Prem Kumar Kaliamoorthi
Cover Designer: Vicky Pearson Esser

Typeset by TNQ Technologies

Contents

9. Solar water desalination system

M. Asaduzzaman Faysal, Nahin Tasmin, M.M. Rashid and Md Hasanuzzaman

List of contributors

Mohsan Uddin Ahmad, Department of Mechanical Engineering, Michigan State University, East Lansing, MI, United States

M. Washim Akram, Department of Mechanical Engineering, Bangladesh Army University of Science and Technology, Saidpur, Rangpur, Bangladesh

Hossain Al Tanjil, Department of Environmental and Resource Management, Brandenburg University of Technology, Cottbus, Germany

M. Asaduzzaman Faysal, Department of Mechatronics Engineering, International Islamic University of Malaysia, Kuala Lumpur, Malaysia

Md. Shariful Alam Emon, Department of Mechanical Engineering, Dhaka University of Engineering & Technology, Gazipur, Bangladesh; Department of Mechanical Engineering, Bangladesh University of Engineering & Technology, Dhaka, Bangladesh

Md Hasanuzzaman, Higher Institution Centre of Excellence (HICoE), UM Power Energy Dedicated Advanced Centre (UMPEDAC), University of Malaya, Kuala Lumpur, Malaysia

Fayaz Hussain, Faculty of Electrical & Electronics Engineering, Ton Duc Thang University, Ho Chi Minh City, Viet Nam

Maliha Kawsary, Department of Enterprise Engineering, The University of Bordeaux, Talence Campus, Talence, France

Zabir Mahmud, Environmental Systems Graduate Program, School of Engineering, University of California Merced, Merced, CA, United States

A.B.M. Abdul Malek, Department of Industrial and Production Engineering, Shah Jalal University of Science and Technology, Sylhet, Bangladesh

Khodadad Mostakim, Department of Mechanical Engineering, Rajshahi University of Engineering and Technology, Rajshahi, Bangladesh

Rafsun Hossain Mridha, Department of Mechanical Engineering, Bangladesh University of Engineering and Technology, Dhaka, Bangladesh

Md Zillur Rahman, Department of Mechanical Engineering, Ahsanullah University of Science and Technology, Dhaka, Bangladesh

M.M. Rashid, Department of Mechatronics Engineering, International Islamic University of Malaysia, Kuala Lumpur, Malaysia

Moham Ed Abdur Razzaq, Department of Mechanical Engineering, Chittagong University of Engineering & Technology, Chattogram, Bangladesh

Nahin Tasmin, Department of Mechanical Engineering, Rajshahi University of Engineering & Technology, Rajshahi, Bangladesh

Chapter 1

Fundamentals of thermal energy and solar system integration

M. Washim Akram[1] and Md Hasanuzzaman[2]
[1]*Department of Mechanical Engineering, Bangladesh Army University of Science and Technology, Saidpur, Rangpur, Bangladesh;* [2]*Higher Institution Centre of Excellence (HICoE), UM Power Energy Dedicated Advanced Centre (UMPEDAC), University of Malaya, Kuala Lumpur, Malaysia*

Chapter outline

1.1 Introduction

The energy within a system that is responsible for its temperature is known as thermal energy. Internal energy of a substance and sensible or latent heat are

Technologies for Solar Thermal Energy. https://doi.org/10.1016/B978-0-12-823959-9.00003-9

two examples of thermal energy in storage or transfer. There are lots of thermal power–driven machineries and appliances in modern life. In most cases, thermal energy performs mechanical action and electricity generation (Alva, Lin & Fang, 2018). According to the 2014 report of the International Energy Association (IEA), worldwide the entire amount of thermal energy stocks retention exists 573 EJ that is equivalent to 13,699 million tonnes of oil and the yearly wreckage exists 394 EJ that is identical to 9425 million tonnes of oil (Alva et al., 2018). Moreover, increase in the world population also indicates the growing demand of energy (e.g., electricity, space heating and cooling, food and consumer products) (Hasanuzzaman, Zubir, Ilham, & Seng Che, 2017; Kamarulzaman, Hasanuzzaman, & Rahim, 2021).

1.2 Foundation of thermodynamics

Thermodynamics is the branch of physical science that deals with the relation between heat and other kinds of energy such as chemical, electrical, mechanical energy. Thermal system engineering is constituted with thermodynamics, heat transfer, and fluid mechanics (Radulescu, 2018; Shyy, 1994). Fig. 1.1 shows the schematic diagram of a thermodynamic system and its corresponding T-S diagram with thermal reservoirs. In thermodynamics, all types of energy along with production of power, cooling, heating, and concerned characteristics of the substances are thoroughly illustrated (Pérez et al., 2017). On the other hand, heat transfer discusses about the three basic modes of transferring heat, namely conduction, convection, and radiation. Lastly, fluid mechanics is concerned with the detail characteristics of different fluids and flow including conservation of mass and conservation of momentum.

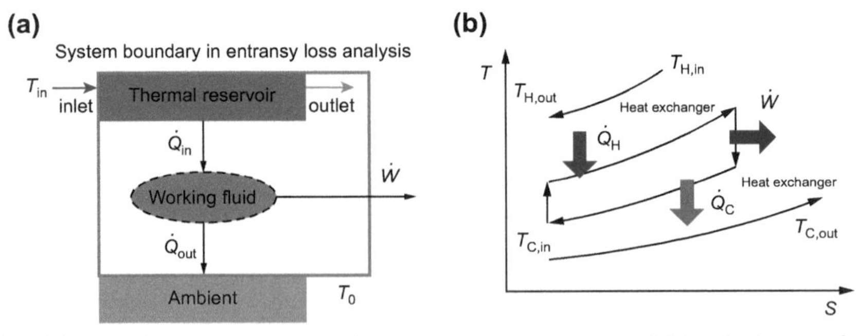

FIGURE 1.1 Schematic diagrams of (A) thermodynamic system and (B) T-S diagram of the thermodynamic cycle with thermal reservoirs (Xu & Guo, 2021).

1.2.1 Basics of thermodynamics

Basic concept of thermodynamics includes sensible heat, latent heat, first, second, and third laws of thermodynamics, zeroth law, the concept of entropy, etc. It also includes the concept of temperature and pressure. The rudiments of thermodynamics are briefly described in the following sections.

System and control volume

A system is a quantity of substance or an area in space which is selected for study. The surroundings is the region or mass outside the system. In other words, everything which interacts with system is known as surroundings (Radulescu, 2018). System and surroundings distinguished by an imaginary surface (either fixed or movable) known as system boundary (Winterbone & Turan, 2015). The system and surroundings both together is known as universe (Pérez et al., 2017). On the other hand, the region of space which is used for observation (of flow processes) is known as control volume. The boundary of control volume is known as control surface (Winterbone & Turan, 2015). For example, the processes occurred in control volume are turbine, boiler pump, condenser, compressor, nozzle, diffuser, etc (Radulescu, 2018).

Properties of a system

Any mathematical combination of measurable or observable characteristics of a system is known as thermodynamic property of the system (Radulescu, 2018). Property value is independent of the path between the end states (Winterbone & Turan, 2015). The most common properties of a system are:

- temperature (T),
- mass (m),
- volume (v), and
- pressure (P).

Also, there are some fewer familiar properties of a system which are thermal conductivity, modulus of elasticity, thermal expansion coefficient, viscosity, electric resistivity, velocity, elevation, etc.

Temperature

The direction of heat transfer is indicated by temperature (Radulescu, 2018). Temperature is measured in different scales in thermometer, e.g., Celsius and Fahrenheit (Pérez et al., 2017). These two scales are based on the melting and boiling point of water (Shyy, 1994). Table 1.1 shows the relation between different temperature scales. Table 1.2 reveals the comparison between different temperature scales. Following equation is used to derive the relation between these two scales (Pérez et al., 2017):

$$\frac{C - \theta_{\text{ice}}}{\theta_{\text{steam}} - \theta_{\text{ice}}} = \frac{F - \theta_{\text{ice}}}{\theta_{\text{steam}} - \theta_{\text{ice}}} \tag{1.1}$$

TABLE 1.1 The relation between different temperature scales (Pérez et al., 2017).

Temperature scale	Relation
Celsius to Fahrenheit	$F = \frac{9}{5}C + 32$
Fahrenheit to Celsius	$C = \frac{5}{9}(F - 32)$
Celsius to Kelvin	$K = C + 273$
Fahrenheit to Rankine	$R = F + 460$
Kelvin to Rankine	$R = \frac{9}{5}K$
Rankine to Kelvin	$K = \frac{5}{9}R$

TABLE 1.2 Comparison of temperature scales (Pérez et al., 2017).

Scale	Fahrenheit (°F)	Centigrade (°C)	Kelvin (K)	Rankine (°R)
Steam point	212	100	375.15	671.67
Triple point	32.02	0.01	273.16	491.69
Ice point	32	0	−273.15	491.67
Absolute zero	−459.67	−273.15	0	0

Pressure

In thermodynamics, pressure is defined as the force (caused by the collisions of the molecules within) exerted per unit area on a system boundary (Winterbone & Turan, 2015). It is measured in terms of newtons per square meter (N/m^2) or Pascal (Pa). The real measure of pressure at a given position, measured relative to absolute vacuum, is termed as absolute pressure. On the other hand, gage pressure is the deviation between absolute pressure and local atmospheric pressure. Now-a-days, sophisticated pressure sensors are used to measure pressures for different purposes (Fig. 1.2). It can be either positive or negative, wherein the negative pressure is called vacuum (below atmospheric). The following equations show the relations among the absolute, gage, and vacuum pressure (Radulescu, 2018).

$$P_{abs} = P_{atm} + P_{gage} \tag{1.2}$$

$$P_{abs} = P_{atm} - P_{vac} \tag{1.3}$$

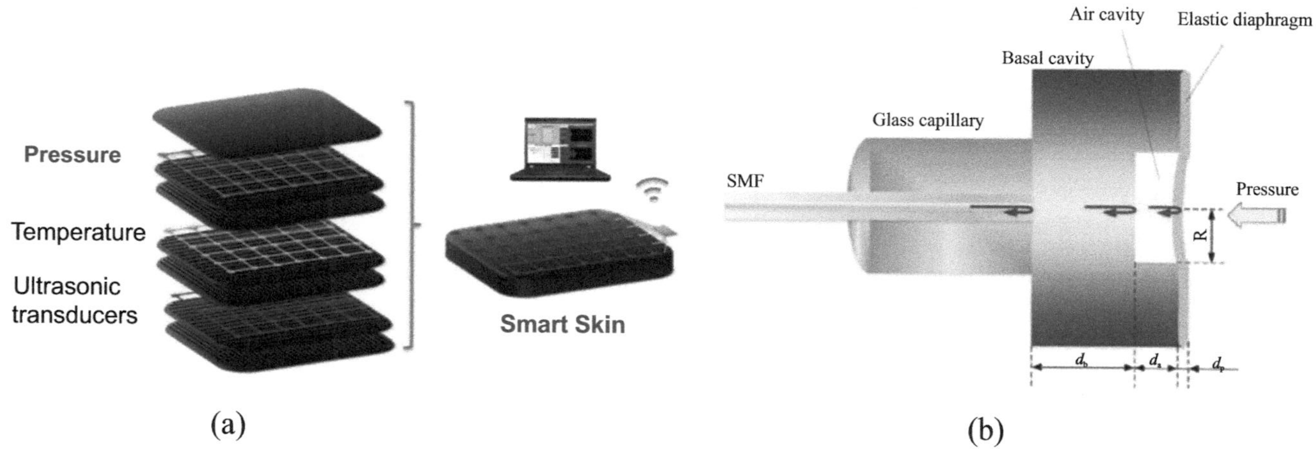

FIGURE 1.2 (A) Stretchable pressure sensor network with smart skin materials (Guo et al., 2016); (B) schematic diagram of a typical pressure sensor (Guo et al., 2019).

1.2.2 Basic energy conversion

The main purpose of energy conversion technologies is to transform the unserviceable form of energy available in nature into useful works. Energy conversion technologies like steam engine or nuclear power stations have brought about revolution in modern human civilization. A list of modern energy conversion technologies is shown in Fig. 1.3 (Islam et al., 2020).

Fossil fuel energy conversion

Fossil fuels (coal, oil, and gas) are hydrocarbon-based nonrenewable source of energy. Different types of heat engines (or turbines) are used to release the chemical energy of fossil fuel and produce heat, thereby this heat is employed to generate steam for running power plant, which is finally converted into electricity through electrical generators. The major fossil fuel energy conversion technologies are discussed below (Islam et al., 2020):

- Heat engine: Fuel chemical energy is converted into work by means of internal combustion (IC) engine. Gas turbines, jet engines, and most of the rocket engines are continuous combustion IC engines.
- Steam turbine: Steam turbines release thermal energy of pressurized steam and perform mechanical work on a rotating shaft (Manushin, 2011). The efficiencies of steam turbine are comparatively higher than the gas turbines as well as IC engines. Besides, the reliability of the steam turbine is greater when high power output is required.
- Gas turbines: Gas turbines are basically run by expanding hot flue gases produced by burning fuels. It is used not only in power generation but also aircraft propulsion systems such as turbojet engine, turboprop engine, turbofan engine, etc.

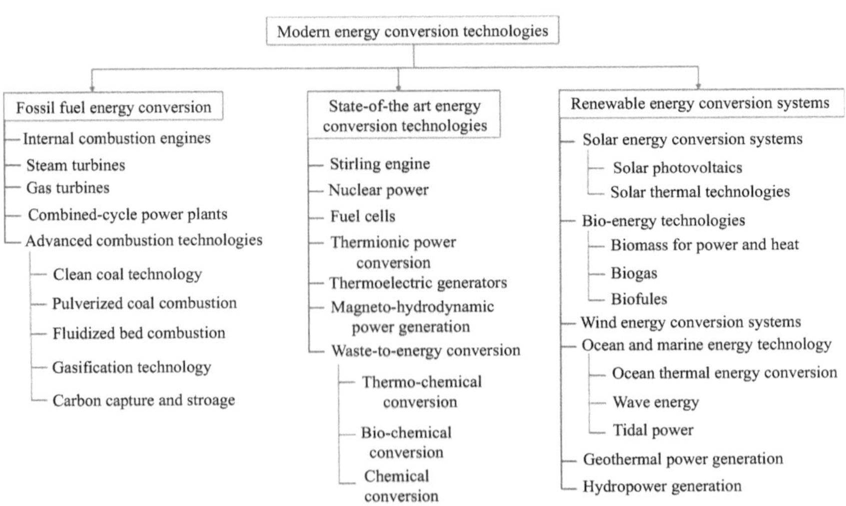

FIGURE 1.3 Modern energy conversion technologies (Islam et al., 2020).

- Modern energy conversion technologies: Apart from the above three major technologies, there are several advanced combustion technologies including fluidized bed combustion, carbon capture and storage, nuclear power, fuel cells, thermionic energy conversion, thermoelectric generator, magnetohydrodynamic generator, waste to energy, etc.

Renewable energy conversion

According to the International Renewable Energy Agency (IRENA), an energy source can be termed as renewable if it is derived from natural processes and replenished at a faster rate than it is being consumed. From this viewpoint, solar energy, wind energy, hydro power, bioenergy, geothermal energy, ocean energy, and hydrogen energy are widely considered as renewables (Kato & Trends, 2002). The sources of renewable energy are briefly discussed as follows:

- Solar energy: Solar energy is the energy directly harvested from solar radiation either in the form of heat (in solar thermal collector, STC) or electricity (in photovoltaic [PV] module). Recently, a hybrid photovoltaic thermal (PVT) modules, a co-generation technology that provides both electricity and heat is getting popular.
- Wind energy: Wind energy is the kinetic energy resulting from pressure gradient in atmosphere. This energy is harvested by means of wind turbines and converted into electricity. The harvested wind power is proportional to the rotor diameter of the turbine and cube of the wind speed (Caduff et al., 2012).
- Hydropower: The power derived from water in motion is known as hydroelectric power briefly called hydropower. The higher-level stored water is allowed to flow down through conducting lines and flow over the turbine vanes. As the water turns the turbine, it runs the generator to produce electricity. Mainly three types of water turbine are used to generate electricity depending on the water head available: for very high heads ($>$1500 m) Pelton wheel, for medium head ($\sim$750 m) Francis turbines, and for low head ($<$100 m) Kaplan turbines are used (International Hydropower Association, 2018).
- Ocean energy: Enormous energy potential is involved in the waves and tides of the ocean that can be harvested for practical applications. Not only the temperature difference between hot surface waters and cold deep waters but also the salinity gradient at river mouths can be encountered for power generation (Kempener & Neumann, 2014). Ocean thermal energy conversion is the best option to convert this huge amount of absorbed ocean energy into electricity, wherein temperature difference between the warm surface water (22–27°C) and very cold water at a depth of 1 km (4–7°C) is used to produce energy (Gupta & Roy, 2007).

- Bioenergy: Bioenergy is derived from recently living organic materials known as biomass, which can be used to produce transportation fuels, heat, and electricity.
- Geothermal: Geothermal refers to the thermal energy derived within the subsurface of the earth. This energy can be used for heating and cooling or be harnessed to generate clean electricity.
- Hydrogen energy: Hydrogen is an energy carrier not an energy source and can deliver or store a tremendous amount of energy. It is used in fuel cells to produce electricity. It is renewable, nontoxic, and when consumed in fuel cell, it produces only water; hence, it is a clean and safe for environment.

1.3 Thermal energy demand and supply

Thermal energy supply depends upon the demand in different sectors, while energy prices affect the consumption pattern (Munasinghe, 1990). Sometimes thermal energy is used as alternative energy form known as process heat in industrial sector. Most of the process heat used in industrial sector comes from the end uses (IAEA, 2006).

1.3.1 Sector wise thermal energy demand

Basically, thermal energy is the most commercial form of energy and widely utilized in industrial, commercial, agricultural, and residential sectors. At present, conventional fossil fuel energy sources supply most of the thermal energy demand for all the sectors with renewable solar energy as well as biomass satisfying a part of it (Wagh & Kulkarni, 2019). Medical sector also needs hot water for patient room, kitchen, and laundry.

Compared to any other sector, the industrial sector consumes more energy. Approximately 74% of industrial energy consumption is used to meet the process heat demand (Fig. 1.4) (IEA-ETSAP and IRENA, 2015). A good portion of this large amount of heat demand in industrial sector can be satisfied by solar energy and solar thermal can be used for various processes like drying, dying, boiling, washing, sterilization, and pasteurization (Schmitt, 2016). On the other hand, steel, glass, and ceramic industries need approximately 13% of the industrial process heat at low temperature thermal energy up to 100°C, whereas 27% process heat required at 200°C and the remaining processes required very high temperature (Farjana et al., 2018). In European industries, about 27% of the overall ending energy demand is expended as heat (Kylili et al., 2018).

Thermal energy demand in different sectors depend upon some factors. Berger et al. revealed three stages to evaluate the connection between thermal energy demand in buildings and climate change with respect to different locations and scales (Berger & Worlitschek, 2019; Saha et al., 2021). In another

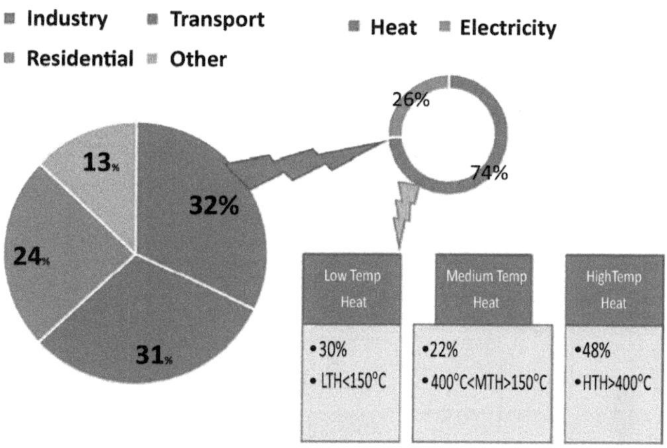

FIGURE 1.4 Statistics of industrial thermal energy consumption by temperature range (Kumar, Hasanuzzaman & Rahim, 2019).

study, Aebischer et al. investigated the impact of climate change on thermal comfort, i.e., the demand of heating and cooling energy in Europe (Aebischer, Catenazzi & Jakob, 2007). The risk of overheating is increased with higher outdoor temperature, particularly in case of internal loads and an absence of (nocturnal) window aeration. The energy requirements of buildings and indoor thermal comfort conditions depend upon the variety of factors such as type of glazing and blinds, thermal insulation levels, appliances and lighting, internal loads from persons, building technology, and operation modes (Aebischer et al., 2007).

1.4 Conventional thermal energy supply technologies

Boiler and combined heat and power (CHP) systems are considered as conventional thermal energy supply technologies.

1.4.1 Boiler

Boilers (also known as steam generator) are closed vessel in which vapor or steam is generated and superheated, under pressure or vacuum, by direct heat from the combustion of fuels or electricity or nuclear energy. The flue gas produced by burning of fossil fuel ducted around tubes containing the water (water tube boiler), or pass through the tubes with the water being in shell (fire tube boiler) (Atinc et al., 2019). The continuous generation of steam is resulted by the absorption of heat by the boiler feed water. Combinedly the steam and feed water are known as working fluid (Sarkar, 2015). The flue gases transfer heat to the working fluid through the heating surfaces and it gets vaporized. The produced saturated steam is further superheated. The flue gas is cooled after passing over the heating surfaces at various zones and finally discharged to the atmosphere through the stack (Sarkar, 2015). Fig. 1.5 shows the schematic diagram of a boiler.

Generally, the function of a steam generator is to produce steam at high pressure. But the modern steam generators have more sophisticated features (Sarkar, 2015). These features include elimination impurities from saturated

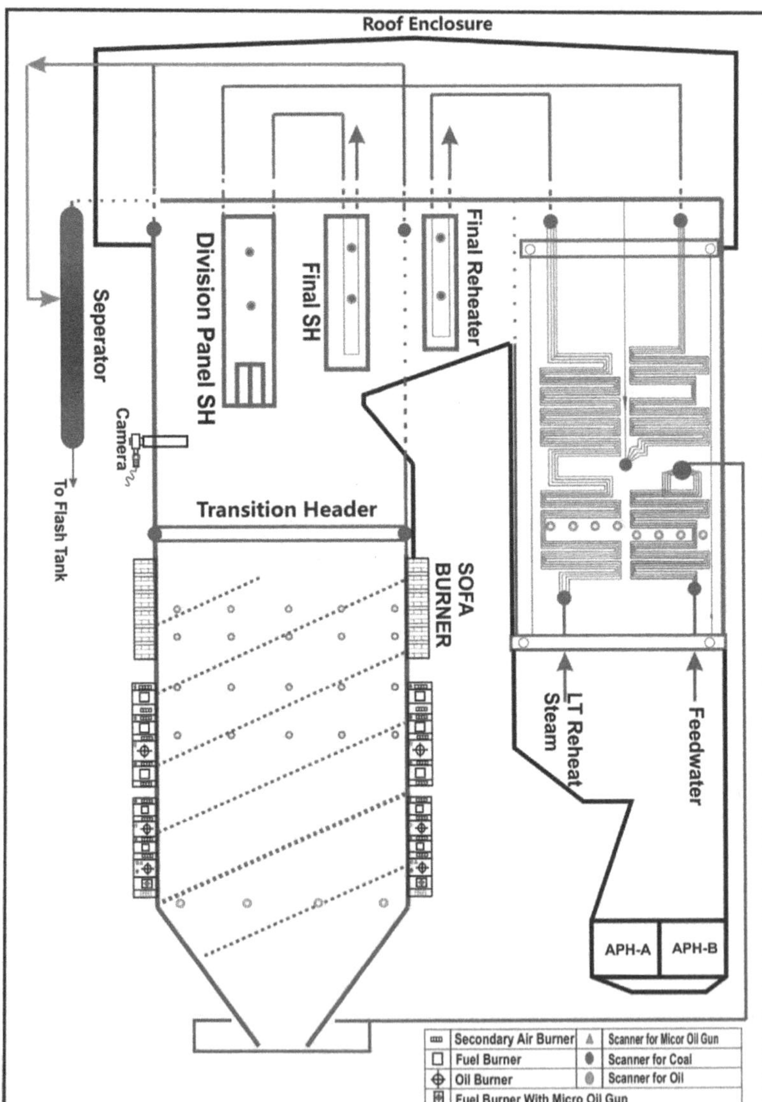

FIGURE 1.5 Schematic diagram of a boiler (Ashraf et al., 2020).

steam and generate high-purity steam, generation of steam at different range of temperatures for different range of load, reheating facility, etc.

Combustion reaction in boiler (Prakash, 2013):

$$\underset{\text{Fuel}}{\boxed{C+H}} + \underset{\text{Air}}{\boxed{O_2+N_2}} \xrightarrow{\text{Combustion}} \underset{\text{Heat}}{\boxed{CO_2+H_2O+N_2}} \tag{1.4}$$

Boiler efficiency is given by (Platzer, 1999):

$$\text{Boiler Efficiency } (\eta) = \frac{\text{Steam flow rate } \times (\text{Steam enthalpy} - \text{Feed water enthalpy})}{\text{Fuel firing rate } \times \text{ Gross calorific value}} \times 100 = \frac{Q \times (H - h) \times 100}{(q \times \text{GCV})} \tag{1.5}$$

where

Q is the quantity of steam generation (kg/h),
q is the quantity of fuel consumption (kg/h),
GCV is the gross calorific value (kCal/kg of fuel),
H is the enthalpy of steam (kCal/kg).,
h is the enthalpy of feed water (kCal/kg).

Classification of steam boiler

The main two types of boiler are:

- Fire-tube boiler
- Water-tube boiler

Fire-tube boiler

In fire tube boiler, water is contained in the shell and the hot gases pass through the tubes. In its general construction, a water tank perforated by tubes that carried hot flue gases from the fire (Fig. 1.6). The shape of the water tank is usually cylindrical, placed either vertically or horizontally. Normally, the fuel is burnt in the furnace under the beginning point of the boiler's shell. The flue gas passes to the end point of the boiler's shell and releases through the stack of the boiler (Sarkar, 2015).

Fire tube boilers are extensively used in stationary engineering field, especially for low pressure steam used as like as heating a building. Since, the steam generation capacity and outlet steam pressure of these types of boiler is limited, this is the reason why they are not able to satisfy the needs of largest units. Another drawback of these type of boilers is that they have higher chances of explosions (Sarkar, 2015).

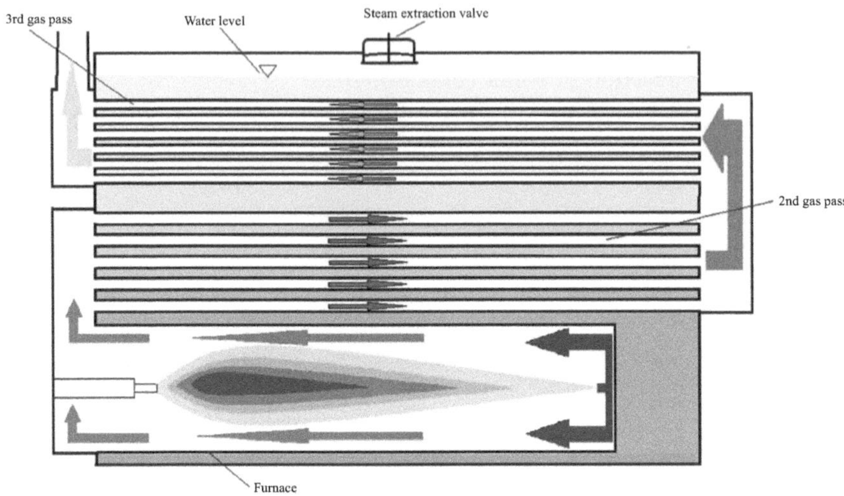

FIGURE 1.6 Fire tube boiler with Stagnation Point Reverse Flow combustor (Tognoli et al., 2019).

Water-tube boiler

The steam generation capacity of water tube boiler is very high compared to the fire tube boiler. The water tube boiler is exorbitantly heavy and can produce high pressure steam. Therefore, the size and weight of water tube boiler is extremely difficult to manage. The water flowing through the tubes is circumvented by the hot flue gases to produce steam (Fig. 1.7). To ensure maximum exposure of the tubes, baffles are installed across the tubes to allow

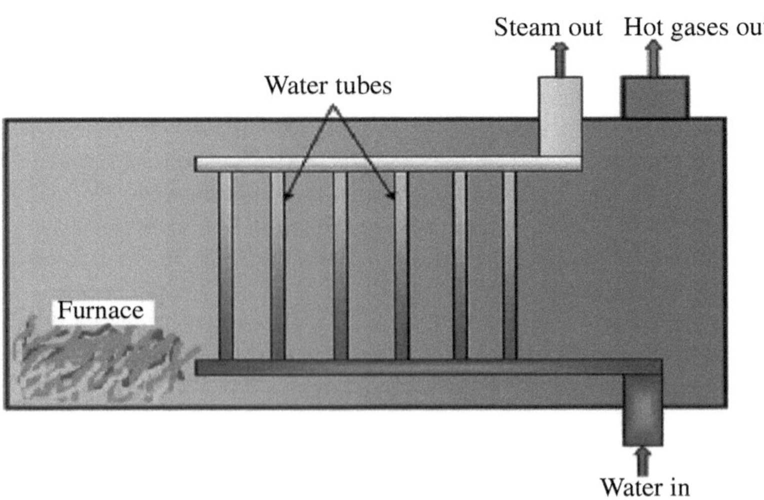

FIGURE 1.7 Schematic view of water tube boiler (Sathish et al., 2021).

cross flow of flue gases. The hot flue gases heat up the water in the steam generating tubes. The cool water at the bottom of the steam drum returns to the feed water drum to further evaporate. Feed water is supplied directly to the steam drum in case of large utility boilers to produce high quantity of steam within very short time. The produced saturated steam enters the superheater to become superheated. In driving turbines or any other machineries, the superheated steam is used.

Modern boiler

There are different types of modified steam generators used now-a-days. According to the arrangement of pressure parts, the modern boiler can be classified as large-volume boiler. Besides, based on the use, the modern boiler can be further classified as utility boilers, recovery boilers as well as heat recovery steam generators (HRSGs). Furthermore, based on the firing methods, the boiler can be further classified as boilers fired by pulverized coal, grate boilers, and fluidized bed boilers (Vakkilainen, 2017a).

Heat recovery steam generators In modern world, HRSGs are so much popular. Most often, they are combined with a diesel generator or a gas turbine so that it can produce additional electricity with the steam. The electricity generating efficiency of this combined cycle units is very high and partial load characteristics is also good. According to the suggestion of different studies in these combined cycle plants, biomass could be used, but they are rarely used commercially. Mainly, there are two types of arrangement. The evaporative tubes are placed horizontally in the horizontal tube HRSG. In large units, this type of arrangement is very popular. The flue gas passage is often square in case of horizontal tube HRSGs. The position of main heat exchanger is vertical in case of vertical type HRSGs. The vertical type HRSG is suitable for small scale plants and are often much higher than they are wide (Vakkilainen, 2017a).

Recovery boilers Concentrated black liquor is burnt in the recovery boiler furnace. The use of inorganic process chemicals is subsequently reduced in kraft pulping. The recovery boiler serves mainly three purposes (Vakkilainen, 2005). Firstly, the organic materials are burnt by it for pulping residue such as black liquor to generate high pressure steam. Secondly, it regenerates and recycles the used chemicals in the black liquor. Lastly, the way of discharging waste streams is environmentally friendly.

Boiler mountings and accessories

Boiler mountings are essentially for economics, safety, and convenience. The three major boiler accessories are: superheaters and reheaters, economizers and air preheaters. On the other hand, boiler accessories are used for measuring boiler operating conditions.

Superheaters and reheaters

In a boiler, the device which is used to convert saturated or wet steam into superheated or dry steam is known as superheater. Superheating steam increases the boiler ideal cycle efficiency and accordingly the specific steam consumption is decreased (Gaffert, 1952).

With the superheat-reheat combined arrangement, the steam is expanded in turbine into two stages. In first stage, the steam coming from the superheater is expanded in high-pressure turbine to some intermediate pressure. Then the low temperature exhaust steam from high-pressure turbine is further reheated and temperature rises to an equal or higher than the superheat steam temperature at constant pressure. The reheated steam is then fed into the intermediate pressure turbine. The superheater or reheater absorbs heat from the products of combustion of flue gases by radiation or convention heat transfer process (Kern, 1997).

Economizers

The temperature of boiler flue gases is generally maintained at about 648–818 K which is above the saturation temperature of the water. The loss of boiler efficiency would be high if the high temperature flue gases exhausted to the atmosphere. These are the reasons why there is a huge possibility to absorb this heat energy of flue gases after leaving from the superheater or reheater by the incoming subcooled feed water in a separate heat transfer surface is known as the economizer. The efficiency of the boiler working fluid increases substantially with this type of additional heat absorption (Hasanuzzaman et al., 2017). The boiler efficiency rises by approximately 1% for each 5.5 K increase in the feed water temperature. Similarly, the efficiency of a conventional boiler increases by 1% for each 22 K drop in flue gas temperature (Kitto & Stultz, 2005; Vakkilainen, 2017b).

Air preheater

The temperature of the economizer exit flue gas is quite high which is 588–698 K. The remaining heat of the economizer exit flue gases is trapped by another heating surface known as air preheater. The heat receiving fluid in the air heater is in ambient temperature, thus the temperature of the flue gases exit from the air heater is reduced to approximately 393–433 K. On the other hand, the ambient air temperature flowing through the air heater raises to 523–613 K prior to the combustion in the boiler (Vakkilainen, 2017b).

The heat from exhaust flue gases from the combustion chamber firstly absorbed by superheater, then the flue gas entered the economizer, lastly the exhaust flue gas from economizer entered into the air heater to absorb heat and improve the overall efficiency of the boiler. These three steps of exhaust heat absorption also reduce the amount of fuel consumption in boiler. The amount of fuel savings is directly proportional to the raise in air temperature at the air heater outlet. If the temperature raise is about 110 K or 280 K, then the typical fuel savings is about 4% or 11% correspondingly (El-Wakil, 1985).

1.4.2 Combined heat and power

The main objective of all the working cycle is to convert heat energy into work. The remaining portion of heat is exhausted as waste. The quality of the exhausted heat is too low that is why it is not suitable for any practical use. In cogeneration, both heat and electricity are produced from same source of energy. The comparable efficiency, i.e., heat loss of both CHP and individually separated heat and power generation is shown in Fig. 1.8. About 30%–40% of fuel energy can be saved by CHP plant compared to the separated power and heat production. Besides, the emission of CHP plant is less compared to the separated generation (Sipilä, 2016).

Gas turbine CHP plants

In gas turbine CHP plants, three equipment are very important (e.g., compressor, combustion chamber, and turbine). Due to the higher flow rates and efficiencies, the axial flow compressors are more common in power generation. To sustain continuous combustion, the compressor draws the air and compress it by the rotation of the shaft. Generator is driven by the remaining shaft power which produce electricity. Almost 55%–65% of turbine power is used to drive the compressor. The gas turbines have multiple stages through which the kinetic energy of gases is extracted and improve the efficiency of the plant. It consists of a gas turbine which generates electricity and a waste heat boiler driven by exhaust gases from the turbine which is used to heat the water. The hot water is supplied to the local district heating service customers (Sipilä, 2016).

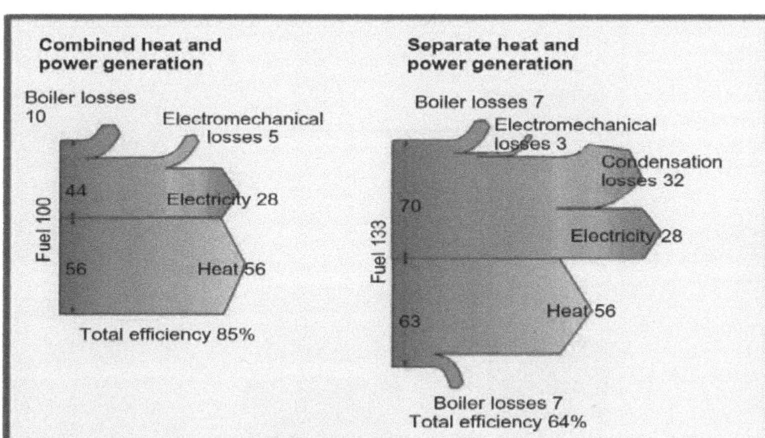

FIGURE 1.8 Energy flow in combined heat and power (CHP) and separated individual heat and power generation (Sipilä, 2016).

Solar energy conversion and integrated CHP plants

Solar energy would be a most effective form of alternative energy due to huge amount of availability in nature. It can be harvested as electrical energy or thermal energy which is shown in Fig. 1.9. The more effective and sustainable solution to convert solar energy into electrical energy is PV cells. The conversion efficiency of photovoltaic cells is comparatively low and it has been found that only 15%–20% solar radiation of the photovoltaic is converted into electric power, rest of the solar radiations being reflected from the surface of the photovoltaic cell (Nahar et al., 2019). These reflected solar radiations or the radiation directly comes from the sun can be converted into thermal energy with the help of STCs (Thirugnanasambandam, Iniyan & Goic, 2010). The most effective solution to obtain both thermal and electrical energy together is the integrated technology such as solar PVT as well as solar concentrated photovoltaic thermal. The electrical and thermal output of these integrated technologies are very high compared to the standalone PV or STCs because of receiving solar radiation is exploited in the interior of the system through energy efficient concentrators.

Electric energy can be calculated by using the following equation (Pearce, 2009):

$$E_T = E_{PV} + E_{CHP} = \sum_{n=1}^{365} s_n \times P_{PV} + \sum_{n=1}^{365} (24 - s_n) \times P_{CHP} \tag{1.6}$$

where

E_{PV} is the electricity from PV,
E_{CHP} is the electricity from CHP,
S_n is the sun hours.

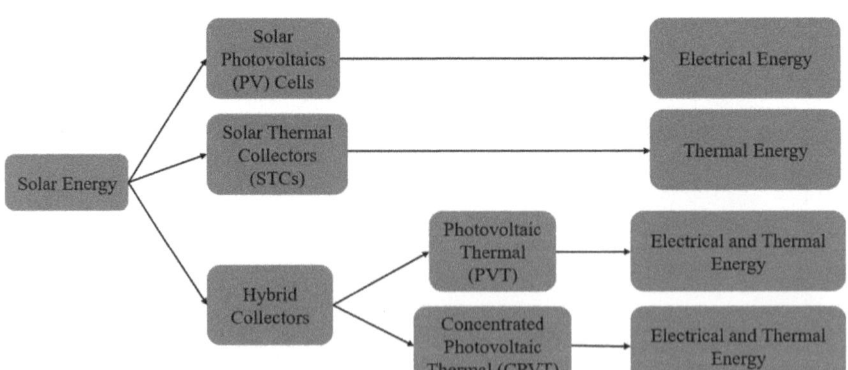

FIGURE 1.9 Solar to electrical as well as solar to thermal energy conversion pathways (Islam et al., 2016; Kumar et al., 2019; Nasrin, Hasanuzzaman, & Rahim, 2018).

Heat generation form the CHP system can be calculated by using the following equation (Pearce, 2009):

$$H_T = \sum_{n=1}^{365} H_{\mathrm{CHP}} \times (24 - s_n) \tag{1.7}$$

where

H_{CHP} is the heat generated by the CHP system.

CHP plants for district heating

The CHP plants generate both heat and power together comparatively efficiently. Steam is superheated in the boiler at higher temperature than the saturation point at constant pressure (Sipilä et al., 2005). When the system is designed such a way that it can produce only power and steam turbine uses cooling water for condensing into water form. If steam is needed as industrial process heat, then an extraction turbine is used in this type of CHP plants. The extraction steam decreases the electrical output of the CHP plant, but the overall efficiency will be still high.

Selection factor for cogeneration systems

There are many factors need to consider for selecting a cogeneration systems. The most common and important selection factors should be considered during selection of a cogeneration plant for a particular industry are as follows (Sipilä, 2016):

- Normal load or maximum/minimum load (thermal and/or electrical) of an industry.
- Power or steam—as emergency backup availability.
- Sudden fluctuation of thermal as well as electricity demand with their time duration and response time
- The step by step rate of increasing the demand of power and steam as the process picks up whether the rise in demand of one utility is more rapid than other, the same or vice-versa under normal process conditions.
- Type of fuel, its availability, and pricing.
- The life span and commercial availability and corresponding outlay for maintenance of various systems.
- Space availability, plant site, raw water availability, soil conditions, infrastructure, and environment.
- Load center, i.e., district heating/cooling for society is near the plant site.
- Cost of project and long-term benefits.

The requirement of heat and power varies from site to site. To match the demand of the CHP plant, i.e., the cogeneration system must be selected with due care and appropriate operating schemes. With the existence of the following, the concept of cogeneration system generally found most attractive (Sipilä, 2016).

- The power to steam output ratio should be consistent to obtain a suitable cogeneration plant.
- The requirement of steam or power of a single industry or group of industries should be enough to permit economies of scale to be achieved.
- Troughs and peaks of demand of steam or power should be managed appropriately.
- To allow optimum techno-economic benefits, there are three main operating regimes of cogeneration plant which are as follows:
- The cogeneration plant is used as base load plant for producing electricity and thermal energy produce for a short fall.
- The cogeneration plant is operated to supply total thermal energy required in the industry, but in case of electric power, it supplies the excess of the industries requirement which may be exported.
- The cogeneration plant is operated to produce thermal energy which is utilized in the industry with the export of surplus heat energy, and the electrical power, with or without export, if possible, to nearby consumers.

By considering all the above-described factors, it is very easy to select an appropriate cogeneration system to produce suitable amount of power and steam. Its low capital cost and highest fuel efficiency make it more acceptable and economical solution.

The potential of CHP in industrial sectors

The use of CHP plant in industrial sector is a viable option for energy conservation in the most effective and economical way. The industries which are moderately large, and the demand of steam is stable that are suitable for cogeneration. The types of fuel used in this type of cogeneration plant are in lower quality, such as furnace oil and lignite, which cannot be directly fired in gas turbines. Energy intensive plants have been traditional investors in CHP facilities. Food processing, pulp and paper, metals, chemicals, and oil refining industries are considered as energy intensive sectors. More than 80% of the total global electric CHP capacities are being represented by these industries (IEA Energy Balances, n.d.). Most of these plants are related with process heat not to weather fluctuations, thus the thermal requirement of these industries is stable. In some industries, the produced waste can be used as low-cost fuel sources of cogeneration plant. Chemical, refinery and wood industries are the big thermal energy consumers.

1.5 Solar thermal integration

Solar thermal system integration is one of the important parts that is included storage tank, heat supply, and process system. Solar thermal integration concept generally divided into as supply level and process level (Muster et al., 2015). Fig. 1.10 shows the solar heat integration concept.

It is a complex task to choose a suitable integration point. Depending upon these aspects, the process, and supply level, solar heat integration concept has been developed.

1.5.1 Integration of supply level

The respective heat transfer medium and integration point directly leads to a possible integration concept for the integration of solar heat at the supply level. Fig. 1.11 shows a schematic diagram of supply level solar heat integration system. The main features of supply level solar heat integration are as follows (Platzer, 2015):

- Solar heat is supplied through the heat distribution network.
- Used in steam networks and high temperature networks where the solar thermal system may deliver preheated feed water or direct high temperature steam.
- It is flexible against process and demand changes.

1.5.2 Integration on process level

Solar thermal energy integration in process level can be defined as the solar energy which is directly used to complete a process such as heating water or air and the heated working fluid is used to maintain the temperature of a certain space. The process level solar thermal heat integration systems are shown in Fig. 1.12. The main features of process level solar thermal heat integration are as follows (Platzer, 2015):

- Solar heat is directly supplied to the process.
- Process level heat integration can be used in the places where low grade (until 100°C) heat is required, for example: washing, hot air drying, heating of industrial baths, cleaning, etc.
- When the heat requirement is restricted to one or two processes, then the process level heat integration is very useful.
- Highly secured control system is required for this for process level solar heat integration.

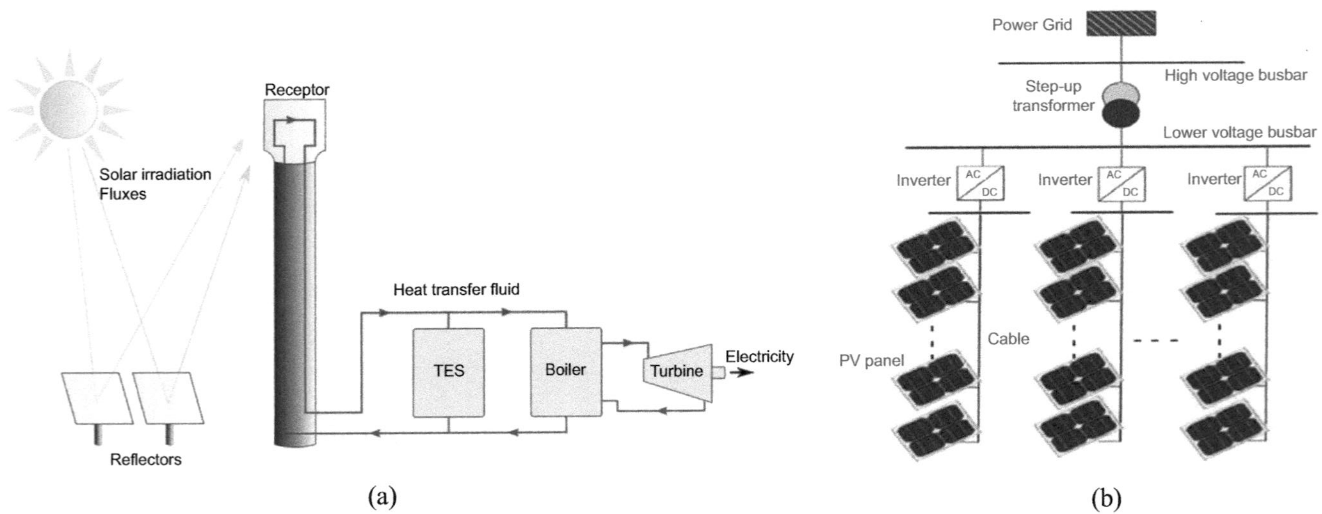

FIGURE 1.10 (A) Concentrated solar power plant with thermal energy storage system (Alnaimat & Rashid, 2019); (B) schematic diagram of a PV power station (Nwaigwe et al., 2019).

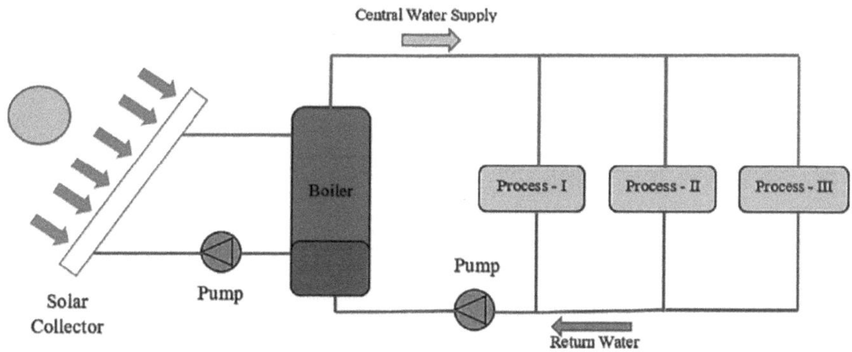

FIGURE 1.11 Schematic diagram of solar energy integration in supply level (Kalogirou, 2004; Kumar et al., 2019).

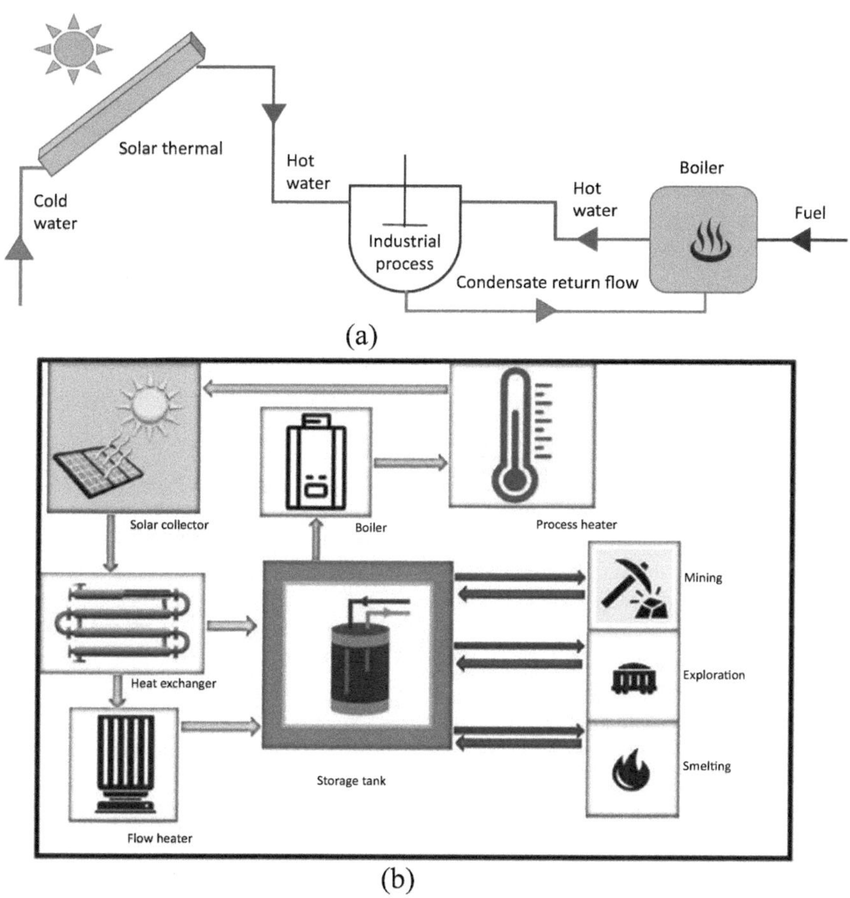

FIGURE 1.12 (A) Principles for solar thermal systems for direct heating to process equipment (Jia et al., 2018); (B) Schematic diagram of solar system integrated with the lead mining process (Farjana et al., 2020).

References

Aebischer, B., Catenazzi, G., & Jakob, M. (2007). Impact of climate change on thermal comfort, heating and cooling energy demand in Europe. In *Proceedings ECEEE* (pp. 859−870).

Alnaimat, F., & Rashid, Y. (2019). Thermal energy storage in solar power plants: a review of the materials, associated limitations, and proposed solutions. *Energies, 12*(21), 41−64. https://doi.org/10.3390/en12214164

Alva, G., Lin, Y., & Fang, G. (2018). An overview of thermal energy storage systems. *Energy, 144*, 341−378.

Ashraf, W., Uddin, G., Arafat, S., Afghan, S., Kamal, A., Asim, M., ... Krzywanski, J. (2020). Optimization of a 660MWe supercritical power plant performance—a case of industry 4.0 in the data-driven operational management part 1. thermal efficiency. *Energies, 13*(21), 55−92. https://doi.org/10.3390/en13215592

Atinc, G., et al. (2019). Boilers. *The Journal of Finance, 11*(1), 128−137. https://doi.org/10.1016/B978-0-12-387785-7.00009-8

IEA Energy Balances. (n.d.). Energy balances of OECD countries, Online database, available at: data. iea. org.

Berger, M., & Worlitschek, J. (2019). The link between climate and thermal energy demand on national level: A case study on Switzerland. *Energy and Buildings, 202*, 109372.

Brain, M. (2020). *Boilers - how steam engines work*. HowStuffWorks. Available at: https://science.howstuffworks.com/transport/engines-equipment/steam2.htm (Accessed 15 November 2020).

Caduff, M., et al. (2012). Wind power electricity: The bigger the turbine, the greener the electricity? *Environmental Science and Technology, 46*(9), 4725−4733. https://doi.org/10.1021/es204108n

El-Wakil, M. M. (1985). *Powerplant technology*. Tata McGraw-Hill Education.

Farjana, S., Mahmud, M. A. P., & Huda, N. (2020). Solar process heat integration in lead mining process. *Case Studies in Thermal Engineering, 22*, 1007−1068. https://doi.org/10.1016/j.csite.2020.100768

Farjana, S. H., et al. (2018). Solar industrial process heating systems in operation−current SHIP plants and future prospects in Australia. *Renewable and Sustainable Energy Reviews, 91*, 409−419.

Gaffert, G. A. (1952). *Steam power stations*.

Guo, Y., Li, Y., Guo, Z., Kim, K., Chang, F., & Wang, S. (2016). Bio-inspired stretchable absolute pressure sensor network. *Sensors, 16*(1). https://doi.org/10.3390/s16010055

Guo, Z., Lv, W., Wang, W., Chen, Q., Zhang, X., Chen, H., & Ma, Z. (2019). Absolute single cavity length interrogation of fiber-optic compound fabry−perot pressure sensors through a white light non-scanning correlation method. *Sensors, 19*(7), 16−28. https://doi.org/10.3390/s19071628

Gupta, H., & Roy, S. (2007). Geothermal energy. In *Geothermal energy*. https://doi.org/10.1016/B978-0-444-52875-9.X5000-X

Hasanuzzaman, M., Zubir, U. S., Ilham, N. I., & Seng Che, H. (2017). Global electricity demand, generation, grid system, and renewable energy polices: A review. *WIREs Energy and Environment, 6*(3), 222. https://doi.org/10.1002/wene.222

IAEA. (2006). *Model for analysis of energy demand (MAED-2) user's manual*. International Atomic Energy Agency.

IEA-ETSAP and IRENA. (2015). Solar heat for industrial processes-technology brief. *Technical Report D*.

International Hydropower Association. (2018). *Hydropower status report: Sector trends and insights*. London: IHA.

Islam, M. M., et al. (2016). Recent progresses and achievements in photovoltaic-phase change material technology: A review with special treatment on photovoltaic thermal-phase change material systems. *Energy Conversion and Management, 126*, 177–204.

Islam, M. M., et al. (2020). Modern energy conversion technologies. In *Energy for sustainable development* (pp. 19–39). Elsevier.

Jia, T., Huang, J., Li, R., He, P., & Dai, Y. (2018). Status and prospect of solar heat for industrial processes in China. *Renewable and Sustainable Energy Reviews, 90*, 475–489. https://doi.org/10.1016/j.rser.2018.03.077

Kalogirou, S. A. (2004). Solar thermal collectors and applications. *Progress in Energy and Combustion Science, 30*(3), 231–295. https://doi.org/10.1016/j.pecs.2004.02.001

Kamarulzaman, A., Hasanuzzaman, M., & Rahim, N. A. (2021). Global advancement of solar drying technologies and its future prospects: A review. *Solar Energy*.

Kato, H., & Trends, G. E. (2002). *World energy outlook 2002, policy modeling for industrial energy use*.

Kempener, R., & Neumann, F. (2014). Ocean thermal energy conversion, technology brief. *IRENA Ocean Energy Technology Brief, 1*.

Kern, D. Q. (1997). *Process heat transfer*. Tata McGraw-Hill Education.

Kitto, J. B., & Stultz, S. C. (2005). *Steam/its generation and use* (41st ed.). USA, Barberton, Ohio: The Babcock & Wilcox Company.

Kumar, L., Hasanuzzaman, M., & Rahim, N. A. (2019). Global advancement of solar thermal energy technologies for industrial process heat and its future prospects: A review. *Energy Conversion and Management, 195*, 885–908.

Kylili, A., et al. (2018). Environmental assessment of solar thermal systems for the industrial sector. *Journal of Cleaner Production, 176*, 99–109.

Manushin, E. A. (2011). Steam turbine. In *A-to-Z guide to thermodynamics, heat and mass transfer, and fluids engineering*. https://doi.org/10.1615/atoz.s.steam_turbine

Munasinghe, M. (1990). Energy demand analysis and forecasting. In *Energy analysis and policy* (pp. 151–185). https://doi.org/10.1016/b978-0-408-05634-2.50018-3

Muster, B., et al. (2015). *Solar process heat for production and advanced applications, Integration Guideline*. IEA SHC Task.

Nahar, A., et al. (2019). Numerical investigation on the effect of different parameters in enhancing heat transfer performance of photovoltaic thermal systems. *Renewable Energy, 132*, 284–295.

Nasrin, R., Hasanuzzaman, M., & Rahim, N. A. (2018). Effect of high irradiation and cooling on power, energy and performance of a PVT system. *Renewable Energy, 116*, 552–569.

Nwaigwe, K. N., Mutabilwa, P., & Dintwa, E. (2019). An overview of solar power (PV systems) integration into electricity grids. *Materials Science for Energy Technologies, 2*(3), 629–633. https://doi.org/10.1016/j.mset.2019.07.002

Pearce, J. M. (2009). Expanding photovoltaic penetration with residential distributed generation from hybrid solar photovoltaic and combined heat and power systems. *Energy, 34*(11), 1947–1954.

Pérez, A., et al. (2017). Basic concepts of thermodynamics. *BMC Public Health, 5*(1), 1–8. Available at: https://ejournal.poltektegal.ac.id/index.php/siklus/article/view/298%0Ahttp://repositorio.unan.edu.ni/2986/1/5624.pdf%0Ahttps://doi.org/10.1016/j.jana.2015.10.005%0Ahttp://www.biomedcentral.com/1471-2458/12/58%0Ahttp://ovidsp.ovid.com/ovidweb.cgi?T=JS&P.

Platzer, W. J. (1999). Energy performance assessment method. In *Proc. Eurosun 1998 2nd international conference on solar heating, Cooling.Portoroz, 14–17- September 1998.*

Platzer, W. J. (2015). *Solar thermal process heat - chances and challenges.*

Prakash, B. (2013). *Combustion optimization in PF boilers.*

Radulescu, L. F. (2018). Introduction & basic concepts of thermodynamics. *A Concise Manual of Engineering Thermodynamics*, 1–13. https://doi.org/10.1142/9789813270855_0001

Saha, S., Ruslan, A. R. M., Morshed, A. K. M. M., & Hasanuzzaman, M. (2021). Global prospects and challenges of latent heat thermal energy storage: A review. *Clean Technologies and Environmental Policy, 23*, 531–559. https://doi.org/10.1007/s10098-020-01997-7

Sarkar, D. K. (2015). Steam generators. *Thermal Power Plant.* https://doi.org/10.1016/b978-0-12-801575-9.00002-0

Sathish, T., Mohanave, V., Afzal, A., Arunkumar, M., Ravichandran, M., Khan, S., … Asif, M. (2021). Advancement of steam generation process in water tube boiler using Taguchi design of experiments. *Case Studies in Thermal Engineering, 27*, 1012–1047. https://doi.org/10.1016/j.csite.2021.101247

Schmitt, B. (2016). Classification of industrial heat consumers for integration of solar heat. *Energy Procedia, 91*, 650–660.

Shyy, W. (1994). Basic concepts of thermodynamics. *Computational Modeling for Fluid Flow and Interfacial Transport, 388*(F 09), 279–298. https://doi.org/10.1016/b978-0-444-81760-0.50015-0

Sipilä, K. (2016). Cogeneration, biomass, waste to energy and industrial waste heat for district heating. In *Advanced district heating and cooling (DHC) systems* (pp. 45–73). Elsevier.

Sipilä, K., et al. (2005). *Small-scale biomass CHP plant and district heating.* Espoo, Finlandia: Technical Research Centre of Finland (VTT), Informe coordinado por Finnish District Heat Association.

Thirugnanasambandam, M., Iniyan, S., & Goic, R. (2010). A review of solar thermal technologies. *Renewable and Sustainable Energy Reviews, 14*(1), 312–322.

Tognoli, M., Najafi, B., Marchesi, R., & Rinaldi, F. (2019). Dynamic modelling, experimental validation, and thermo-economic analysis of industrial fire-tube boilers with stagnation point reverse flow com. *Applied Thermal Engineering, 149*, 1394–1407. https://doi.org/10.1016/j.applthermaleng.2018.12.087

Vakkilainen, E. (2005). *Kraft recovery boilers—principles and practice.* Valopaino Oy.

Vakkilainen, E. K. (2017a). Principles of steam generation. In *Steam generation from biomass* (pp. 1–17). https://doi.org/10.1016/b978-0-12-804389-9.00001-0

Vakkilainen, E. K. (2017b). Thermal design of boiler parts. *In Steam Generation from Biomass*, 115–143. https://doi.org/10.1016/b978-0-12-804389-9.00005-8

Wagh, M. M., & Kulkarni, V. V. (2019). Thermal energy demand fulfillment of Kolhapur through modeling and optimization of integrated renewable energy systems. *Renewable Energy Focus, 29*, 114–122.

Winterbone, D. E., & Turan, A. (2015). Introduction and revision. *Advanced Thermodynamics for Engineers*, 1–12. https://doi.org/10.1016/b978-0-444-63373-6.00001-0

Xu, S., & Guo, Z. (2021). Entransy transfer analysis methodology for energy conversion systems operating with thermodynamic cycles. *Energy, 224*, 1201–1289. https://doi.org/10.1016/j.energy.2021.120189

Chapter 2

Solar thermal energy conversion

Md. Shariful Alam Emon[1,2], Mohsan Uddin Ahmad[3] and Md Hasanuzzaman[4]

[1]*Department of Mechanical Engineering, Dhaka University of Engineering & Technology, Gazipur, Bangladesh;* [2]*Department of Mechanical Engineering, Bangladesh University of Engineering & Technology, Dhaka, Bangladesh;* [3]*Department of Mechanical Engineering, Michigan State University, East Lansing, MI, United States;* [4]*Higher Institution Centre of Excellence (HICoE), UM Power Energy Dedicated Advanced Centre (UMPEDAC), University of Malaya, Kuala Lumpur, Malaysia*

Chapter outline

Technologies for Solar Thermal Energy. https://doi.org/10.1016/B978-0-12-823959-9.00011-8

2.1 Introduction

Energy is one of the basic needs for automation, modernization, and economic development. Fossil fuels and renewable resources (e.g., solar, wind, water, geothermal, and biomass sources) are the major sources of energy. Solar power is a green form of renewable energy. It is huge, plenty, and free of charge. Today, energy demand is rising daily due to the growth of technology and industry. Furthermore, due to the decrease in fossil fuel resources and pollution, it is important to utilize alternative and clean sources. More emphasis should now be placed on pure and environmentally friendly fuels such as solar energy. However, all current energy resources can't be replaced by solar power. Nearly 60%–70% of our present energy requirements may be taken care of by utilizing solar energy. Photovoltaic (PV) devices/solar cells or solar collectors are the main technologies to capture solar power.

2.2 The geometry of solar radiation

2.2.1 Latitude

It is the angular measurement of the position of a point or location on the surface of the earth which the location or point how much north or south remains with respect to the plane of the equator. It is indicated by, ϕ and ranges from 0° to 90°. It is 0° at the equator and 90° at the poles. From the equator to the north pole, latitude has a positive value, whereas from the equator to the south pole, it has a negative sign, as illustrated in Fig. 2.1.

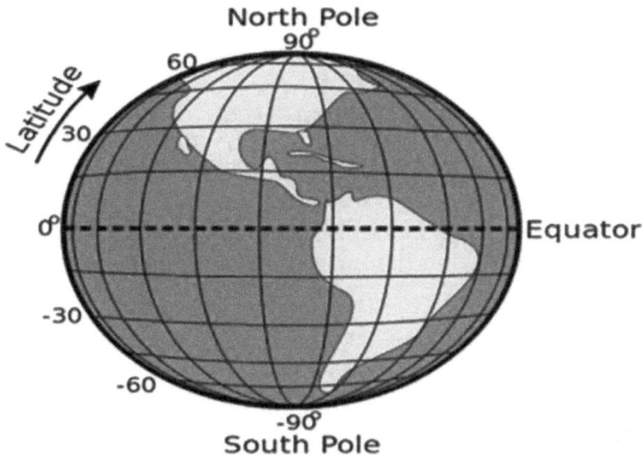

FIGURE 2.1 Illustration of latitude (Awasthi et al., 2020).

2.2.2 Declination angle

It is the angle produced between the plane of the Equator and the line traced from the center of the earth to the center of the sun, shown in Fig. 2.2. It refers to a certain day of the year and it is indicated by δ and ranges from -23.45 to $+23.45$ degrees. The formula for calculating the declination angle (δ) is as follows (Batayneh, Bataineh, Soliman, & Hafees, 2019):

$$\delta = \left(\frac{180}{\pi}\right) * (0.006918 - 0.399912\cos(B) + 0.070257\sin(B) \\ -0.006758\cos(2B) + 0.000907\sin(2B) \\ -0.002697\cos(3B) + 0.00148\sin(3B)) \tag{2.1}$$

where the coefficient B is figured out by (Batayneh et al., 2019):

$$B = (n-1) * \frac{360}{365} \tag{2.2}$$

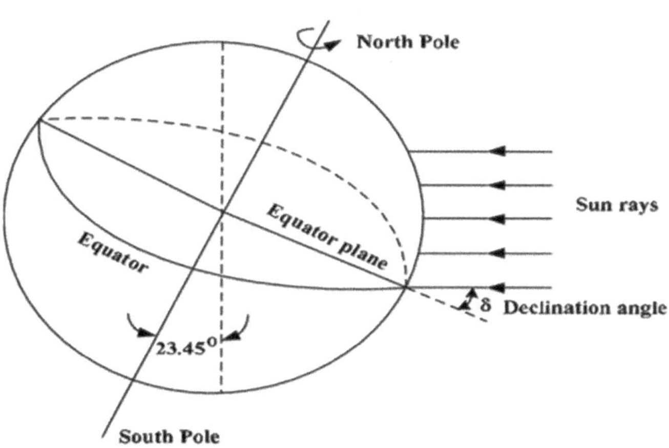

FIGURE 2.2 Schematic of the declination angle (Awasthi et al., 2020).

The outcome of the declination angle equation is in degree units. Furthermore, n is the integer number of consecutively counted days in a year ($0 \leq n \leq 365$), i.e., $n = 1$ for January 1.

It is calculated by (Stanciu & Stanciu, 2014):

$$\delta = 23.45 \sin\left[2\pi \frac{(n + 284)}{365}\right] \tag{2.3}$$

where n denotes the year's nth day and it is an integer value.

2.2.3 Solar noon

The sun begins to radiate at sunrise in the morning and continues to do so until dusk. On a given day, it is the moment for the sun to reach its peak in the sky. It fluctuates from day to day throughout the year.

2.2.4 Hour angle

The earth rotates about its axis and it creates night and day on the earth. The cycle from morning to evening is due to the east-west apparent motion of earth. During the day, the earth rotates 15 degrees an hour. The angle of time is the transformation from day to solar noon. In terms of solar noon, it is symmetrical. The Hour Angle (ω) is provided by the following equation (Batayneh et al., 2019):

$$\omega = 15^\circ * (\text{Local solar time} - 12) \tag{2.4}$$

The hour angle is normally 0° at solar noon, positive in the afternoon, and negative before noon.

2.2.5 Elevation/altitude angle

It denotes the angle produced by the sun's beams and the horizontal plane or axis and is indicated by (Batayneh et al., 2019):

$$\sin(\alpha) = [\cos(\phi)\cos(\delta)\cos(\omega) + \sin(\phi)\sin(\delta)] \tag{2.5}$$

Every minute, as well as every day, the elevation/altitude angle varies. It reaches its highest during the day at solar noon. When the sunlight is perpendicular to the horizontal plane or axis, the angle it forms is 90°.

2.2.6 Zenith angle (θ_z)

It denotes the angle produced by the sun's beams and the vertical plane or axis and is indicated by (Batayneh et al., 2019) that is shown in Fig. 2.3:

$$\cos(\theta_z) = \cos(\phi)\cos(\delta)\cos(\omega) + \sin(\phi)\sin(\delta) \tag{2.6}$$

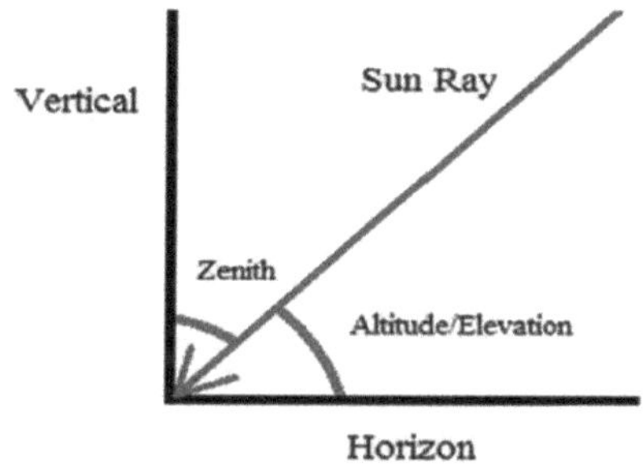

FIGURE 2.3 Schematic of the elevation and Zenith angle (Awasthi et al., 2020).

It reaches its lowest value during the day at solar noon. It is the elevation angle's counterpart angle that can be calculated as follows:

$$\theta_z = 90^\circ - \alpha \tag{2.7}$$

2.2.7 Sunset hour angle (ω_s) and daylight hours

When the elevation/altitude angle (α) reaches zero or the zenith angle (θ_z) reaches 90°, sunset happens. The equation for calculating the sunset hour angle is (Batayneh et al., 2019):

$$\cos(\omega_s) = -\left[\frac{\sin(\phi)\sin(\delta)}{\cos(\phi)\cos(\delta)}\right] = -\tan(\phi)\tan(\delta) \tag{2.8}$$

Another angle is the sunrise hour angle (ω_{sr}) which is identical to the sunset hour angle (ω_s) but has the opposite sign. Thus, the number of daylight hours between sunrise and sunset can be found using the given equation (Batayneh et al., 2019):

$$N = \frac{2}{15}\cos^{-1}[-\tan(\phi)\tan(\delta)] \tag{2.9}$$

It can be written from the previous equation as:

$$N = \frac{2}{15}\omega_s \tag{2.10}$$

2.2.8 Solar azimuth angle (γ_S)

In the spherical coordinate system, it is a form of angular measurement. The angle produced by a reference line from the center of the Earth to the north and a sunray projection from a horizontal plane is called the solar azimuth angle, as shown in Fig. 2.4. In the traditional method, the line joining the center of the

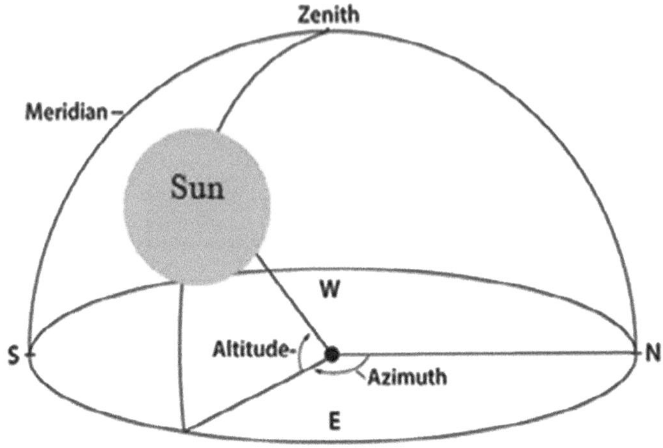

FIGURE 2.4 Schematic of the azimuth and altitude angle (Awasthi et al., 2020).

earth and the south is taken as the reference line, but the reference line along the north is a common and acceptable practice. The angle is determined from the clockwise direction, i.e., West-270°, South-180°, East-90°.

$$\gamma_S = \text{sign}(\omega)\left[\cos^{-1}\left(\frac{\cos(\theta_z)\sin(\phi) - \sin(\delta)}{\sin(\theta_z)\cos(\phi)}\right)\right] \tag{2.11}$$

Therefore, above equation separately can be written as below:

$$\gamma_S = \begin{cases} +\cos^{-1}\left(\dfrac{\cos(\theta_z)\sin(\phi) - \sin(\delta)}{\sin(\theta_z)\cos(\phi)}\right) & \text{if } \omega > 0 \\ -\cos^{-1}\left(\dfrac{\cos(\theta_z)\sin(\phi) - \sin(\delta)}{\sin(\theta_z)\cos(\phi)}\right) & \text{if } \omega < 0 \end{cases} \tag{2.12}$$

In the morning and before the solar noon, the angle fluctuates between 0 and 180 degrees, while in the evening and after the solar noon, it changes between 180 and 360 degrees.

2.2.9 Tilt angle or inclination angle (β)

If the collector or solar PV panel is horizontal, it may not make efficient use of solar radiation as the sun's direction varies seasonally and daily. To absorb more radiation, it is required to follow the motion of the sun, so that the tracking or following surface may be tilted or inclined to the horizontal plane or axis, and the angle is called tilt or inclination angle (β), as illustrated in Fig. 2.5. The angle should be changed from time to time or optimized for a certain location to use the solar potential effectively.

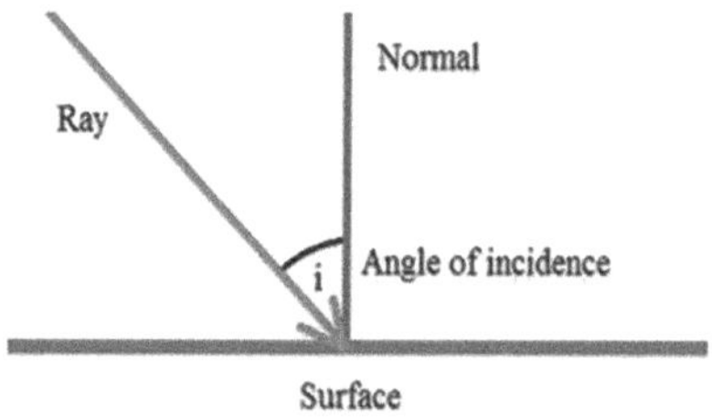

FIGURE 2.5 Schematic of angle of incidence (Awasthi et al., 2020).

2.2.10 Surface azimuth angle (γ)

To track the sun and capture solar radiation, the tracking plane should be focused on the sun, so that the plane should be rotated into the horizontal plane. In this case, surface azimuth angle plays an important role. It is the angle produced by the line connecting the center of the earth to the north, and the projection of the normal vector of the tracking surface onto the horizontal surface, which refers to the surface azimuth angle as shown in Fig. 2.6.

2.2.11 Angle of incidence (θ)

It is a special type of inclination angle. It's the angle formed by the tracking plane's normal line and the incident sun's rays.

$$\begin{aligned}\cos(\theta) = {} & \sin(\delta)\sin(\phi)\cos(\beta) - \sin(\delta)\cos(\phi)\sin(\beta)\cos(\gamma) \\ & + \cos(\delta)\cos(\phi)\cos(\beta)\cos(\omega) + \cos(\delta)\sin(\phi)\sin(\beta)\cos(\gamma)\cos(\omega) \\ & + \cos(\delta)\sin(\beta)\sin(\gamma)\sin(\omega)\end{aligned} \tag{2.13}$$

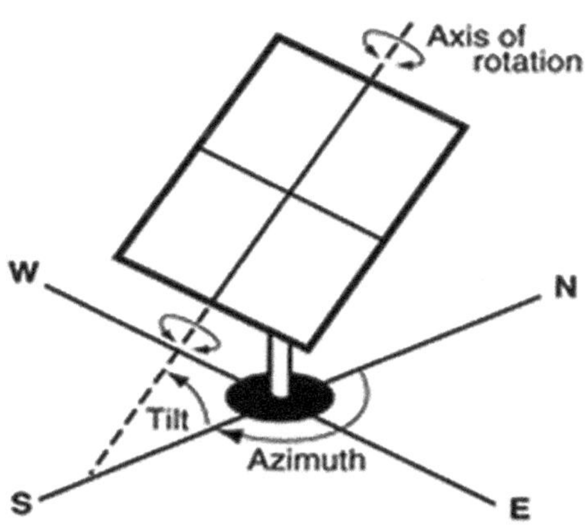

FIGURE 2.6 Schematic of inclination or tilt angle and surface azimuth angle (Awasthi et al., 2020).

The angle of incidence can be calculated by the alternate equation given below:

$$\cos(\theta) = \cos(\theta_z)\cos(\beta) + \sin(\theta_z)\sin(\beta)\cos(\gamma_S - \gamma) \quad (2.14)$$

Here, γ is the tracking surface azimuth angle and β is the tilt angle of the tracking surface. The incident angle plays a significant role in solar tracking since it is utilized to find out the optimal location of the tracking surface with respect to direct radiation and, therefore, to put the tracking surface in the best position to capture the greater solar radiation.

2.2.12 Geometric factor (R_b)

At a given time, it is the ratio of the amount of solar radiation hitting an inclined surface to that hitting a horizontal surface. The ratio of slanted surface total radiation to horizontal surface total radiation is also known as this. The geometric factor can be utilized in the evaluation and comparison of the tilted panel's performance. It is calculated by the given formula (Batayneh et al., 2019):

$$R_b = \frac{\cos(\theta)}{\cos(\theta_z)} = \frac{I_{b,t}}{I_b} \quad (2.15)$$

where $I_{b,t}$ is the tilted surface incident direct radiation and I_b is the horizontal surface incident direct radiation.

2.3 Components and types of solar radiation

Measurement of solar radiation is usually done in two parameters:

2.3.1 Irradiance

It is the solar radiation per unit time and per unit area that is indicated as G (W/m^2). For extraterrestrial radiations, it is denoted as G_o.

2.3.2 Irradiation or insolation

It is the solar radiation captured per unit area at a certain time. Its unit is J/m^2. It is measured on hourly, daily, and monthly average daily basis and is denoted as I, H, and $\overline{H}$, respectively.

The entire radiation on the earth's surface, or global radiation, is split into two solar radiation components:

Direct radiation: It comes from the sun's disc and refers to radiation that is not dispersed in the atmosphere, and so has a fixed direction. The term "beam radiation" or "direct beam radiation" is also used to describe it.

Diffuse radiation: The remainder of the radiation is spread throughout the atmosphere and molecules, aerosols, as well as clouds, scatter it in all directions. This scattered component is known as diffuse radiation.

The presence of an atmosphere and other particles creates several obstacles before solar radiation reaches the surface of the planet. Scattering, reflection, and absorption all reduce the intensity of the light. As a result, terrestrials may divide radiation into two kinds:

Extraterrestrial radiation: It is the solar radiation striking the surface of the earth if there is no atmosphere present. It is the solar radiation striking the surface of the earth outside of the atmosphere. It can be calculated without the influence of atmosphere.

Terrestrial or global radiation: It is solar radiation striking the surface of the earth while the atmosphere is present. It changes depending on the weather and the state of the sky. It can be measured or estimated from the database.

2.4 Calculation of extraterrestrial radiation

Figs. 2.7 and 2.8 shows the flowchart of calculation of extraterrestrial radiation.

The equation below (Batayneh et al., 2019) is used to compute extraterrestrial radiation falling on a plane that is normal to the radiation (G_{on}):

$$G_{on} = G_{SC}(1.000110 + 0.034221\cos(B) + 0.001280\sin(B) + 0.000719\cos(2B) + 0.000077\sin(2B)) \tag{2.16}$$

where the value of G_{SC}, solar constant is 1367 W/m^2 and the value of B is calculated by Eq. (2.2). It can be converted into the extraterrestrial radiation

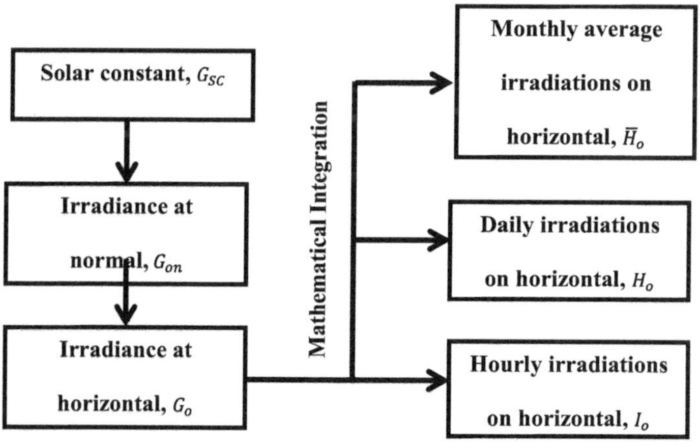

FIGURE 2.7 Flowchart of calculation of extraterrestrial radiation.

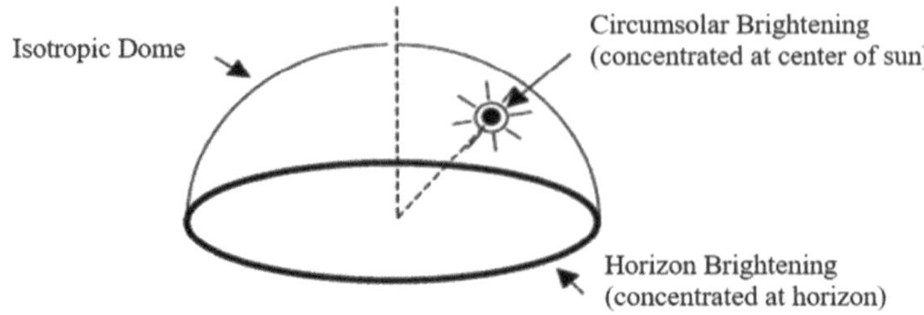

FIGURE 2.8 Schematics of sky depicting the component of solar radiance distribution (Yadav & Chandel, 2013).

incident on a plane that is horizontal, G_o by the formula given by (Batayneh et al., 2019):

$$\begin{aligned} G_o = G_{SC}[(1.000110 + 0.034221\cos(\mathrm{B}) + 0.001280\sin(\mathrm{B}) \\ + 0.000719\cos(2\mathrm{B}) + 0.000077\sin(2\mathrm{B}))]\cos(\theta_z) \end{aligned} \tag{2.17}$$

An alternative relationship of the aforementioned equation can be expressed as (Hawlader, 1984):

$$G_o = G_{SC}\left(1 + 0.033\cos\left(\frac{360\,n}{365}\right)\right)\cos\theta_z \tag{2.18}$$

The aforementioned equation can be combined to the zenith angle Eq. (2.6) such as:

$$G_o = G_{SC}\left(1 + 0.033\cos\left(\frac{360\,n}{365}\right)\right) \times (\cos\phi\cos\delta\sin\omega_s + \sin\phi\sin\delta) \tag{2.19}$$

Now, the equation can be integrated to obtain the hourly extraterrestrial radiation over a certain time using the aforementioned connection.

$$\begin{aligned} I_o = & \frac{24 \times 3600\, G_{SC}}{\pi}\left(1 + 0.033\cos\left(\frac{360\,n}{365}\right)\right) \\ & \times\left(\cos\phi\cos\delta(\sin\ \omega_2 - \sin\omega_1) + \frac{\pi \times (\omega_2 - \omega_1)}{180}\sin\phi\sin\delta\right) \end{aligned} \tag{2.20}$$

The daily extraterrestrial radiation on a horizontal plane integrating Eq. (2.19) from sunrise to sunset is as follows:

$$\begin{aligned} H_o = & \frac{24 \times 3600\, G_{SC}}{\pi}\left(1 + 0.033\cos\left(\frac{360\,n}{365}\right)\right) \\ & \left(\cos\phi\cos\delta\sin\omega_s + \frac{\pi\omega_s}{180}\sin\phi\sin\delta\right) \end{aligned} \tag{2.21}$$

Now, the monthly average daily extraterrestrial radiation on a horizontal surface can be calculated from Eq. (2.21) by averaging the days of a particular month as follows:

$$\overline{H}_o = \frac{24 \times 3600\, G_{SC}}{\pi}\left(1 + 0.033 \cos\left(\frac{360\, \overline{n}}{365}\right)\right)\left(\cos\phi \cos\delta \sin\omega_s + \frac{\pi\omega_s}{180}\sin\phi \sin\delta\right) \tag{2.22}$$

where $\overline{n}$ is the average day of a particular month.

2.5 Calculation of terrestrial radiation

Terrestrial radiation can usually be found in three ways.

- Instrument measurement
- Information from databases such as TMY, NASA, SSE, etc.
- Estimation from different correlations

2.5.1 Measurement of terrestrial radiation

Total irradiance can be measured by pyranometer. Diffuse radiation can be measured using a shade ring pyranometer. Pyrheliometer measures the direct sunlight or beam radiation. Irradiance of the beam may also be evaluated by difference in the pyranometer readings with and without shadow band.

Nowadays, solar radiation is measured using two methods such as thermoelectric and photoelectric effects. It operates by combining thermocouples of various metals. The temperature differential between them generates a very small current. By appropriately assembling the thermopiles and painting them in a dull black finish where may be employed as nearly excellent blackbody detectors. A thin metal plate is connected to the heated connection. The plate's opposing side is blackened to absorb solar radiation. The cold junction of the instrument is exposed. The output is electrically adjusted for the cavity temperature. The amount of insolation is proportional to the temperature raised by the hot junction and the resulting electromagnetic force. Due to the linearization and calibration of the output voltage, it is simply translated to radiative flux.

Photoelectric sensors are simpler, faster, and more stable. The photoelectric effect happens when solar radiation hits a light-sensitive detector, absorbing some of the photons' energy. The atoms emit electrons in this excited state, which can travel through a conductor to create an electrical current. The current is proportional to the detector's energy. The main flaw is a nonuniform spectral response in the solar band.

There are one or two glass domes underneath the sensor in pyranometer. The sensor may be thermoelectric or photoelectric type. The glass is designed specifically for the transmission of solar light through a range of wavelengths and is thermally separated from the sensor. The pyranometer has been designed for usage in the horizontal position that it is permanently mounted for calibration.

Pyrheliometer allows only to see light from the sun directly. To avoid dispersed light, the detector only collects radiation from a narrow cone of sky. As a result, only approximately normal incident radiation reaches the sensor, which is placed at the tube's base. This computer-controlled instrument automatically tracks the sun.

2.5.2 Terrestrial radiations databases

The NASA Surface Meteorology and Solar Energy (SSE) database, accessible online, may give monthly average total daily irradiance on horizontal surfaces.

2.5.3 Terrestrial irradiation estimation from correlation

The following correlation can be written according to the most often used Angstrom–Prescott regression model (Angstrom, 1924; Robaa, 2009):

$$\frac{\overline{H}}{\overline{H}_o} = a + b\,\frac{\overline{n}}{\overline{N}} \tag{2.23}$$

where $\overline{H}$ is monthly average daily global radiation, $\overline{H}_o$ denotes monthly average daily extraterrestrial radiation, $\overline{n}$ denotes monthly average daily hours of bright sunshine, $\overline{N}$ is the day length of the average day of the month, and a and b denote empirical coefficients.

2.6 Calculation of beam and diffuse radiation on horizontal plane

The full contribution of extraterrestrial radiation may not be available due to the cloud cover of the area or location, and it varies from time to time, from day to day, and from month to month. It is necessary to understand the average number of clouds on the site. It is represented by an index named the cloudiness index or clearness index. The clearness index has a value of one, indicating that the sky is clear for the day or hour. There are three common types of cloudiness index:

- Monthly average daily clearness index
- Daily clearness index
- Hourly clearness index

2.6.1 The monthly average daily clearness index

It is the proportion of the total or global radiation on a monthly average daily basis to the monthly average daily extraterrestrial radiation for a definite location, defined as (Liu & Jordan, 1960):

$$\overline{K}_T = \frac{\overline{H}}{\overline{H}_o} \tag{2.24}$$

2.6.2 The hourly clearness index

It is the proportion of the total or global radiation falling on a horizontal surface at a certain hour to the extraterrestrial radiation for the given hour at a particular location, given as (Liu & Jordan, 1960):

$$K_T = \frac{I}{I_o} \tag{2.25}$$

2.6.3 The daily clearness index

It is the proportion of the daily total or global radiation falling on a horizontal surface to the daily extraterrestrial radiation for the particular location, given as (Liu & Jordan, 1960):

$$K_T = \frac{H}{H_o} \tag{2.26}$$

The correlation developed by Erbs, which is a function of K_T (the hourly clearness ratio), can be used to estimate diffuse radiation in proportion to total radiation. It was achieved through the collection of meteorological data for a particular location over years of measurements. The relations are given by (Erbs, Klein, & Duffie, 1982):

$$\frac{I_d}{I} = \begin{cases} 1.0 - 0.09\,K_T, & K_T \le 0.22 \\ 0.9511 - 0.1604\,K_T + 4.388\,K_T^2 - 16.638\,K_T^3 + 12.336\,K_T^4, & 0.22 < K_T < 0.80 \\ 0.177, & K_T \ge 0.80 \end{cases} \tag{2.27}$$

Orgill and Holland (1977) suggested the above relationship in another form that is given below:

$$\frac{I_d}{I} = \begin{cases} 1 - 0.249\,K_T, & K_T \le 0.35 \\ 1.557 - 1.84\,K_T, & 0.35 < K_T < 0.75 \\ 0.177, & K_T \ge 0.75 \end{cases} \tag{2.28}$$

From this correlation, the diffuse component of the total radiation can be estimated at different hours of the day in various months of a year. Then, the direct component of total radiation can therefore be calculated by the relationship (Orgill & Hollands, 1977).

$$I = I_b + I_d \tag{2.29}$$

Collares-Pereira and Rabl (1979) suggested the relationship of the diffuse component of the daily irradiation on the horizontal surface shown below:

$$\frac{H_d}{H} = \begin{cases} 0.99, & K_T \le 0.17 \\ 1.188 - 2.272\,K_T + 9.473\,K_T^2 - 21.865\,K_T^3 + 14.648\,K_T^4, & 0.17 < K_T < 0.75 \\ -0.54\,K_T + 0.632, & 0.75 < K_T < 0.8 \\ 0.2, & K_T \ge 0.8 \end{cases} \tag{2.30}$$

Collares-Pereira and Rabl (1979) suggested another correlation of monthly average daily diffuse irradiation given below:

$$\frac{H_d}{H} = 0.775 + 0.347 \times \frac{(\omega - 90)\pi}{180} - \left[0.505 + 0.261 \times \frac{(\omega - 90)\pi}{180}\right] \times \cos\left(\frac{360(K_T - 0.9)}{\pi}\right) \tag{2.31}$$

Hourly total irradiation from daily irradiation on horizontal

For any midpoint of an hour (ω):

$$I = r_t\,H \tag{2.32}$$

According to Collares- Pereira and Rabl (1979):

$$r_t = \frac{\pi}{24}(a + b\cos\omega)\left[\frac{(\cos\omega - \cos\omega_S)}{\left(\sin\omega_S - \frac{\pi\,\omega_S}{180}\cos\omega_S\right)}\right] \tag{2.33}$$

where:

$$a = 0.409 + 0.5016\sin(\omega_S - 60) \tag{2.34}$$

$$b = 0.6609 - 0.4767\sin(\omega_S - 60) \tag{2.35}$$

Hourly diffuse irradiations from daily diffuse irradiation on horizontal

For any midpoint of an hour (ω) (Collares-Pereira & Rabl, 1979):

$$I_d = r_d H_d \tag{2.36}$$

From Liu and Jordan (1960):

$$r_d = \frac{\pi}{24}\left[\frac{(\cos\omega - \cos\omega_S)}{\left(\sin\omega_S - \frac{\pi\,\omega_S}{180}\cos\omega_S\right)}\right] \tag{2.37}$$

2.7 Radiations on a tilted plane

The difficulty of total radiation estimate is related to the extremely changeable cloud conditions that are difficult to anticipate. The diffused components are made up of three parts (Mamun, Islam, Hasanuzzaman, & Selvaraj, 2021):

- **Isotropic component (iso):** It is the diffuse radiation obtained equally from all across the sky dome.
- **Circumsolar brightening (cs):** It is the intensely focused radiation in the sky surrounding the solar globe which is produced by the spread of sun rays ahead.
- **Horizon brightening (hz):** It is the more intense and localized radiation, closer to the horizon, that may be seen in clear sky conditions.

The magnitude of the angular scattering of the diffuse radiation part is greatly influenced by ground reflectance. To accommodate changing reflectance with various weather scenarios, an estimate of $\rho_g = 0.5$ is employed for dry tropical regions (Fan, Yang, & Hu, 2018). From the above components of diffuse radiation, the diffuse sky assumption that can be categorized into two types:

- **Isotropic sky (iso):** This assumption only considers the isotropic component and exclude the other two parts.
- **Anisotropic sky (iso + cs, iso + cs + hz):** The anisotropic model is designed to accommodate the influence of circumsolar diffusion on inclined surfaces together with the horizon brightening elements (Liu & Jordan, 1960). There have been many anisotropic models.

2.7.1 Calculation of radiations on a tilted plane

It is necessary to have the following facts in order to compute radiations on an inclined plane:

(a) Tilt or inclination angle of the tracking surface
(b) At least two items of total, direct, and diffusion radiations on horizontal plane
(c) Assumptions of diffuse sky (isotropic or anisotropic)
(d) The model for calculation

2.7.2 Models for calculation of radiations on a tilted plane

Liu and Jordan model (iso, γ = 0, I) (Liu & Jordan, 1960)

$$I_T = I_b R_b + I_d \left(\frac{1 + \cos \beta}{2} \right) + I \rho_g \left(\frac{1 - \cos \beta}{2} \right) \tag{2.38}$$

$$R_b = \frac{G_{bT}}{G_b} = \frac{G_{bn} \cos\theta}{G_{bn} \cos\theta_z} = \frac{\cos \theta}{\cos \theta_z} = \frac{\cos(\phi - \beta)\cos \delta \cos \omega + \sin (\phi - \beta) \sin \delta}{\cos \phi \cos \delta \cos \omega + \sin \phi \sin \delta} \tag{2.39}$$

where

ρ_g (albedo) = 0 (dull black surface) to 1 (white shiny surface) [Recommended value 0.2].

Liu and Jordan model (iso, γ = 0, H) (Liu & Jordan, 1960)

$$\overline{H}_T = \overline{H}_b \overline{R}_b + \overline{H}_d \left(\frac{1 + \cos \beta}{2} \right) + \overline{H} \rho_g \left(\frac{1 - \cos \beta}{2} \right) \tag{2.40}$$

where, for northern hemisphere, surface facing south:

$$\overline{R}_b = \left[\frac{\cos (\phi - \beta) \cos \delta \cos \omega'_S + \left(\frac{\pi \, \omega'_S}{180} \right) \sin(\phi - \beta) \sin \delta}{\cos \phi \cos \delta \sin\omega_S + \left(\frac{\pi \, \omega_S}{180} \right) \sin \phi \sin \delta} \right] \tag{2.41}$$

$$\omega'_S = \min \left[\begin{array}{c} \cos^{-1}(-\tan \phi \tan \delta) \\ \cos^{-1}(-\tan(\phi - \beta)\tan \delta) \end{array} \right] \tag{2.42}$$

where, for southern hemisphere, surface facing north:

$$\overline{R}_b = \left[\frac{\cos (\phi + \beta) \cos \delta \cos \omega'_S + \left(\frac{\pi \, \omega'_S}{180} \right) \sin (\phi + \beta) \sin \delta}{\cos \phi \cos \delta \sin \omega_S + \left(\frac{\pi \, \omega_S}{180} \right) \sin \phi \sin \delta} \right] \tag{2.43}$$

$$\omega'_S = \min\begin{bmatrix} \cos^{-1}(-\tan\phi\tan\delta) \\ \cos^{-1}(-\tan(\phi+\beta)\tan\delta) \end{bmatrix} \tag{2.44}$$

ρ_g (albedo) = 0 (dull black surface) to 1 (white shiny surface) [Recommended value 0.2].

Then, the direct component of daily total radiation can therefore be calculated as follows:

$$H_b = H - H_d \tag{2.45}$$

Hay, Davies, Klucher and Reindl model (HDKR model) (iso + cs + hz, γ = 0, I) (Yadav & Chandel, 2013)

$$I_T = (I_b + I_d A_i)R_b + I_d(1-A_i)\left(\frac{1+\cos\beta}{2}\right)\left[1+f\sin^3\left(\frac{\beta}{2}\right)\right] + I\rho_g\left(\frac{1-\cos\beta}{2}\right) \tag{2.46}$$

$$R_b = \frac{G_{bT}}{G_b} = \frac{G_{bn}\cos\theta}{G_{bn}\cos\theta_z} = \frac{\cos\theta}{\cos\theta_z} = \frac{\cos(\phi-\beta)\cos\delta\cos\omega + \sin(\phi-\beta)\sin\delta}{\cos\phi\cos\delta\cos\omega + \sin\phi\sin\delta} \tag{2.47}$$

where, A_i is the anisotropy index and f is the correction factor to account cloudiness, both are calculated by the equations are given below:

$$A_i = \frac{I_b}{I_o} \tag{2.48}$$

$$f = \sqrt{\frac{I_b}{I}} \tag{2.49}$$

ρ_g (albedo) = 0 (dull black surface) to 1 (white shiny surface) [Recommended value 0.2].

2.8 Solar power conversion

The two most common ways of conversing the solar energy into electricity are:

- PV systems
- Solar thermal

2.8.1 PV systems

The PV cell, often known as a solar cell, is a device that converts solar radiation into electrical energy. Silicon similar semiconductors are found in solar cells. A common method is to dope the top layer with boron, which serves as a p-type semiconductor, and the bottom layer with phosphorous, which serves as a pentavalent semiconductor (n-type semiconductor). A PN junction occurs when two kinds of semiconductors are placed together. Holes gather near the surface, causing electrons to migrate from the n-type side to the p-type side and holes to move from the p-type layer to the n-type layer (Fan et al., 2018). The sunlight hits the solar cells, transferring photon energy to the silicon electrons, subsequently freeing them from the silicon parent atom. These freed electrons go into the n-type silicon layer. When an electron goes from an n-type layer to a p-type layer, via the use of electrodes, an electric current is generated (Stapleton, 2017). PV modules are one of the most efficient, long-lasting, and environmentally friendly systems. Only a small percentage of the solar energy that falls on these modules is transformed into electricity. The remaining irradiation is converted into thermal energy, which warms the PV module and reduces its output. The basic equations of heat transfer from the module's top surface are given below.

The top of the module's energy consumption is calculated by the equation given below (Dubey & Tay, 2013; Dubey & Tiwari, 2008; Nahar, Hasanuzzaman, & Rahim, 2017a; Nahar, Hasanuzzaman, Rahim, & Parvin, 2019; Tiwari, Sodha, Chandra, & Joshi, 2006):

$$E_{ab} = \tau_g \alpha_{sc} p_{sc} \mathrm{GWA} \tag{2.50}$$

where τ_g is the transmissivity of glass, α_{sc} is the absorptivity of solar module, p_{sc} is the packing factor of solar module, G is the incident irradiation (W/m^2), W is the width of the solar cell, A is the PV area. Packing factor is the ratio of the total solar cell area to the entire module area. It is determined by the geometry of the solar cells utilized in the module. It influences the output and operating temperature of the module.

At the top of the module, the energy loss due to convection is represented by (Dubey & Tay, 2013; Dubey & Tiwari, 2008; Nahar et al., 2017a; Tiwari et al., 2006):

$$E_{ctop} = U_{sca}(T_{sc} - T_a)\ \mathrm{WA} \tag{2.51}$$

U_{sca} is overall heat transfer coefficient through glass cover from module top surface to ambient (W/m^2 K), is the solar module top surface temperature (°C), is the ambient temperature (°C).

The energy transferred to the PV module's bottom surface is indicated by (Dubey & Tay, 2013; Dubey & Tiwari, 2008; Nahar, Hasanuzzaman & Rahim, 2017b; Tiwari et al., 2006):

$$E_b = U_t(T_{sc} - T_b)\ \mathrm{WA} \tag{2.52}$$

where U_t is the overall heat transfer coefficient from module top surface to tedlar (polyvinyl fluoride) back surface of the module (W/m^2 K), the tedlar (polyvinyl fluoride) back surface temperature of the module (°C).

The entire energy transformed into electricity is shown by the given equation (Dubey & Tay, 2013; Dubey & Tiwari, 2008; Rahman, Hasanuzzaman, & Rahim (2015); Tiwari et al., 2006):

$$E_e = \eta_{sc} p_{sc} \mathrm{GWA} \tag{2.53}$$

where η_{sc} is the electrical efficiency of solar module.

The equation of energy balance on the top of the module is represented by the following equation (Rahman et al., 2015; Rahman, Hasanuzzaman, & Rahim, 2017):

$$E_{ab} = E_{ctop} + E_b + E_e \tag{2.54}$$

The temperature of the solar cell is obtained from the above energy balancing equation as (Rahman et al., 2015, 2017):

$$T_{sc} = \frac{p_{sc} G(\tau_g \alpha_{sc} - \eta_{sc}) + (U_{sca} T_a + U_t T_b)}{(U_{sca} + U_t)} \tag{2.55}$$

Also, the solar cell efficiency is determined by the equation as below:

$$\eta_{sc} = \frac{\text{Electrical energy}}{\text{Incident solar energy}} = \frac{E_e}{p_{sc} \mathrm{GWA}} \tag{2.56}$$

2.8.2 Solar thermal power production

Solar energy can be converted into thermal energy by using solar thermal collectors which capture the radiation and transfer it to the fluid in the collector tubes. Fig. 2.9 shows the schematics of solar thermal power conversion (Kumar, Hasanuzzaman, & Rahim, 2019). The temperature of the collector fluid increases and the thermal energy of the fluid may be stored in storage and can be used for different industrial processes and applications.

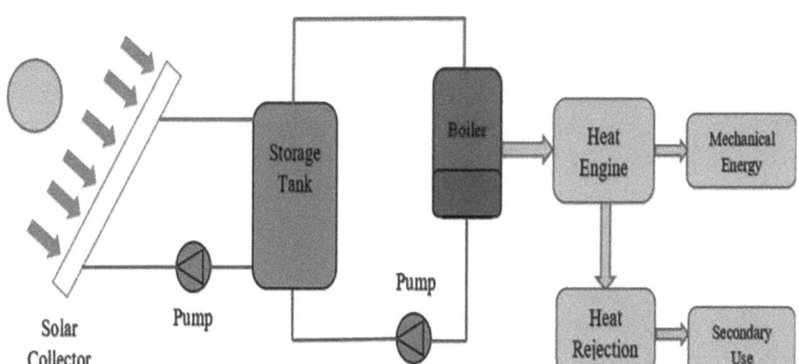

FIGURE 2.9 Schematics of solar thermal power conversion (Kumar et al., 2019).

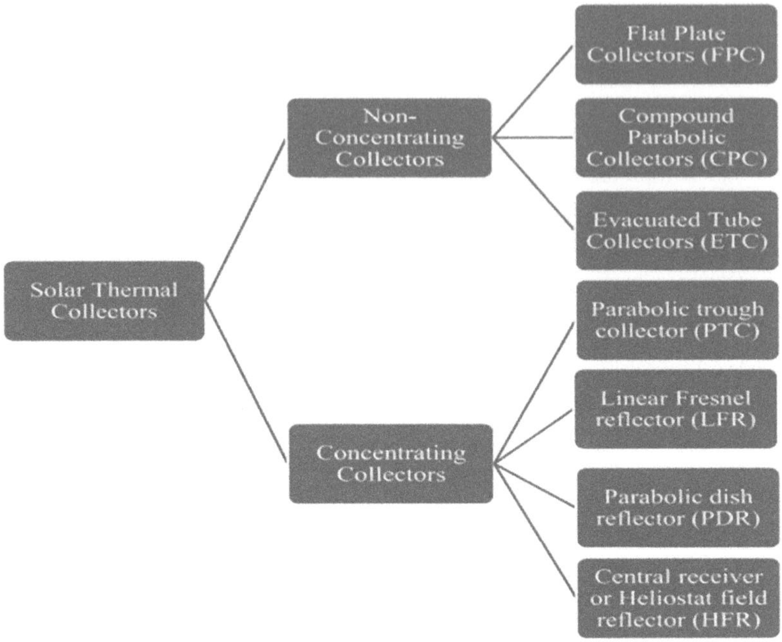

FIGURE 2.10 Types of solar thermal collectors (Kumar et al., 2019).

Nowadays, solar photovoltaic thermal is very popular technology where the solar thermal collector is integrated with PV panel and it gives more power output. There are several types of solar collectors and listed below that is shown in Fig. 2.10.

For a flat plate collector (FPC), the simplest case, the basic equations of thermal energy conversion can be written as follows. The heat gain, Q_u under the steady state condition by the collector (FPC) (Kumar et al., 2019):

$$Q_u = A_c\left[\tau\alpha G_t - U_L\left(T_p - T_a\right)\right] = mC_p(T_o - T_i) \tag{2.57}$$

where A_c is the total aperture area of the collector, α is the absorptivity, τ is the transmissivity, G_t is the total radiation on the collector aperture $\left(\mathrm{W}/\,m^2\right)$, U_L is the solar collector heat transfer loss coefficient $\left(\mathrm{W}\,/m^{2\circ}\mathrm{C}\right)$, T_p is the average temperature of the absorbing surface (°C), T_a is the ambient temperature (°C), C_p is the specific heat of the fluid at constant pressure (J/kg K), T_i is the temperature of fluid entering the collector (°C), and T_o is the temperature of fluid leaving the collector (°C).

In Eq. (2.57), T_p is replaced by T_i and a correction factor, F_R is incorporated as follows (Kumar et al., 2019):

$$Q_u = A_c F_R[\tau\alpha G_t - U_L(T_i - T_a)] \tag{2.58}$$

The factor, F_R is called the heat removal factor and it is found by the given equation:

$$F_R = \frac{mC_p}{A_c U_L}\left(1 - \frac{U_L A_c F'}{mC_p}\right) \tag{2.59}$$

The new factor, F' in the above equation can be calculated as follows:

$$F' = \frac{U_o}{U_L} \tag{2.60}$$

Now, the efficiency of the collector can be calculated as heat gain divided by incident solar energy shown in the following equation (Kumar et al., 2019):

$$\eta = \left[\tau\alpha - \frac{U_L(T_i - T_a)}{G_t}\right] \tag{2.61}$$

2.8.3 Installations of PV modules or thermal collectors

It is extremely essential that the PV modules or thermal collectors should be correctly placed to provide maximum sunshine and to prevent partial shading. Incorrect PV modules or collector orientation may lead to reduced power production. To capture the maximal intensity of the radiation module, the incident sunlight must be normal (Hafez, Soliman, El-Metwally, & Ismail, 2017). The use of a solar tracking device for continually following the sun's route across the sky is the finest way to capture the maximum solar power. But the trackers of the module are complicated, costly devices that need maintenance and repair, thus not necessarily appropriate for general usage (Nann, 1990; Sungur, 2009). Thus, fixed type PV installations with properly designed tilt angle remain popular in the PV sector (Mehleri, Zervas, Sarimveis, Palyvos, & Markatos, 2010).

2.8.4 Fixed type PV installations with an optimum tilt angle

The optimal performance of a PV panel relies on how much solar radiation incidence occurs. A panel must thus be sloping at such an angle that the top surface is intercepted by the most sunlight. Optimal tilt determination relies on installation methods, land topography, and climate conditions (Babatunde, Abbasoglu, & Senol, 2018). PV modules are typically oriented at regional latitude. However, it is a prerequisite to allow the maximum irradiation of the PV system. The tilt and the azimuth angle of tilt are two major angles used to describe the orientation of the array (Bakirci, 2012). The optimal tilting angle for a certain installation may be achieved by capturing the maximum radiation intensity in the given surrounding environment. For optimized system tilt angles, which incorporates the sun's effect of on the solar collectors and panels are positioned from the horizon to the equator to get the most benefit from the sun. Location of latitude and climatic conditions influence the tracking angles.

2.9 Solar tracking technology

There has been a huge increase in the use of solar energy due to its vast availability and environmental friendliness. But although efficiency is still an issue in many cases, that is especially true in many applications. Solar tracking devices are used to help minimize this issue.

The first solar tracker was created by C. Finster in 1962, who developed a mechanical technique for solar trackers. Though the Finster solar tracker was insignificant in terms of increasing overall PV system efficiency, years of research and testing allowed the development of better tracking systems, including concentrators and nonconcentrators, resulting in a distinct increase in the PV system's overall efficiency. Solar cells have improved and, as a result, a solar tracking system has come into use instead of having complexity, greater costs, and continuous maintenance.

2.9.1 Classification of solar tracking

It can be categorized according to the control system used, the drivers utilized, and the extent of movement freedom shown in the system, which is described in the diagram. Fig. 2.11 shows the types of solar tracking technologies (Awasthi et al., 2020).

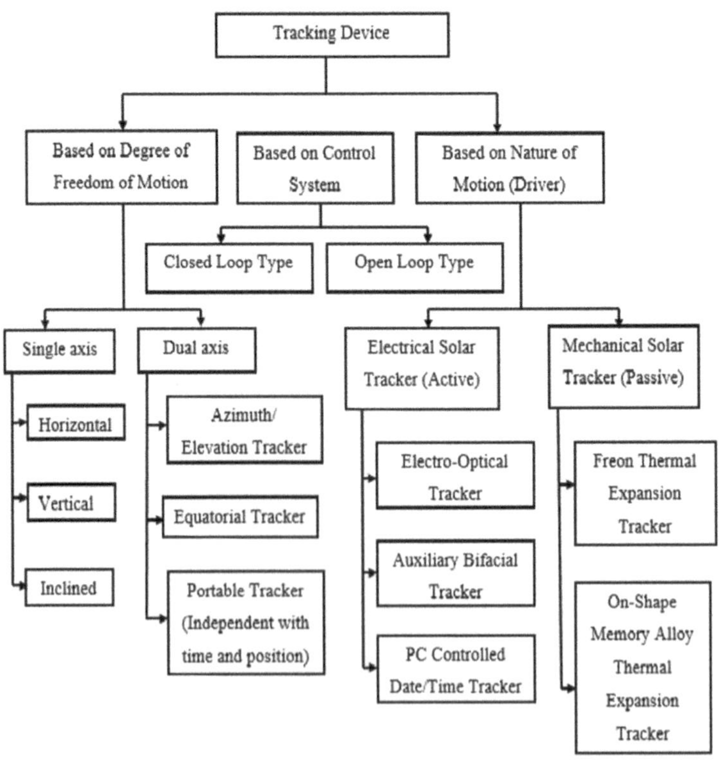

FIGURE 2.11 Types of solar tracking technologies (Awasthi et al., 2020).

Tracking system can be categorized as follows (Awasthi et al., 2020) based on the control system used:

- Closed loop tracking system
- Open loop tracking system
- Passive solar trackers
- Single axis tracking (SAT)
- Double axis tracking (DAT)

2.9.2 Closed loop tracking system

It is based on feedback control system concepts. Sensors are employed in this system to detect the location of the sun, which is then sent back to the system. The comparator or microcontroller may then identify the error and send the necessary actuation signal to the motors to fix it. Fig. 2.12 shows a schematic diagram of closed loop tracking system (Seme, Štumberger, Hadžiselimović, & Sredenšek, 2020).

2.9.3 Open loop tracking system

This system employs a controller that utilizes data received from sensors and operating algorithms to provide the driving signal to the motor, depending only on current data inputs. There is no component of observation and evaluation when it comes to the output of the data to arrive at the specified result. There is no correction procedure involved. As a result, it is more cost-effective and easier to implement; therefore, it is used in most instances. Fig. 2.13 shows a schematic diagram of open loop tracking system (Seme et al., 2020).

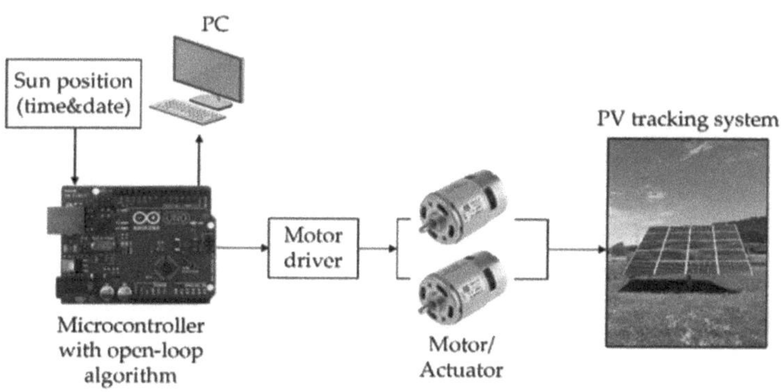

FIGURE 2.12 Schematic diagram of closed loop tracking system (Seme et al., 2020).

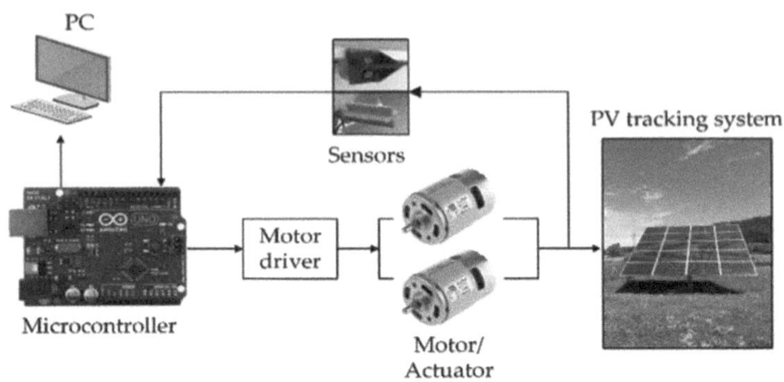

FIGURE 2.13 Schematic diagram of open loop tracking system (Seme et al., 2020).

2.9.4 Passive solar trackers

There are no mechanical devices to direct the panel toward the radiation of the sun. Instead, it utilizes pressurized gas fluid or shape memory alloys as actuators at low boiling points. The solar panel provides angular mobility by inducing heat expansion into expansive gases or memory alloys in the form. It is less complicated and efficient, but at low temperatures, it doesn't achieve great efficiency. Fig. 2.14 shows a schematic of solar tracking: (A) Passive (B) Active (Seme et al., 2020).

2.9.5 Single axis tracking

Only a single axis of rotation of the panel is associated with the tracking system. The axis of rotation of the solar tracking system may be vertical, horizontal, or inclined. These are cheaper and less complicated, but less effective than the solar tracking system with DAT. PV single-axis tracking systems are classified into three categories. These include a single-axis

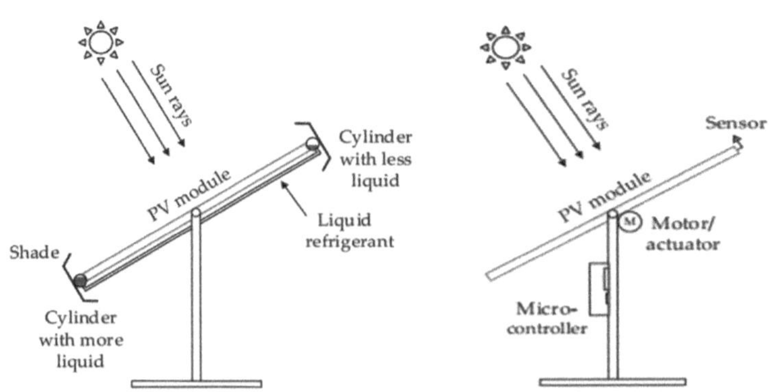

FIGURE 2.14 Schematic of solar tracking: (A) Passive (B) Active (Seme et al., 2020).

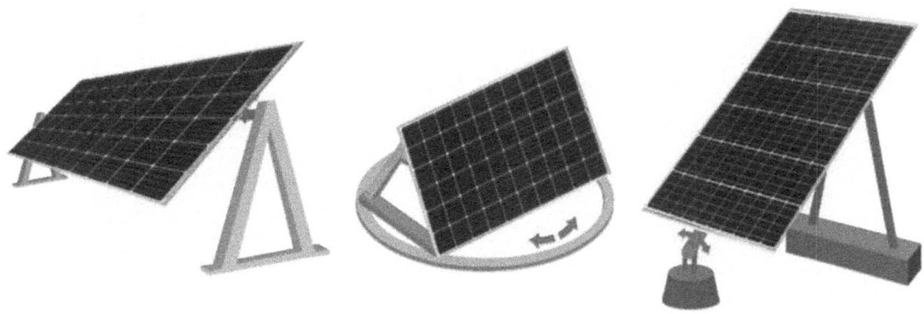

FIGURE 2.15 Schematics of single-axis PV tracking systems: (A) HSAT, (B) VSAT, and (C) TSAT (Seme et al., 2020).

horizontal system (HSAT), a single-axis vertical system (VSAT), and a tilted single-axis system for tracking (TSAT). The HSAT rotating axis is horizontal to the floor. The axis of rotation of the VSAT is vertical to the floor. These systems rotate over the day from east to west. The TSAT's rotating axis is tilted to the ground. Fig. 2.15 shows a schematic of single-axis PV tracking systems: (A) HSAT, (B) VSAT, and (C) TSAT (Seme et al., 2020).

2.9.6 Double axis tracking

The sun is tracked on two separate axes, utilizing two pivot points, so the panel may be turned around a full 360° in this method. In this kind of solar tracker system, the horizontal and vertical axes are typically both present. Although it needs a sophisticated control system, it is better than SAT. The PV dual-axis systems are characterized by the azimuth of their principal axes with relation to their ground into two different categories. The azimuth-altitude dual axis tracking (AADAT) system and the tip-tilt dual axis tracking (TTDAT) system are two typical types. The TTDAT system has its primary axis horizontally to the ground, whereas the secondary axis is perpendicular to the primary axis. The primary axis of an AADAT is perpendicular to the ground, whereas the secondary axis is perpendicular to the primary axis. Fig. 2.16 shows dual axis PV tracking systems: (A) TTDAT and (B) AADAT (Seme et al., 2020).

In the case of the single-axis tracking system, it can track the sun's daily path, but it can't keep track of the whole year's movement. This is one of the biggest demerits of the SAT. But in typical tracking systems, the initial cost of the actuators and mechanisms and the additional complexity are concerning. Several researchers have thus taken a novel method to sun tracking named discrete solar tracking. Essentially, the main idea is to lower the starting cost of two motors, as well as their excessive power use, by measuring each day.

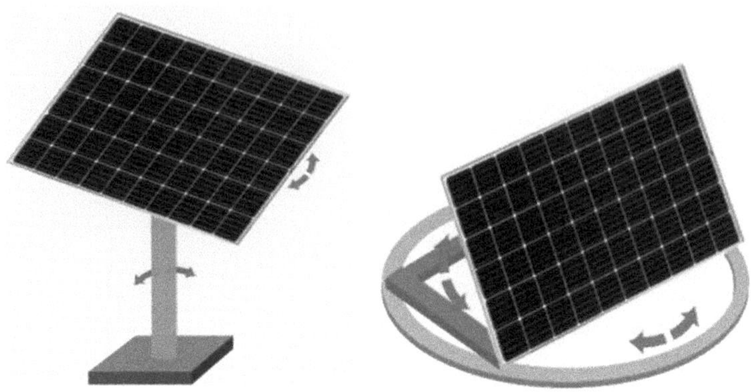

(a) (b)

FIGURE 2.16 Dual axis PV tracking systems: (A) TTDAT and (B) AADAT (Seme et al., 2020).

2.10 Worked out problems

Example 2.1. Find out the extraterrestrial solar irradiance on the horizontal surface for latitude, $\phi = 3.139\ N$, Kuala Lumpur, Malaysia, at 11:00 a.m. on February 25.

Solution. For March 19, $n = 56$ and at 11.00 a.m., the hour angle, $\omega = -15$ (as per hour 15°).

So, the declination angle can be calculated from the equation:

$$\begin{aligned}\delta &= 23.45 \sin\left[2\pi\ \frac{(n\ + 284)}{365}\right] \\ &= 23.45 \sin\left[2\pi\ \frac{(56\ + 284)}{365}\right] \\ &= -9.78\end{aligned}$$

Now, from the relationship of the zenith angle:

$$\begin{aligned}\cos(\theta_z) &= \cos(\phi)\cos(\delta)\cos(\omega) + \sin(\phi)\sin(\delta) \\ &= \cos(3.139)\cos(-9.78)\cos(-15) + \sin(3.139)\sin(-9.78) \\ &= 0.94\end{aligned}$$

The well-known solar constant, $G_{SC} = 1367\ \mathrm{W/m^2}$.

The extraterrestrial solar irradiance on the horizontal surface, from the equation:

$$G_o = G_{SC}\left(1 + 0.033\cos\left(\frac{360\,n}{365}\right)\right)\cos\theta_z$$

$$= 1367 \times \left(1 + 0.033\cos\left(\frac{360 \times 56}{365}\right)\right) \times 0.94$$

$$= 1309.16\,\mathrm{W/m^2}$$

Example 2.2. Find out the solar radiation on the horizontal surface if there is no atmosphere at latitude, $\phi = 3.139\,N$, Kuala Lumpur, Malaysia, between 11:00 a.m. and 1:00 p.m. on February 25.

Solution. If there is no presence of an atmosphere, the radiation belongs to extraterrestrial solar radiation.

From the previous example, it is seen that the declination angle, $\delta = -\,9.78°$ and for February 25, n = 56. For 11:00 a.m., $\omega = -15°$ and for 1:00 p.m., $\omega = 15°$.

Now, using the equation:

$$I_o = \frac{24 \times 3600\, G_{SC}}{\pi}\left(1 + 0.033\cos\left(\frac{360\,n}{365}\right)\right)$$
$$\times\left(\cos(\phi)\cos(\delta)(\sin(\omega_2) - \sin(\omega_1)) + \frac{\pi \times (\omega_2 - \omega_1)}{180}\sin(\phi)\sin(\delta)\right)$$

Now, inserting the known values:

$$I_o = \frac{24 \times 3600 \times 1367}{\pi}\left(1 + 0.033\cos\left(\frac{360 \times 56}{365}\right)\right)$$
$$\times\left(\cos(3.139)\cos(-9.78)(\sin(15) - \sin(-15))\right.$$
$$\left.+\frac{\pi \times (15 - (-15))}{180}\sin(3.139)\sin(-9.78)\right)$$
$$= 38.096 \times 10^6\,\mathrm{J/m^2}$$

Example 2.3. What is the extraterrestrial daily solar irradiation on the horizontal surface on February 25 in Kuala Lumpur, Malaysia (latitude, $\phi = 3.139\,N$)?

Solution. From the previous example, the declination angle, $\delta = -9.78°$ and for February 25, $n = 56$.

Now, the sunset hour angle can be determined by the relationship as:

$$\cos(\omega_s) = -\tan(\phi)\tan(\delta) = -\tan(3.139)\tan(-9.78) = 9.45 \times 10^{-3}$$

The sunset hour angle $\omega_s = 89.46°$

$$H_o = \frac{24 \times 3600\, G_{SC}}{\pi}\left(1 + 0.033\cos\left(\frac{360\, n}{365}\right)\right) \times \left(\cos(\phi)\cos(\delta)\sin(\omega_s) + \frac{\pi\omega_s}{180}\sin(\phi)\sin(\delta)\right)$$

$$= \frac{24 \times 3600 \times 1367}{\pi}\left(1 + 0.033\cos\left(\frac{360 \times 56}{365}\right)\right) \times \left(\cos(3.139)\cos(-9.78)\sin(89.46) + \frac{\pi \times 89.46}{180}\sin(3.139)\sin(-9.78)\right)$$

$$= 37.13 \times 10^6\ \mathrm{J/m^2}$$

References

Angstrom, A. (1924). Solar and terrestrial radiation. Report to the international commission for solar research on actinometric investigations of solar and atmospheric radiation. *Quarterly Journal of the Royal Meteorological Society, 50*(210), 121–126. https://doi.org/10.1002/qj.49705021008

Awasthi, A., Shukla, A. K., Murali Manohar, S. R., Dondariya, C., Shukla, K. N., Porwal, D., et al. (2020). Review on sun tracking technology in solar PV system. *Energy Report, 6*, 392–405. https://doi.org/10.1016/j.egyr.2020.02.004

Babatunde, A. A., Abbasoglu, S., & Senol, M. (2018). Analysis of the impact of dust, tilt angle and orientation on performance of PV Plants. *Renewable and Sustainable Energy Reviews, 90*, 1017–1026. https://doi.org/10.1016/j.rser.2018.03.102

Bakirci, K. (2012). General models for optimum tilt angles of solar panels: Turkey case study. *Renewable and Sustainable Energy Reviews, 16*(8), 6149–6159. https://doi.org/10.1016/j.rser.2012.07.009

Batayneh, W., Bataineh, A., Soliman, I., & Hafees, S. A. (2019). Investigation of a single-axis discrete solar tracking system for reduced actuations and maximum energy collection. *Automation in Construction, 98*, 102–109. https://doi.org/10.1016/j.autcon.2018.11.011

Collares-Pereira, M., & Rabl, A. (1979). The average distribution of solar radiation-correlations between diffuse and hemispherical and between daily and hourly insolation values. *Solar Energy, 22*(2), 155–164. https://doi.org/10.1016/0038-092X(79)90100-2

Dubey, S., & Tay, A. A. O. (2013). Testing of two different types of photovoltaic–thermal (PVT) modules with heat flow pattern under tropical climatic conditions. *Energy for Sustainable Development, 17*(1), 1–12. https://doi.org/10.1016/j.esd.2012.09.001

Dubey, S., & Tiwari, G. N. (2008). Thermal modeling of a combined system of photovoltaic thermal (PV/T) solar water heater. *Solar Energy, 82*(7), 602–612. https://doi.org/10.1016/j.solener.2008.02.005

Erbs, D. G., Klein, S. A., & Duffie, J. A. (1982). Estimation of the diffuse radiation fraction for hourly, daily and monthly-average global radiation. *Solar Energy, 28*(4), 293–302. https://doi.org/10.1016/0038-092X(82)90302-4

Fan, S., Yang, W., & Hu, Y. (2018). Adjustment and control on the fundamental characteristics of a piezoelectric PN junction by mechanical-loading. *Nanomaterials and Energy, 52*, 416–421. https://doi.org/10.1016/j.nanoen.2018.08.017

Hafez, A. Z., Soliman, A., El-Metwally, K. A., & Ismail, I. M. (2017). Tilt and azimuth angles in solar energy applications – a review. *Renewable and Sustainable Energy Reviews, 77*, 147–168. https://doi.org/10.1016/j.rser.2017.03.131

Hawlader, M. N. A. (1984). Diffuse, global and extra-terrestrial solar radiation for Singapore. *International Journal of Ambient Energy, 5*(1), 31–38. https://doi.org/10.1080/01430750.1984.9675406

Kumar, L., Hasanuzzaman, M., & Rahim, N. A. (2019). Global advancement of solar thermal energy technologies for industrial process heat and its future prospects: A review. *Energy Conversion and Management, 195*, 885–908. https://doi.org/10.1016/j.enconman.2019.05.081

Liu, B. Y. H., & Jordan, R. C. (1960). The interrelationship and characteristic distribution of direct, diffuse and total solar radiation. *Solar Energy, 4*(3), 1–19. https://doi.org/10.1016/0038-092X(60)90062-1

Mamun, M. A. A., Islam, M. M., Hasanuzzaman, M., & Selvaraj, J. (2021). Effect of tilt angle on the performance and electrical parameters of a PV module: Comparative indoor and outdoor experimental investigation. *Energy and Built Environment.* https://doi.org/10.1016/j.enbenv.2021.02.001

Mehleri, E. D., Zervas, P. L., Sarimveis, H., Palyvos, J. A., & Markatos, N. C. (2010). Determination of the optimal tilt angle and orientation for solar photovoltaic arrays. *Renewable Energy, 35*(11), 2468–2475. https://doi.org/10.1016/j.renene.2010.03.006

Nahar, A., Hasanuzzaman, M., & Rahim, N. A. (2017a). Numerical and experimental investigation on the performance of a photovoltaic thermal collector with parallel plate flow channel under different operating conditions in Malaysia. *Solar Energy, 144*, 517–528. https://doi.org/10.1016/j.solener.2017.01.041

Nahar, A., Hasanuzzaman, M., & Rahim, N. A. (2017b). A three-dimensional comprehensive numerical investigation of different operating parameters on the performance of a photovoltaic thermal system with pancake collector. *Journal of Solar Energy Engineering, 139*(3). https://doi.org/10.1115/1.4035818

Nahar, A., Hasanuzzaman, M., Rahim, N. A., & Parvin, S. (2019). Numerical investigation on the effect of different parameters in enhancing heat transfer performance of photovoltaic thermal systems. *Renewable Energy, 132*, 284–295. https://doi.org/10.1016/j.renene.2018.08.008

Nann, S. (1990). Potentials for tracking photovoltaic systems and V-troughs in moderate climates. *Solar Energy, 45*(6), 385–393. https://doi.org/10.1016/0038-092X(90)90160-E

Orgill, J. F., & Hollands, K. G. T. (1977). Correlation equation for hourly diffuse radiation on a horizontal surface. *Solar Energy, 19*(4), 357–359. https://doi.org/10.1016/0038-092X(77)90006-8

Rahman, M. M., Hasanuzzaman, M., & Rahim, N. A. (2015). Effects of various parameters on PV-module power and efficiency. *Energy Conversion and Management, 103*(0), 348–358. https://doi.org/10.1016/j.enconman.2015.06.067

Rahman, M. M., Hasanuzzaman, M., & Rahim, N. A. (2017). Effects of operational conditions on the energy efficiency of photovoltaic modules operating in Malaysia. *Journal of Cleaner Production, 143*, 912–924. https://doi.org/10.1016/j.jclepro.2016.12.029

Robaa, S. M. (2009). Validation of the existing models for estimating global solar radiation over Egypt. *Energy Conversion and Management, 50*(1), 184–193. https://doi.org/10.1016/j.enconman.2008.07.005

Seme, S., Štumberger, B., Hadžiselimović, M., & Sredenšek, K. (2020). Solar photovoltaic tracking systems for electricity generation: A review. *Energies, 13*(16), 4224. Retrieved from: https://www.mdpi.com/1996-1073/13/16/4224.

Stanciu, C., & Stanciu, D. (2014). Optimum tilt angle for flat plate collectors all over the world – a declination dependence formula and comparisons of three solar radiation models. *Energy Conversion and Management, 81*, 133–143. https://doi.org/10.1016/j.enconman.2014.02.016

Stapleton, A. (2017). How solar cells turn sunlight into electricity. *Cosmos magazine, 2017*. https://cosmosmagazine.com/technology/how-solar-cells-turn-sunlight-into-electricity/ (Accessed 20 August 2021).

Sungur, C. (2009). Multi-axes sun-tracking system with PLC control for photovoltaic panels in Turkey. *Renewable Energy, 34*(4), 1119–1125. https://doi.org/10.1016/j.renene.2008.06.020

Tiwari, A., Sodha, M. S., Chandra, A., & Joshi, J. C. (2006). Performance evaluation of photovoltaic thermal solar air collector for composite climate of India. *Solar Energy Materials and Solar Cells, 90*(2), 175–189. https://doi.org/10.1016/j.solmat.2005.03.002

Yadav, A. K., & Chandel, S. S. (2013). Tilt angle optimization to maximize incident solar radiation: A review. *Renewable and Sustainable Energy Reviews, 23*, 503–513. https://doi.org/10.1016/j.rser.2013.02.027

Chapter 3

Heat exchanger for solar thermal energy

Rafsun Hossain Mridha[1] and Md Hasanuzzaman[2]

[1]*Department of Mechanical Engineering, Bangladesh University of Engineering and Technology, Dhaka, Bangladesh;* [2]*Higher Institution Centre of Excellence (HICoE), UM Power Energy Dedicated Advanced Centre (UMPEDAC), University of Malaya, Kuala Lumpur, Malaysia*

Chapter outline

Technologies for Solar Thermal Energy. https://doi.org/10.1016/B978-0-12-823959-9.00009-X

3.1 Basic concept of heat exchanger

3.1.1 Concept and definition

Heat exchanger is an important equipment which is widely used in the most of the energy-intensive industries. Enhancing heat transfer, discovering a new type of heat exchanger, design and control of new heat exchangers are always a matter of concern for engineers and researchers. Heat exchangers play a vital role in using solar energy at the time of storing and releasing heat. In this chapter, solar thermal energy is linked up with different types of heat exchanger where a detailed discussion has been made on their basic concepts, design process, and performance analysis.

Heat exchanger is a thermomechanical equipment which acts as a medium while exchanging heat between two or more separated fluids. The major performing criteria of a heat exchanger is to transfer heat between the fluids. In order to transfer heat, there must be temperature difference between the fluids. In short, heat exchanger is a thermofluid system where heat is transferred among the fluids due to temperature difference.

3.1.2 Classification of heat exchangers

Heat exchangers can be classified based on the following categories:

- Heat transfer processes
- Heat exchanger physical construction
- Heat exchanger flow arrangement
- Heat transfer mechanism
- Heat exchanger compactness

Classification based on transfer processes

Direct contact: In such kind of heat exchangers, direct contact of fluid occurs where one phase can dissolve in the other. Good example of the direct contact is cooling tower that is shown in Fig. 3.1.

Indirect contact: In this type of heat exchangers, there occurs no contact between the fluids directly. Good example of indirect contact heat exchanger is shell and tube heat exchanger that is shown in Fig. 3.2.

Classification based on physical construction

Tubular heat exchanger: Fluids are contained in several tubes and indirect heat exchange occurs among the fluids. The tubular heat exchanger can be classify as following:

- Double pipe heat exchanger
- Shell and tube heat exchanger
- Helically coiled heat exchanger (Fig. 3.3)

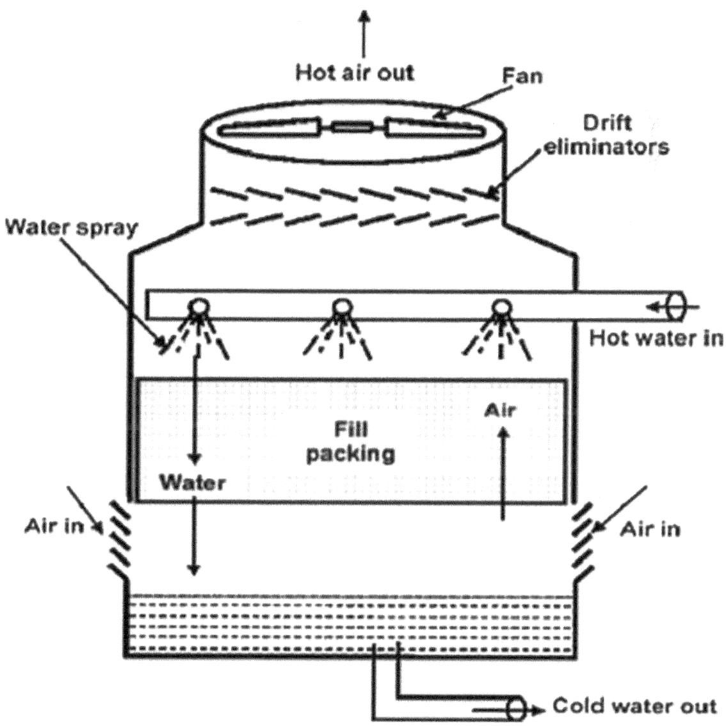

FIGURE 3.1 Schematic diagram of a cooling tower and working principle (Tyagi, Pandey, Pant, & Tyagi, 2012).

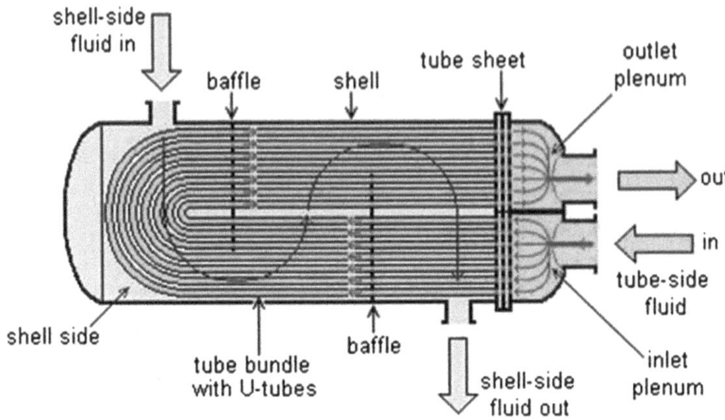

FIGURE 3.2 Schematic of a shell and tube heat exchanger with vertical baffles (Ali, Ul-Hamid, Alhems, & Saeed, 2020).

FIGURE 3.3 Schematic diagram of a helically coiled heat exchanger (Joshi & Tiwari, 2018).

Plate heat exchanger: Fluids are passed through several plates. The plates can be of two types as follows:

- Gasket plate heat exchanger
- Spiral plate heat exchanger (Fig. 3.4)

Classification based on flow arrangement

Parallel flow: The flow direction of both the fluids is same.

Counter flow: Both the fluids flow in the direction opposite to each other.

Cross flow: The flow direction of the fluids is perpendicular to each other (Fig. 3.5).

Compact heat exchanger

A compact heat exchanger is an equipment typically used for efficient heat transfer from one fluid to another. It is characterized by high heat transfer area

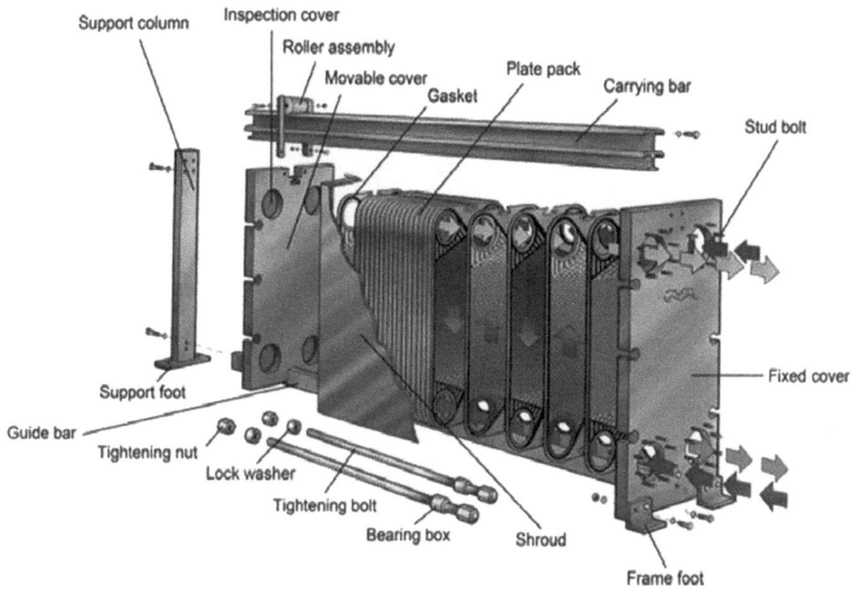

FIGURE 3.4 Schematic diagram of a plate heat exchanger (Pandya, Shah, Molana, & Tiwari, 2020).

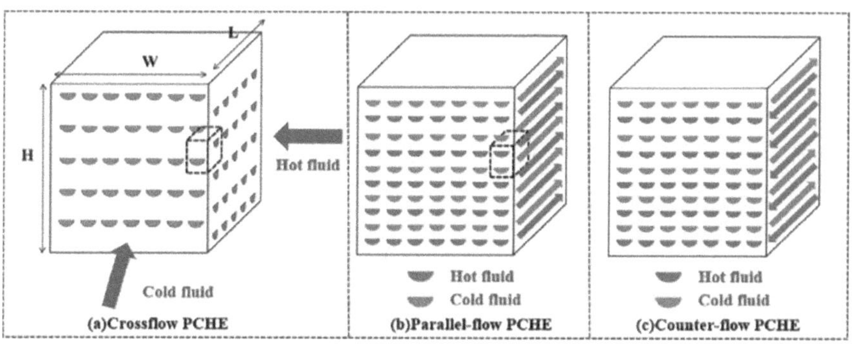

FIGURE 3.5 Schematic diagrams printed circuit heat exchanger (A) cross flow; (B) parallel flow; (C) counter flow (Huang et al., 2019; Kim, Baik, Jeon, Jeon, & Byon, 2017).

to volume ratio, large heat transfer coefficients, small flow passages, and laminar flow (Fig. 3.6).

Heat exchangers coupled with solar thermal technology

- Heat exchanger using phase change materials (PCM) along with prilling or granulation concept
- Earth water heat exchanger

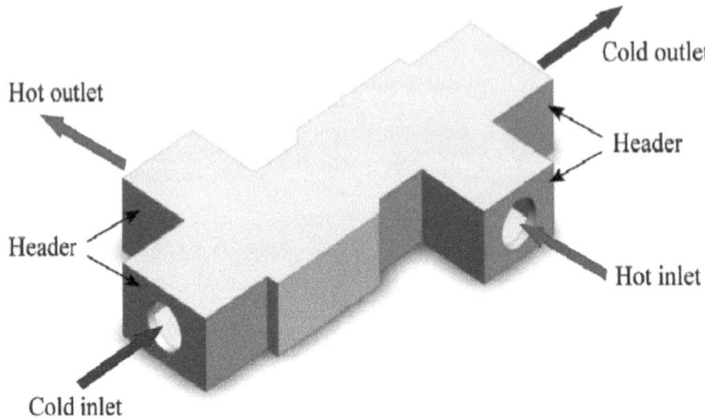

FIGURE 3.6 Schematic diagram of a compact heat exchanger (Sarmiento et al., 2020).

- Multi tank modular heat storage
- Earth air heat exchanger (EAHE)
- Single slope solar still integrated with a heat exchanger
- Vertical straight and helical coiled pipe heat transfer
- Solar pond heat transfer
- Thermoplate conical heat transfer
- Roll-bond heat exchanger
- Multi-pass shell and tube heat exchanger
- Submerged heat exchanger for solar absorption
- Fluidized bed (FB) heat exchanger

3.1.3 Common heat exchanger technologies for solar thermal energy

There are some heat exchanger technologies which are frequently used in several fields of solar energy. In this segment of the chapter, the most common heat exchanger has been discussed.

Salt-steam heat exchanger system

It consists of a boiler, reheater, and superheater. Hot salt enters at 565°C and leaves at 290°C. Both the boiler and superheater are kept at a pressure of 170 bar. On the other hand, the reheater is kept at a pressure of 30 bar. Metal halide salts are used as PCM, as the enthalpy of fusion and phase transition temperature of these salts are significantly higher. Moreover, prilling or granulation concept is closely associated with this type of heat exchangers (Figs. 3.7 and 3.8).

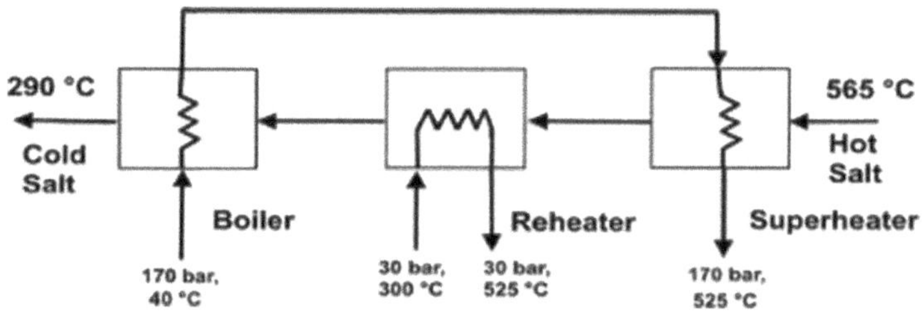

FIGURE 3.7 Schematic of salt-steam heat exchanger system (Badenhorst, 2016).

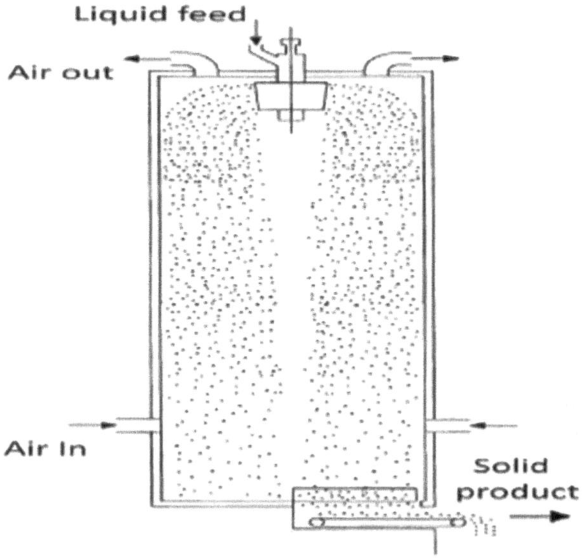

FIGURE 3.8 Prilling process (Badenhorst, 2016).

Integrated photovoltaic thermal solar system with earth water heat exchanger

Integrated photovoltaic thermal solar (IPVTS) is generally coupled with EWHE system which experiences climatic conditions which are semidry, dry, and tolerable. The following diagram illustrates IPVTS integrated with the heat exchanger. There are two different arrangements. Both of the setups use monocrystalline silicon (Si) PV module which are set on a frame of steel (mild). The first scheme is called IPVTS which is a surrounding constructed on rear side of the PV module. The fluid flows uniformly through a pipe in aforementioned system (Fig. 3.9).

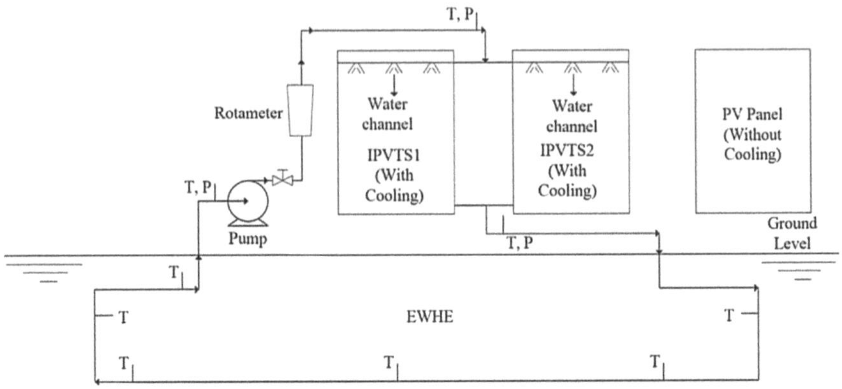

FIGURE 3.9 Schematic diagram of IPVTS coupled with EWHE (Jakhar, Soni, & Gakkhar, 2017).

The second scheme is known as PV and it performs without any external cooling source. The enclosure of the IPVTS is made of sheets of galvanized iron (GI). These sheets are set on the rear side of photovoltaic module which uses sealant of silicon that creates a channel that is broad in size. Holes, made in the pipe which carries the fluid, are uniform that makes fluid available in the channel. The enclosure boundaries are sealed so that they can sustain pressure difference. Additionally, for the outlet of fluid, a pipe is provided separately at the bottom.

Multitank modular heat storage for solar thermal systems

A heat storage system that is sensible consists of a bed made of pebble or rocks which is typically packed. Heat transfer fluid (HTF) flows over this packed bed. If the HTF is hotter than the bed material, then it gets heated. After that, the heat is recovered cold HTF over packed bed. A cubic-shaped block (made of concrete) with appropriate heat transfer properties is generally used as the bed material. A table is presented below that shows the dimensions of concrete storage block (CSB) and insulated storage tank. The CSB is located within storage tank that is insulated (Table 3.1).

TABLE 3.1 Dimensions of concrete storage block and insulated storage tank (Singh, Sharma, & Bhattacharya, 2020).

Item	Dimensions
Concrete storage block (CSB)	50 mm × 50 mm × 50 mm
Insulated storage tank	100 mm × 100 mm × 100 mm

In charging and discharging processes, the modular system has better exergetic efficiency. In order to design a solar thermal energy storage, particularly with a view to eliminating effects of temperature fluctuations, the multitank storage system can undoubtedly increase effectiveness by eradicating the losses. The packed bed notion is more beneficial for power producing thermal units, for instance, the central receiver tower technology.

Earth air heat exchanger

The schematic sketches illustrate two configurations of a system that includes building integrated photovoltaic thermal (BIPVT) compound EAHE that is shown in Fig. 3.10. Both of the configurations consist of two mode: heating and cooling. In mode of heating, the first configuration (configuration A), the outdoor air that is cold enters the HE. Afterward, it is heated with the help of soil. Then, the air enters the collector of BIPVT. After that it is again heated by the thermal energy of the photovoltaic modules. It cools the photovoltaic modules, as a result, electrical efficiency increases. In the cooling mode, the hot air of outdoor transfers heat to the surrounding soil as well as gets precooled. Besides, the exhaust air of building passes through duct which is under the PV modules. Thus, it decreases temperature and increases efficiency.

In the heating mode of the second configuration B, the air of outdoor enters the collector of BIPVT. After that it flows through the heat exchanger system. The air which passes the collector is cooler in A than B. In addition, in configuration B, the modules are cooled in a better manner. On the other hand, both soil temperature and the difference of temperature between entering air in heat exchanger system are less in configuration B. It indicates low efficiency of heat exchanger system in B configuration. Moreover, cooling mode is same for both the configurations. Both configurations produce electricity. Some part of it is absorbed by the fans to cycle air using BIPVT and EAHE systems. Finally, rest of the generated electricity can cover electricity demand of the building (Fig. 3.11).

Single slope solar still integrated with helically coiled heat exchanger

Fig. 3.12 illustrates the schematic diagram of a conventional active solar distillation system coupled with a heat exchanger which is helically coiled. The heat exchanger is made of copper. N identical thermal collectors are fixed up in series and basin of the still is connected with outlet.

The still has:

- a basin which is made of fiber reinforced plastic
- a glass that is floating and
- a mild steel used in case of supporting purpose.

The near and lower parts are made black to ensure maximum intensity being consumed and transmitted to water which is in the solar still basin.

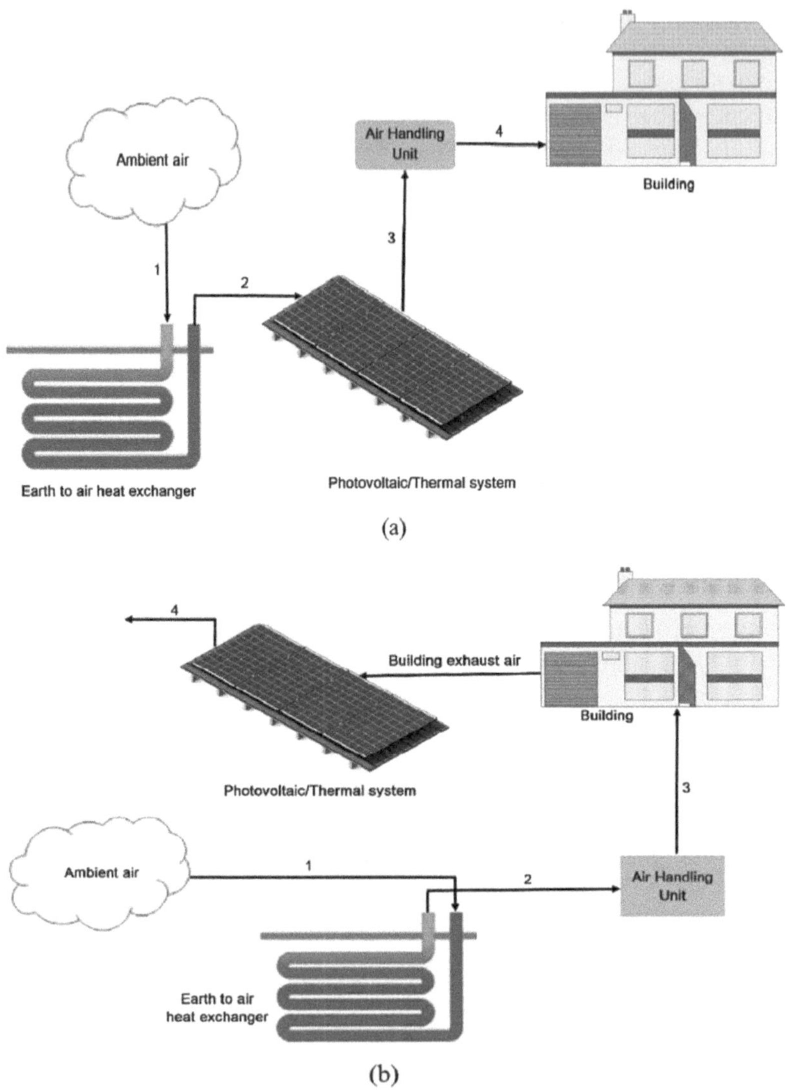

FIGURE 3.10 Working principle: (A) heating mode and (B) cooling mode of configuration (Afrand, Shahsavar, Sardari, Sopian, & Salehipour, 2019).

The HTF is water which flows through the heat exchanger. Heat energy is transferred to the impure water which is in basin of the still. The system is designed in such a way that the water that is hot flows inside the tubes underneath the upper layer. The collectors are connected in series connection, for example, outlet of the previous collector is inlet for the collector which is next. This assists to gain higher temperature of water. Therefore, N identical

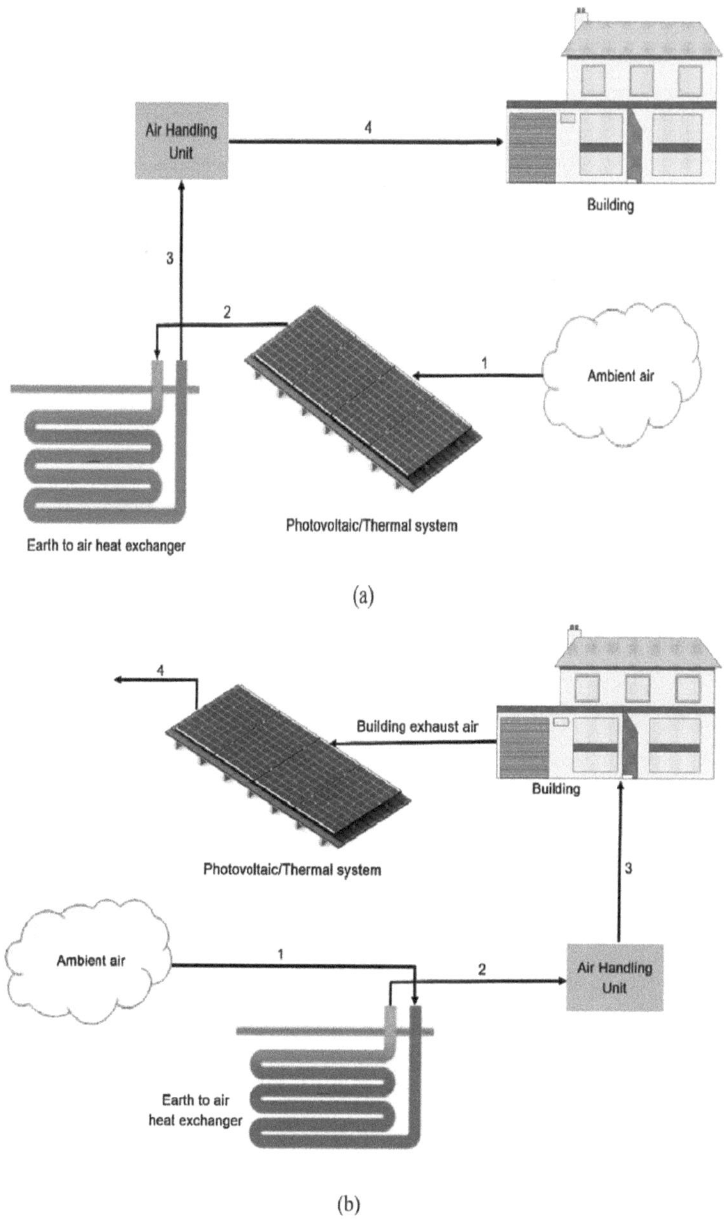

FIGURE 3.11 Working principle: (A) heating mode and (B) cooling mode of configuration (Afrand et al., 2019).

collectors are kept in a closed loop. The loop completes the circuit of an active solar distillation system. A direct current (DC) motor is attached to the system of distillation to allow water for overcoming pressure drop. Moreover, it helps

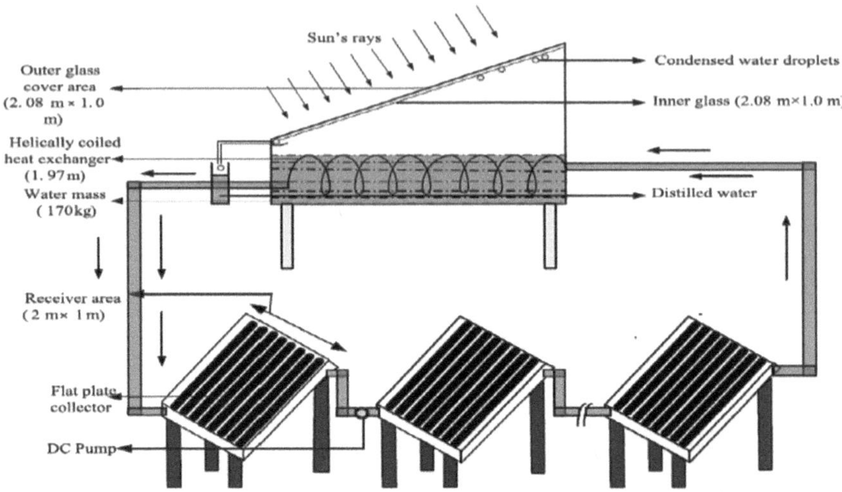

FIGURE 3.12 Schematic diagram of a conventional active solar distillation system coupled with a helically coiled heat exchanger (Joshi & Tiwari, 2018).

water to circulate having high velocity compared to circulation naturally. Generally, electricity is provided to the DC motor.

Vertical straight and helical coiled pipe heat exchanger

A two cylindrical storage tank that is co-axial which contain the charging and discharging pipe is typically used. For decreasing heat loss in the inner tank, insulator is filled between the two tanks which is made of microfiber. The vertical pipe is composed of steel (stainless). The fluid is generally pumped by a pump (Fig. 3.13).

The pump is a positive displacement one and attached to a motor of variable speed in case of charging. Opposite in case of discharging. Various flow rates are used to notice the heat transfer rate from the fluid to the storage. Oil is used as the fluid of variable speed. It is extracted from a reservoir which is charged continuously by a heater.

Solar pond heat exchanger

The heat supply system consists of two heat exchangers kept at the bottom and lateral side of the solar pond. The system is composed of four solar thermal collectors. The orientation of the solar collectors is directly toward the equator that faces south. The tilt angle is generally identical to latitude of the Solvay facilities that is 41°S (Fig. 3.14).

Heat exchangers are connected to the solar collectors by virtue of PVC pipes. All the external pipes are thermally insulated by neoprene. The heat is

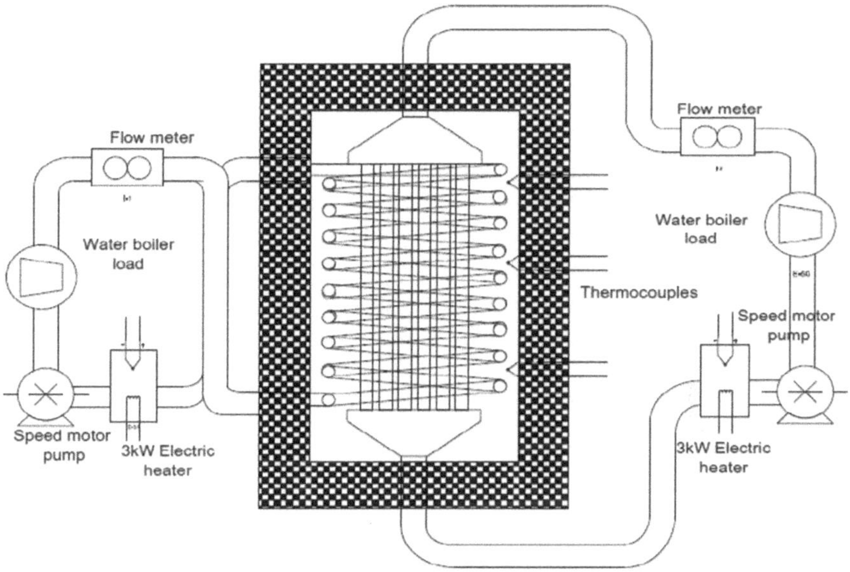

FIGURE 3.13 The whole system of heat energy storage with the vertical straight pipe and the helical coiled pipe (Vaivudh, Rakwichian, & Chindaruksa, 2008).

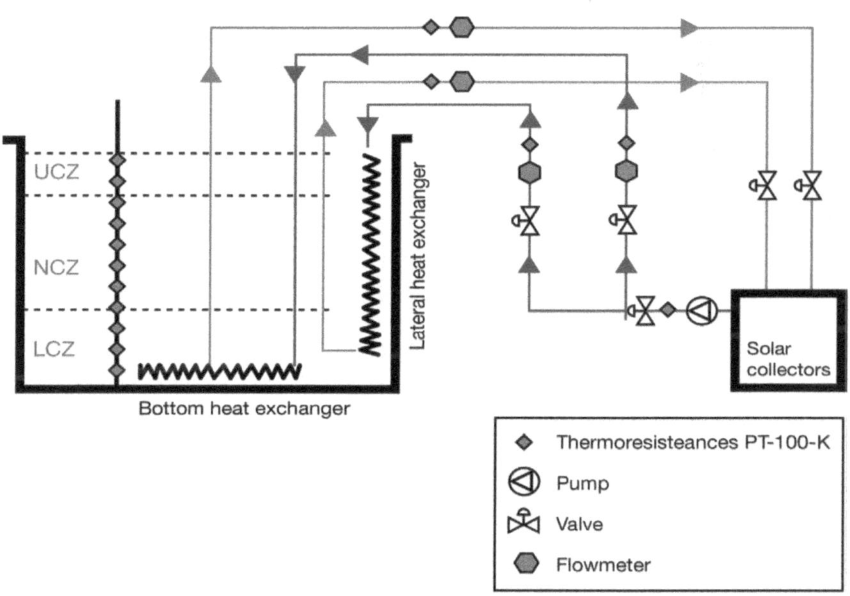

FIGURE 3.14 Heat supply of solar collectors to the solar pond using lateral and bottom heat exchanger (Alcaraz et al., 2018).

provided to a specific location of the pond in heat extraction system, as it depends on which HE is used (bottom or lateral). To measure the flow rate and temperature of the working fluid, a sensor and flow meter are set at inlet along with outlet. Two temperature sensors (Pt1000) are installed for the proper functioning of the heat supply system. One is installed at the top of the lower convective zone (LCZ) which possess the maximum temperature of the system. The other sensor is located at the exit of pipe of the solar collectors. The water that is hot is pumped from solar collectors to the heat exchanger that is situated at the bottom of the pond while the system is supplying heat to the LCZ. The HTF transmits energy to storage zone. After that it returns to a temperature close to the LCZ temperature which is heated with the help of solar thermal collectors. The hot water transfers heat through the non-convective zone (NCZ) and LCZ by flowing to bottom of the pond. Finally, the heated water is sent back to the pond to begin a fresh cycle. If the temperature difference of the two sensors is more than 15°C, the cycle will continue.

Shell and tube heat exchangers

The probable design for molten salt and oil heat exchanger which is suitable in case of concentrated solar power. This type is a U-tube design having two shell passes and two tube passes. A baffle that is longitudinal separates the entire shell into two sections and it defines the two shell passes. There is a specific number of vertical baffles that are equally spaced having a baffle cut that is vertical. They allow the fluid of shell side to flow through a route that is "s-shaped" (Fig. 3.15).

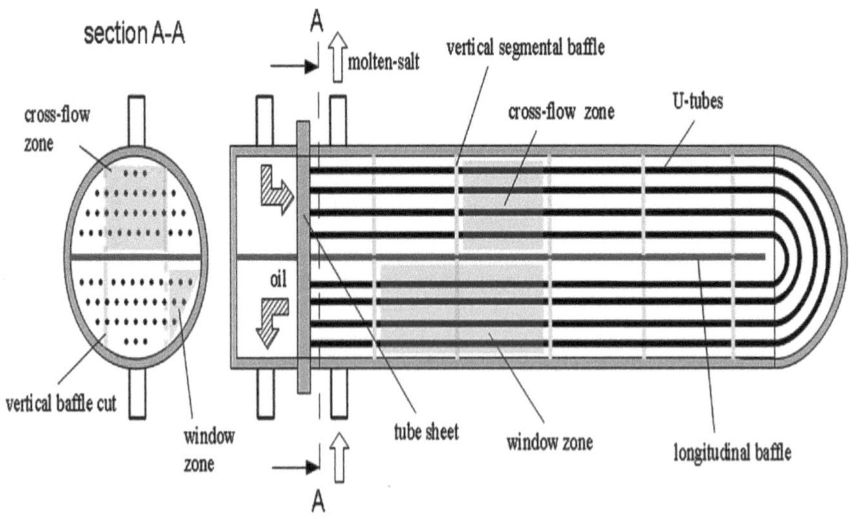

FIGURE 3.15 Front schematic view of a shell and tube heat exchanger with vertical segmental baffles and a baffle cut (Zaversky, Sánchez, & Astrain, 2014).

The fluid passes through three segments:

- Begins inlet nozzle
- Passes through shell side passes and
- Ends at the exit nozzle.

To increase heat transfer coefficient of shell side, the "s-shaped" route is required. The coefficient of heat transfer has the maximum value for cross-flow over tubes.

The vertical baffle cut as well as installed baffles assist to identify two bundle regions namely:

- The window zone
- The cross-flow zone

Coefficient of heat transfer reaches its peak value within the cross-flow zone, as it occurs over the tubes. In contrast, in the window area, flow direction in the shell side is changed by 180 degrees and flow conditions vary. Along the tubes, there is pure parallel flow and angle of flow is less than 90 degrees (90 degrees is considered ideal cross-flow). The flow of tube side is simpler than the "s-shaped" flow. The flow of tube side enters the HE through the nozzle of inlet. After that it flows through inlet channel. Finally, the distribution occurs over a specific number of tubes of U shape in the tube sheet. For each and every tube, the fluid flows through the straight tube section at the upper side, passes the bend, flows to the straight tube section at the lower side and finally, before entering outlet nozzle, it gets mixed in the outlet channel.

Submerged heat exchanger used for solar absorption

Water-filled window with submerged heat exchanger is a trending multiglazing solar-absorption system. At the same time, it can heat up services water and reduce consumption of room air conditioning. Distilled water is used to fill the whole cavity that is between two glass panes. The solar heat absorbed from water is extracted by secondary water flow without direct contact. It is the thermal process of the system (Fig. 3.16).

Solar radiation incident includes:

- Reflection
- Transmission
- Absorption

The values of these parts depend on thermal and optical properties of the material. At all the four glass surfaces, convective heat transfer occurs. The glasses remain in contact with air or water. These surface temperatures of the glasses generally vary. 3D thermal conduction occurs at all components of the window. The distilled water ascends and descends with glass surface at higher and lower temperature, respectively, due to the buoyancy force.

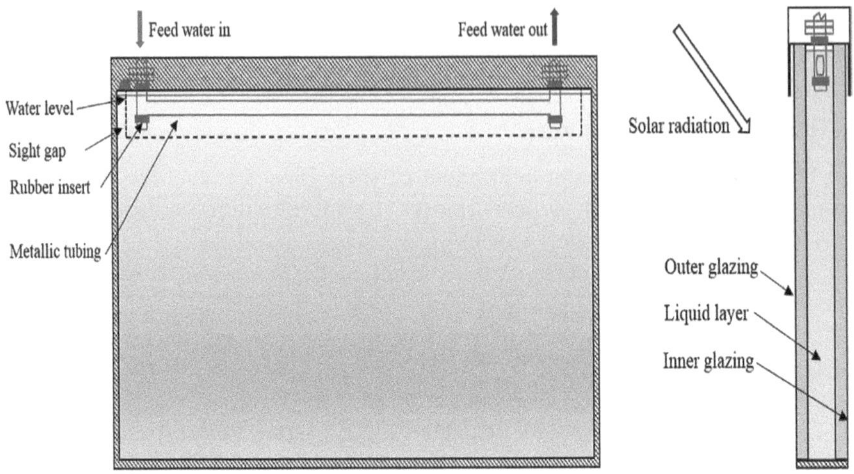

FIGURE 3.16 Schematic view of water-filled double glazing with metallic tubing that is submerged (Liu & Chow, 2020).

Therefore, inside the cavity, a flow is formed that is cellular. The flow becomes stronger by removing heat continuously from the submerged heat exchanger. The Rayleigh number defines the mode of heat transfer. If the Rayleigh number is smaller (e.g., lower than 103), then thermal conduction dominates. If the Rayleigh number increases, free convection occurs. The glazing temperature decreases because the cold stream removes thermal energy from window cavity. Thus, the water-filled window saves power in services involved in water heating along with air conditioning. The pump-operated feed water supply helps to minimize the friction loss in the whole circuit in a hot water system. In order to lower the water flow velocity and friction loss, hot water system coupled with water flow windows that are arranged in parallel, where additional pumping power becomes negligible.

Fluidized bed heat exchanger

Fig. 3.17 illustrates the schematic view of the solar-based coupled power plant system. The system is composed of two thermodynamic cycles:

- Brayton cycle
- Rankine cycle

FB heat exchanger is used in Rankine cycle. The water of the heat exchanger extracts heat from solid particles and evaporates, superheats, reheats as per the cycle. The steam gets superheated and enters the steam turbine that is at high pressure. After that it returns to the fluidized bed that is at low pressure. In the end, the outlet water leaves steam turbine and goes to the open feed water heater (OFWH). The OFWH preheats the water and

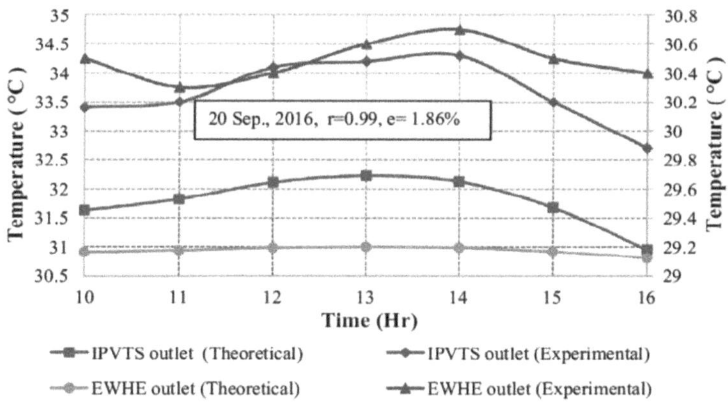

FIGURE 3.17 Schematic view of the solar based coupled power plant system (Farsi & Dincer, 2019).

delivers it to the fluidized bed heat exchanger that reduces the irreversibilities associated with generation of steam and therefore, augments the thermodynamic performance of the entire system (Fig. 3.18).

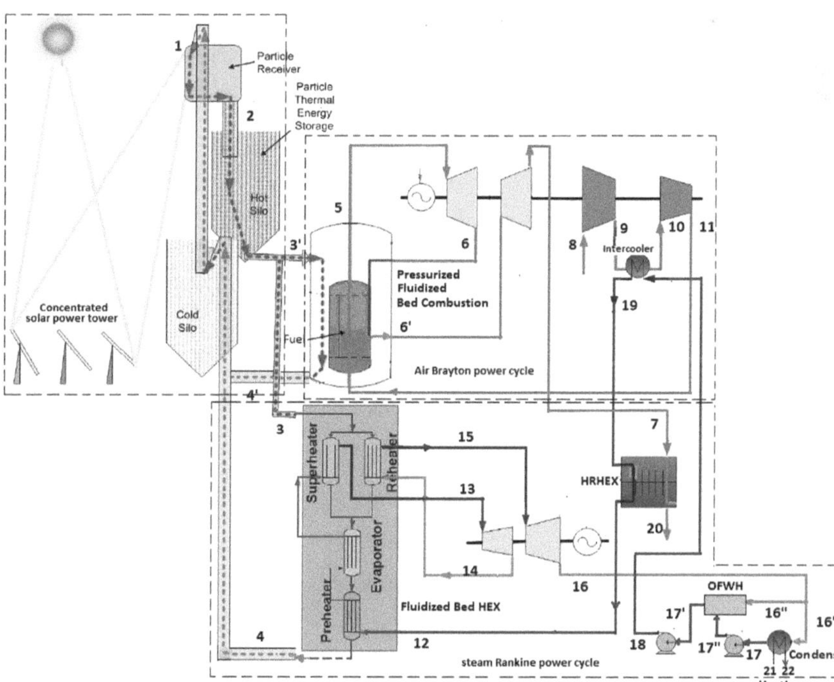

FIGURE 3.18 Basic thermal resistance concept for a pipe (Sarkar et al., 2014; Chen et al., 2017).

3.2 Design of heat exchanger

Designing a heat exchanger basically includes:

- Analysis of heat transfer
- Analysis of pressure drop
- Rating and sizing
- Economical aspects

In this segment of this chapter, mathematical and numerical modeling of heat exchangers associated with solar thermal energy will be briefly discussed.

3.2.1 Mathematical modeling of earth water heat exchanger coupled with IPVTS (Jakhar et al., 2017)

In order to model the EWHE system, three heat transfer modes are generally considered which include:

1. Convection in the pipe's inner surface which is buried
2. Conduction in the pipe that is buried and
3. Conduction between earth and the external surface of the pipe.

Following assumptions are implemented:

- 1D heat transfer
- Isotropic soil and homogenous thermal conductivity
- Temperature of soil is considered as average ambient temperature
- Circular cross-sectional area of pipe is uniform
- State that is quasi steady

Pipe thickness and thermal resistance of the soil annulus can be calculated as follows (Fig. 3.19):

$$R_{\text{soil}} = \frac{\ln\left(\frac{r_s}{r_o}\right)}{2\pi l_p k_s} \tag{3.1}$$

$$R_{\text{pipe}} = \frac{\ln\left(\frac{r_o}{r_i}\right)}{2\pi l_p k_p} \tag{3.2}$$

Here, r_s = soil annulus radius (m)
$R_{i/o}$ = inner and outer radius of HE pipe (m)
$k_{s/p}$ = thermal conductivity of soil and pipe (W/mK)
l_p = length of pipe (m)

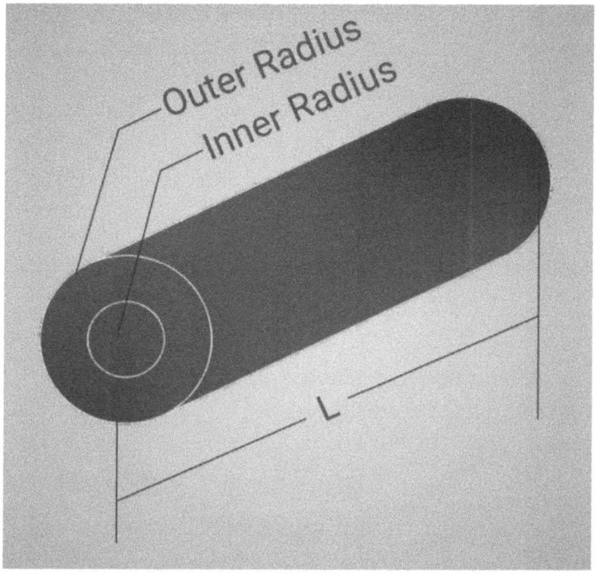

FIGURE 3.19 Tube bundle model scheme (Zaversky et al., 2014).

The Nusselt number (Nu) and Reynolds number (Re) are given by:

$$\mathrm{Re} = \frac{\rho v_w 2r_i}{\mu} \tag{3.3}$$

For laminar flow: Nu = 4.36, for Re < 2300.
For turbulent flow in a circular pipe:

$$\mathrm{Nu} = \frac{(f/8)(\mathrm{Re} - 1000)\mathrm{pr}}{1 + 12.7(f/8)^{0.5}(\mathrm{pr}^{0.66} - 1)} \tag{3.4}$$

where $0.5 < \mathrm{pr} < 2000$ and $2300 < \mathrm{Re} < 5 \times 10^6$, pr = Prandtl number.

f = friction factor. f is determined from Petukhov's relation for smooth pipes.

$$f = (0.79\ln\mathrm{Re} - 1.64)^{-2} \tag{3.4}$$

Now the thermal resistance between flowing water and inner surface of buried pipe due to convective heat transfer can be calculated as:

$$R_{\mathrm{Convec}} = \frac{1}{2\pi r_i l_p h} \tag{3.5}$$

where h = convective heat transfer coefficient and can be determined as:

$$h = \frac{\mathrm{Nu}k_w}{2r_i} \tag{3.6}$$

k_w = thermal conductivity of water (W/mK)

The overall thermal resistance between soil and flowing water is estimated from:

$$R_{total} = R_{soil} + R_{pipe} + R_{Convec} \tag{3.7}$$

The overall heat transfer coefficient is:

$$U = \frac{1}{R_{total}} \tag{3.8}$$

For a constant temperature pipe, the effectiveness is calculated as:

$$\varepsilon = 1 - e^{\left(-\frac{U}{m_w C_w}\right)} \tag{3.9}$$

Here, C_w = specific heat capacity of water (J/kg K)
m_w = mass flow rate of water (kg/s)
Temperature of outlet of the HE is:

$$T_{wo1} = T_{wi1} - (T_{wi1} - T_s) \tag{3.10}$$

Here, T_{wi1} = EWHE temperature for inlet
T_{wo1} = EWHE temperature for outlet
T_s = Temperature of soil from water
The cooling potential of the HE can be calculated by:

$$Q = m_w C_w (T_{wi1} - T_{wo1}) \tag{3.11}$$

The exergy input and output in total per unit flow rate of mass of the HE system is calculated from:

$$\Psi_{in} = h_{in} - h_a - T_a(s_{in} - s_a) \tag{3.12}$$

$$\Psi_{out} = h_{out} - h_a - T_a(s_{out} - s_a) \tag{3.13}$$

where $\Psi_{in/out}$ = specific exergy at inlet and outlet (kJ/kg)
$S_{in/out/a}$ = specific entropy of inlet, outlet and ambient (kJ/kgK)
The exergy destruction rate of EWHE can be calculated as:

$$\dot{E}\chi_{des,\ EWHE} = \dot{m}_w(\Psi_{in} - \Psi_{out}) - Q_g\left(1 - \frac{T_o}{T_{soil}}\right) + \frac{T_a \times \dot{m}_w \times \Delta P}{\rho \times T_{w,avg1}} \tag{3.14}$$

$\dot{E}\chi_{des,EWHE}$ = exergy destruction rate
Q_g = heat transferred from EWHE to soil (W)
ΔP = pressure drop (Pa)
ρ = density (Kg/m^3)
$T_{w,\ avg1}$ = average temperature of water of EWHE

$$Q_g = \dot{m}_w(T_{wi1} - T_{wo1}) \tag{3.15}$$

The second law exergetic efficiency of the HE can be determined from:

$$\eta_{\text{II law, EWHE}} = \left(1 - \frac{\dot{E}\chi_{\text{des, EWHE}}}{\dot{m}_w \Psi_{\text{in}}}\right) \tag{3.16}$$

The overall second law efficiency is as follows:

$$\eta_{\text{II law, coupled system}} = \left[1 - \frac{(\dot{E}\chi_{\text{loss}} + \dot{E}\chi_{\text{des}})_{IPVTS} + \dot{E}\chi_{\text{des, EWHE}}}{\dot{E}\chi_{\text{input, total}} + \dot{E}\chi_{\text{pump}}}\right] \tag{3.17}$$

3.2.2 Mathematical modeling of Earth Air Heat Exchanger system (Afrand et al., 2019)

Ɛ-NTU process is generally used to determine and estimate the performance of the HE system. Ɛ can be obtained by dividing actual heat transfer ($\dot{Q}_{\text{EAHE}}$) by maximum possible heat transfer ($\dot{Q}_{\text{EAHE, max}}$).

$$\varepsilon = \frac{\dot{Q}_{\text{EAHE}}}{\dot{Q}_{\text{EAHE, max}}} = \frac{T_{\text{out,EAHE}} - T_{\text{in, EAHE}}}{T_{\text{soil}} - T_{\text{in, EAHE}}} \tag{3.18}$$

Here, T = temperature of air, in = inlet, out = outlet.
The effectiveness can be calculated as:

$$\varepsilon = 1 - \exp(-NTU) \tag{3.19}$$

where NTU = number of transfer units which is:

$$NTU = \frac{hA}{\dot{m}_f c_p} \tag{3.20}$$

In this equation, c_p = specific heat capacity of air (J/kg K)
$\dot{m}_f$ = air mass flow rate (kg/s)
h = convective heat exchange coefficient (W/K m^2)
A = surface area of heat transfer which is:

$$A = \pi D_{i,\text{i, EAHE}} L_{\text{EAHE}} \tag{3.21}$$

where $D_{i,\text{ EAHE}}$ and L_{EAHE} are inner diameter and length, respectively.
The convective heat exchange coefficient can be evaluated as:
If Reynolds number of EAHE, $\text{Re}_{\text{EAHE}} < 2300$:

$$h = 3.66 \frac{k}{D_{i,\text{EAHE}}} \tag{3.22}$$

k = thermal conductivity (W/m K)

If Reynolds number of EAHE, $2300 \leq \mathrm{Re}_{\mathrm{EAHE}} < 5 \times 10^6$:

$$h = \frac{k}{D_{i,\mathrm{EAHE}}}\left[\frac{(\xi/8)(\mathrm{Re}_{\mathrm{EAHE}} - 1000)\mathrm{Pr}}{1 + 12.7\sqrt{\frac{\xi}{8}}\left(\mathrm{Pr}^{2/3} - 1\right)}\right] \tag{3.23}$$

where

$$\xi = (1.82\log\mathrm{Re}_{\mathrm{EAHE}} - 1.64)^{-2} \tag{3.24}$$

and Pr = Prandtl number.

To calculate the outlet air temperature, the following expression can be used:

$$T_{\mathrm{out,\,EAHE}} = T_{\mathrm{in,\,EAHE}} + \varepsilon(T_{\mathrm{soil}} - T_{\mathrm{in,\,EAHE}}) \tag{3.25}$$

3.2.3 Mathematical modeling of vertical straight and helical coiled pipe heat exchanger (Vaivudh et al., 2008)

Vertical straight pipe equations

When highly viscous fluid flows through a small diameter tube, laminar flow is induced. Thermal entry length is counted for $10 < L/D < 400$ in order to determine Nusselt number from the following expression:

$$\mathrm{Nu} = 0.036\mathrm{Re}^{0.8}\mathrm{Pr}^{0.33}\left(\frac{D}{L}\right)^{0.055} \tag{3.26}$$

Here, Re = Reynolds number

Pr = Prandtl number

D = diameter of the pipe

L = length of vertical straight pipe

Nusselt number for turbulent flow (fully developed) in smooth tubes is determined from Eqs. (3.4–3.5).

The average coefficient (h_o) of free convection heat transfer is determined from:

$$h_o = \left(\frac{k}{L}\right)0.10(Gr_L\mathrm{Pr})^{1/3} \tag{3.27}$$

k = thermal conductivity (W/m^{2}K)

Gr_L = Grashof number

Helical coiled pipe equations

The coefficient of convective heat transfer depends on Reynolds number (Re) and Dean number (D_n).

Dean number is the division of the tube diameter by coil diameter.

$$D_n = \mathrm{Re}(D/d)^{1/2} \tag{3.28}$$

Here, d = coiled diameter (m)

There are three regions of D_n which give Nu as:

$$D_n < 20 \quad \mathrm{Nu} = 1.7\left(D_n^2 \mathrm{Pr}\right)^{1/6} \tag{3.29}$$

$$20 < D_n < 100 \quad \mathrm{Nu} = 0.9\left(\mathrm{Re}^2 \mathrm{Pr}\right)^{1/6} \tag{3.30}$$

$$D_n > 100 \quad \mathrm{Nu} = 0.7\mathrm{Re}^{0.47}\mathrm{Pr}^{1.6}\left(\frac{D}{d}\right)^{0.07} \tag{3.31}$$

3.2.4 Mathematical modeling of tube bundle of a shell and tube heat exchanger (Zaversky et al., 2014)

The basic element of tube bundle is a single metal pipe which contains the HTF. For this reason a dynamic pipe model composed of steel wall is formed as the base model (Fig. 3.20).

The flow is modeled based on the method of finite volume which formulates the balance of momentum in steady state. The wall of the pipe is considered as a cylindrical 1D conduction model which has a temperature node toward the radial direction. Capacitance of the wall is considered as lumped. It is assumed that there is no heat conduction toward the flow direction.

However, in flow direction, the model is parted into several segments in such a way that the wall segment can be coupled to each control volume. Each section gives a wall surface temperature of characteristics for every control volume (Fig. 3.21). The left side of the aforementioned figure depicts a segment of the model. Squares in red indicate a Modelica, for instance, limit between submodels.

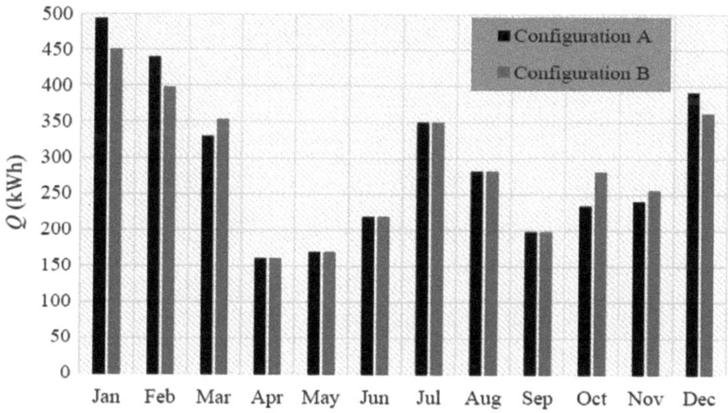

FIGURE 3.20 Experimental and simulated values of IPVTS along with EWHE inlet and outlet temperature during test (Jakhar et al., 2017).

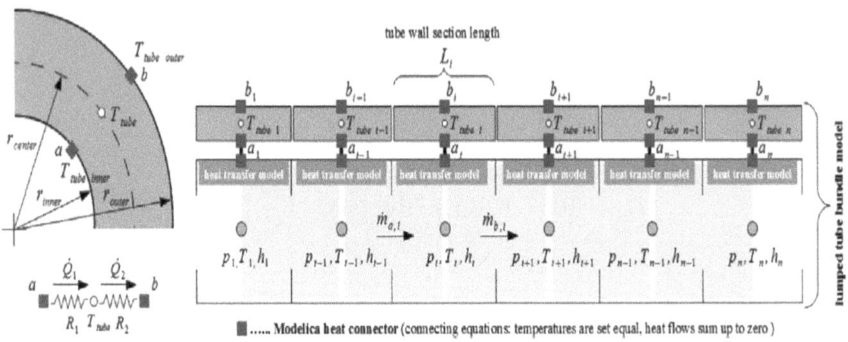

FIGURE 3.21 Total destruction of exergy and exergetic efficiency of the system (Jakhar et al., 2017).

At connecting points:

- Summation of flows of heat is zero
- Equal temperature

A "single layer cylinder" model is taken from Modelica Buildings Library (Wetter et al., 2014).

For every discrete longitudinal section (i), the governing equations are:

$$R_1 = \frac{\ln\left(\frac{r_{\text{center}}}{r_{\text{inner}}}\right)}{2\pi L_i k_{\text{tube}}} \tag{3.32}$$

$$R_2 = \frac{\ln\left(\frac{r_{\text{outer}}}{r_{\text{center}}}\right)}{2\pi L_i k_{\text{tube}}} \tag{3.33}$$

$$\dot{Q}_1 = \frac{T_{\text{tube inner } i} - T_{\text{tube } i}}{R_1} \tag{3.34}$$

$$\dot{Q}_2 = \frac{T_{\text{tube } i} - T_{\text{tube outer } i}}{R_2} \tag{3.35}$$

$$C_{\text{tube } i} \cdot \frac{dT_{\text{tube } i}}{dt} = \dot{Q}_1 - \dot{Q}_2 \tag{3.36}$$

Here, r_{center} = center radius of cylindrical conduction model (m)

r_{inner} = inner radius of cylindrical conduction model (at heat connector a) (m)

r_{outer} = outer radius of cylindrical conduction model (at heat connector b) (m)

R = thermal resistance of the model (K/W)
$\dot{Q}$ = the control volume boundary heat flow (W)
k = thermal conductivity (W/mK)
L = tube length (m)
$T_{\text{tube } i}$ = node temperature of the tube for discrete segment i (K)
$T_{\text{tube inner } i}$ = inner surface temperature of tube for discrete segment i (K)
$T_{\text{tube outer } i}$ = outer surface temperature of tube for discrete segment i (K)
$C_{\text{tube } i}$ = capacity of heat for discrete tube section i (J/K)
The heat transfer coefficient is determined from:

- Nusselt number (Nu) correlations and
- Newton's law of convective heat transfer

$$\dot{Q}_{\text{net } i} = h_{\text{fluid } i} A_{\text{tube inner } i}(T_{\text{tube inner } i} - T_i) \tag{3.37}$$

where $\dot{Q}_{\text{net } i}$ = the control volume boundary net heat flow (W)
h_{fluid} = heat transfer coefficient (W/m^{2}K)
$A_{\text{tube inner } i}$ = each discrete tube segment's inner area (m^2)
T_i = temperature of bulk fluid within control volume i (K)
The pipe flow is considered turbulent, as it occurs while ideal operations. Therefore, for smooth pipes Nusselt number correlation is:

$$\text{Nu}_{\text{fluid}} = \frac{h_{\text{fluid}} D_{\text{inner}}}{k} = \frac{(f/8)(\text{Re} - 1000)\text{Pr}}{1 + 12.7(f/8)^{0.5}(\text{pr}^{2/3} - 1)} \left[1 + \left(\frac{D_{inner}}{L}\right)^{2/3}\right] \left[\frac{\text{Pr}}{\text{Pr}_{wall}}\right]^{0.11} \tag{3.38}$$

Eq. (3.38) is valid for Reynolds numbers between 2300 and 1×10^6 $(2300 < \text{Re} < 1 \times 10^6)$ and Prandtl numbers (Pr) between 0.6 and 1×10^5.
Here, D_{inner} = tube inner diameter (m)
f = friction factor which can be determined from Eq. (3.25)
L = tube length (m)

3.2.5 Hydrodynamics of fluidized bed heat exchanger (Farsi & Dincer, 2019)

The fluidized gas velocity has a great domination on the behavior of the FB. The FB pressure drop highly affects the fluidizing gas compression power. Therefore, a hydrodynamic analysis plays a vital role in the design of FB. The pressure drop of the bed per unit length of the chamber and the lowest velocity of fluidization are major factors for hydrodynamic analysis and can be estimated from the following expressions:

$$\frac{\Delta P}{L} = 150\frac{(1 - \varepsilon_{mf})^2}{\varepsilon_{mf}^3} \frac{\mu u_{mf}}{(\Phi_s D_{par})^2} + 1.75\frac{(1 - \varepsilon_{mf})}{\varepsilon_{mf}^3} \frac{\rho_g u_{mf^2}}{(\Phi_s D_{par})} \tag{3.39}$$

$$\mathrm{Ar} = \frac{1.75}{\varepsilon_{mf}^3 \Phi_s}\mathrm{Re}_{mf}^2 + \frac{150(1-\varepsilon_{mf})}{\varepsilon_{mf}^3 \Phi_s^2}\mathrm{Re}_{mf} \tag{3.40}$$

Here, Re = Reynolds number and Ar = Archimedes number. These numbers are expressed as:

$$\mathrm{Re}_{mf} = \frac{D_{par} u_{mf} \rho_g}{\mu_g} \tag{3.41}$$

$$\mathrm{Ar} = \frac{g {D_{par}}^3 \rho_l (\rho - \rho_g)}{{\mu_g}^2} \tag{3.42}$$

where ε_{mf} = the void fraction (Ɛ) at the lowest fluidization (mf) condition

ΔP = pressure drop (Pa)

L = height of the fluidized chamber (m)

u_{mf} = velocity at minimum fluidization (m/s)

D_{par} = particle diameter (m)

ρ_g = density of gas (kg/m^3)

ϕ = the sphericity of the particles

μ = fluid viscosity (Ns/m^2)

g = gravitational acceleration (m^2/s)

3.2.6 Heat recovery heat exchanger in hybrid particle-based concentrated solar power plant (Farsi & Dincer, 2019)

In order to preheat the water of Rankine cycle, exhaust gas of the turbine is utilized in HRHE. The following expressions are thermodynamic balance correlations for the HRHE:

$$\dot{Q}_{\mathrm{HRHE}} = \dot{m}_{\mathrm{fluegas}}(h_7 - h_{20}) = \dot{m}_{\mathrm{steam}}(h_{12} - h_{19}) \tag{3.43}$$

$$\dot{m}_7 ex_7 + \dot{m}_{19} ex_{19} = \dot{m}_{20} ex_{20} + \dot{m}_{12} ex_{12} + \dot{E}\chi_{des,\mathrm{HRHE}} \tag{3.44}$$

$$\dot{E}\chi_{des,\mathrm{HRHE}} = \dot{Q}_{\mathrm{HRHE}}\left(1 - \frac{T_0}{T_{20-7}}\right) \tag{3.45}$$

where $\dot{Q}_{\mathrm{HRHE}}$ = net heat transfer rate of HRHE (kW)

$\dot{m}_{\mathrm{fluegas}}$ = flow rate of mass (kg/s)

h = specific enthalpy (kJ/kg)

ex = specific exergy (kJ/kg)

$\dot{E}\chi_{des,\mathrm{HRHE}}$ = rate of exergy destruction (kW)

T = temperature (°C, K)

3.3 Heat exchanger performance analysis

Performance of a system dictates that how efficiently the system is functioning. It is required to implement some adjustments if necessary. Moreover, it indicates the system's durability and reliability. Therefore, in this part of the chapter, performance analysis of heat exchangers will be discussed which are used in solar thermal systems.

3.3.1 Performance analysis of IPVTS system with EWHE

In the following figure, the theoretical and experimental inlet and outlet temperatures of the IPVTS coupled with EWHE are presented. Calculations are based on segment 3.2.1 (Fig. 3.17).

In the coupled system, the outlet temperature of the heat exchanger is considered as the temperature of inlet of IPVTS. In contrast, the temperature of the outlet of the IPVTS is considered as the temperature of inlet of the HE. The outlet temperature of the EWHE during the experiment is more than the theoretical value, as there is transfer of heat between connecting pipes along with the surroundings. Additionally, the outlet temperature of IPVTS of the experiment is greater than theoretical temperature. The maximum outlet temperature of IPVTS during experiment is determined to be 34.3°C. The temperature of inlet of the IPVTS varies between 30.3 and 30.7°C. The outlet temperatures of the IPVTS largely depend on the intensity of solar insolation. Analysis is done statistically shows a better relation within experimental and theoretical values, where:

e = percent deviation of root mean square and r = coefficient of correlation

Fig. 3.22 illustrates total exergy destruction and exergetic efficiency of the combined scheme.

It can be declared that the exergetic efficiency in overall of the integrated scheme is the highest from 1 PM to 3 PM in comparison with the other operating hours due to the following reasons:

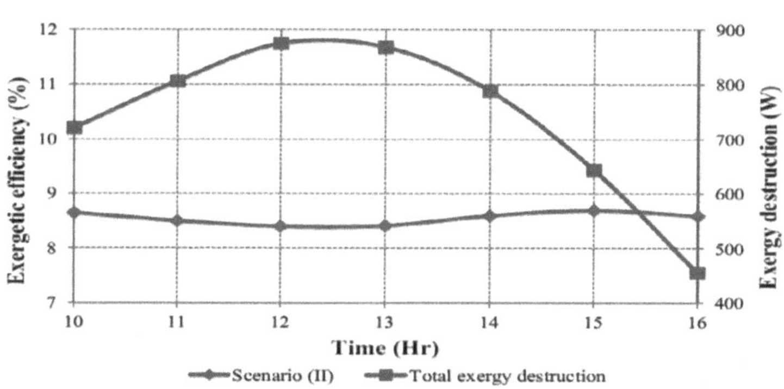

FIGURE 3.22 Thermal power that is gained monthly from the different configurations of coupled system (Afrand et al., 2019).

- High exergy input
- Low destructions and losses of exergy

Finally, overall exergetic efficiency of the combined system varies from 8.39% to 8.69%. In the above-mentioned figure, Scenario (II) is a condition to overcome the deficiencies due to the derivation of the second law efficiency.

3.3.2 Performance analysis of BIPVT integrated with earth air heat exchanger

Fig. 3.23 depicts the rate of thermal energy received monthly with the help of air which is obtained from the combined system illustrated in Figs. 3.4 and 3.5.

Both configurations show the identical performance in the mode of cooling. On the contrary, in the mode of heating, the configuration A shows more improved performance except for two months: March and October, as during these months, the following factors are higher than the other months:

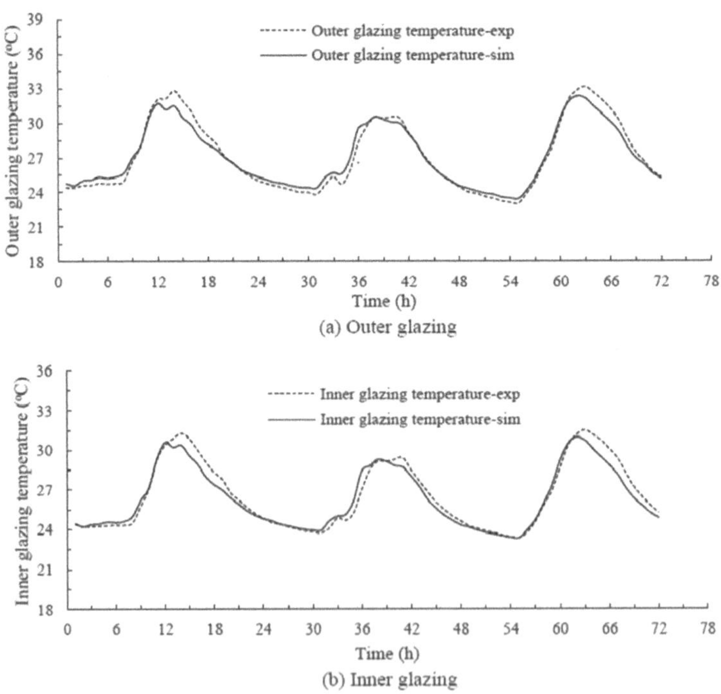

FIGURE 3.23 Temperatures of HTF of 1 kW load by the vertical pipes at several flow rates (Vaivudh et al., 2008).

- Air temperature of the ambient
- Intensity of solar radiation
- Temperature of the PV panel

The maximum rates of heat energy for the configurations are: 493.62 (A) and 449.63 (B) kWh, respectively, that occurs in January. However, the smallest rate of heat energy is 160.02 kWh for A and B which belongs to month April. The heat energy gained by both configurations based on yearly rate are 3499.59 and 3468.16 kWh, respectively. Finally, it is observed that heat transfer of A is slightly higher (0.91%) than configuration B.

3.3.3 Performance analysis of vertical straight and helical coiled pipe HE

Typically, for discharging experiments, thermal oil is used as HTF. Heat is drawn to a load measuring 1 kW from storage tank which consists of a boiler which possesses water of around 14.3 kg in order to absorb heat from the pipe of discharging. It is recommended to increase the average temperature about 1°C within a minute. The temperature of storage decreases from 300°C, as heat loss occurs to ambient when HTF flows through the pipe. In Fig. 3.24, the temperature of HTF increases for approximately 50 min. After that it declines to 100°C which is a minimum temperature of the load. The storage and HTF temperature begin at 300°C, where the ambient temperature is 29°C. In Fig. 3.25, the storage temperatures decline at several flow rates of HTF for about 2 h.

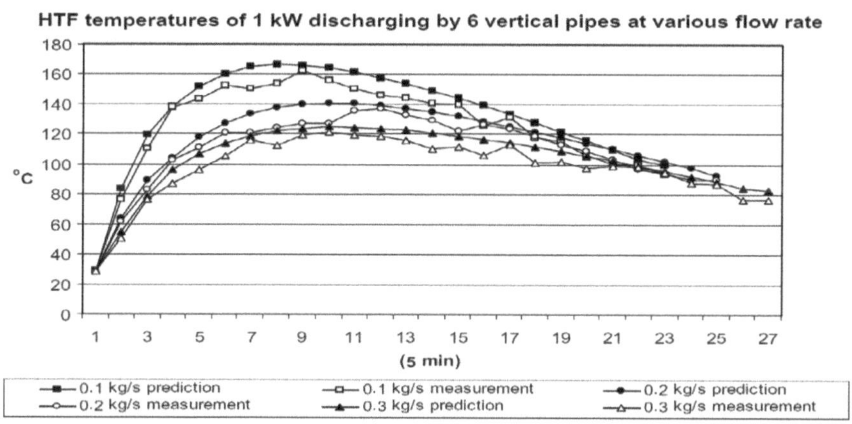

FIGURE 3.24 Temperatures of storage of 1 kW load by the vertical pipes at different flow rates (Vaivudh et al., 2008).

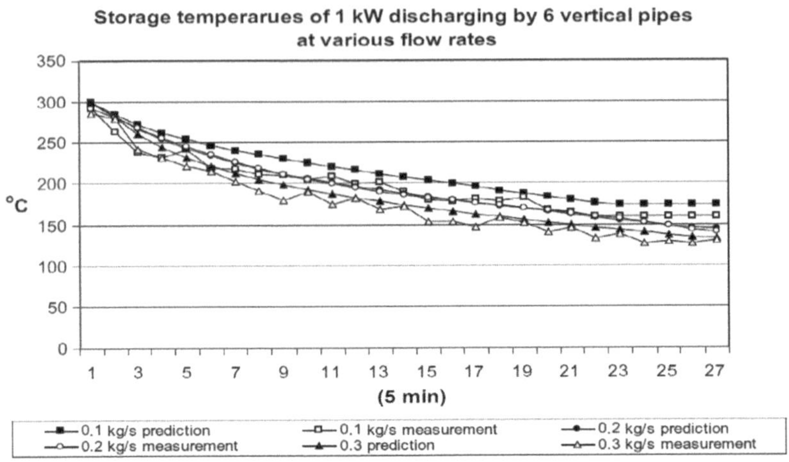

FIGURE 3.25 Comparison (A) outer glazing temperature and (B) inner glazing temperature (Liu & Chow, 2020).

3.3.4 Performance analysis of submerged heat exchanger for solar absorption

In Figs. 3.23 and 3.26, the comparison between water-in-cavity and glazing temperatures are shown which are obtained from experimental and numerical studies. It is observed that the numerical values are adjacent to typical data which were tested.

The relative errors (REs) are within ±8.0%, where for the days 1 to 3, the relative mean errors of the glazing of outer are 1.8%, 1.4%, and 1.7%, respectively. In case of water layer, these are 1.7%, 1.8%, and 1.7%, however, in case of the glazing of inner, results are 1.5%, 1.4%, and 1.7%.

As REs and relative mean errors are much smaller in values, the deviations between numerical and experimental results are quite acceptable.

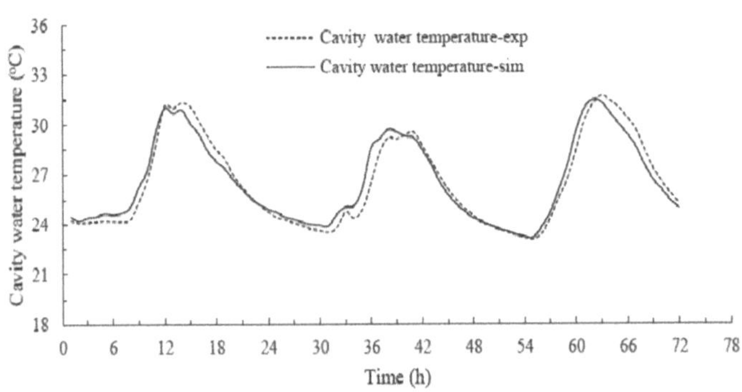

FIGURE 3.26 Comparison of temperature for the water-in-cavity (Liu & Chow, 2020).

3.3.5 Performance analysis of fluidized bed heat exchanger

In Fig. 3.27, the variations of the exergy and energy efficiencies for the entire system including Brayton, Rankine cycle, and the temperature of FB that is pressurized, are depicted. The pressure ratio of the compressor is decreased from a value of 6 to a lower value of 5 with the lowest and highest temperature of combustion, respectively, for reliable and safe operation. The thermodynamic performance increases with the increase of the pressurized fluidized bed combustion (PFBC) temperature. To achieve high PFBC temperatures, the pressure and fuel flow must be increased which results in producing high temperature flue gas. Therefore, higher power is produced by the turbine. In addition, if the gas turbine temperature is increased, mass flow rate and exhaust gas temperature will also increase. Additionally, the temperature of water in Rankine cycle may augment. Exergy and energy efficiencies of the whole system augments with the increase of temperature of PFBC.

Finally, the exergy and energy efficiencies of entire system reaches 50% and 53.8%, respectively, at maximum PFBC temperature that is 1500°C.

In Fig. 3.28, exergy destruction rates are shown for the components of Brayton (except PFBC) and Rankine cycle. Among the depicted components, FB HEX has the highest exergy destruction rate where the value is around 440 KW. The heat recovery heat exchanger has also higher value of exergy destruction rate that is about 210 KW.

The rate of exergy destruction in components of solar tower is higher because of the temperature differences within the source (sun) of heat and temperatures of operation.

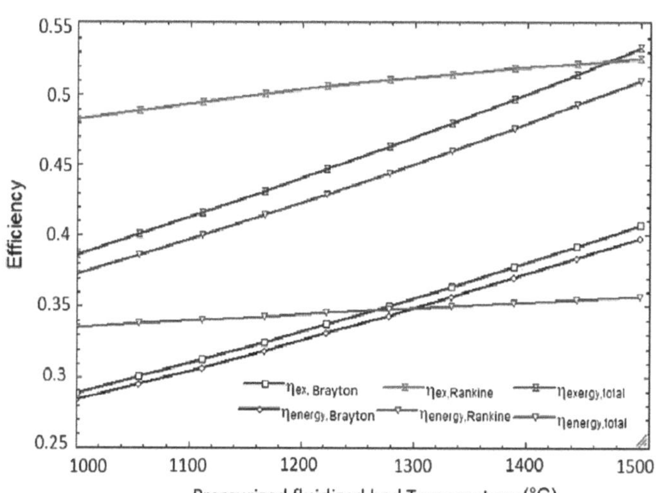

FIGURE 3.27 Outcome of the PFB temperature on exergy and energy efficiencies of Rankine cycle, Brayton cycle, and the entire system (Farsi & Dincer, 2019).

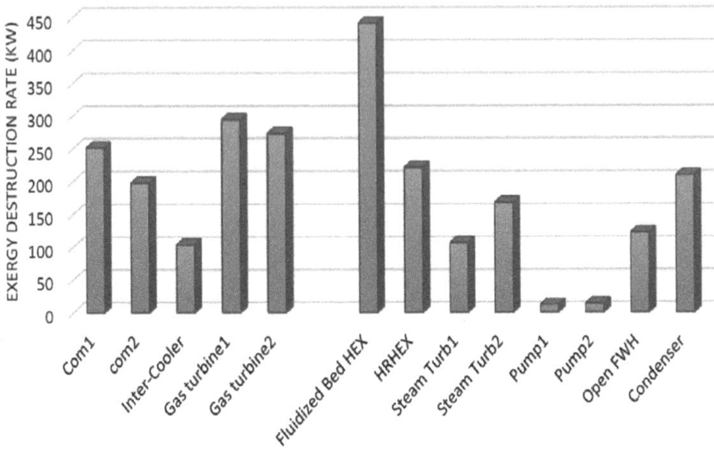

FIGURE 3.28 Rate of exergy destruction for the Brayton cycle (except PFBC) and Rankine cycle components (Farsi & Dincer, 2019).

3.4 Problems and solutions on HEs

In this part of the chapter, some example problems will be solved and discussed.

Example 3.1. Oil is needed to be cooled in a double tube counter flow heat exchanger. The inner tubes have a diameter of 3 cm with negligible thickness. The inner diameter of the outer tube (the shell) is 4 cm. Water passes the tube at a flow rate of 0.6 kg/s. The oil flows through the shell at a rate of 0.9 kg/s. The average temperatures of the water and the oil are 45 and 80°C, respectively, estimate the overall coefficient of heat transfer of this heat exchanger.

The properties of water at 45°C are:

$\rho = 990.1\ \text{kg/m}^3$ $\quad$ $\text{Pr} = 3.91$

$k = 0.637\ \text{W/Mk}$ $\quad$ $\nu = \mu/\rho = 0.602 \times 10^{-6}\ \text{m}^2/\text{s}$

The properties of oil at 80°C are:

$\rho = 852\ \text{kg/m}^3$ $\quad$ $\text{Pr} = 499.3$

$k = 0.138\ \text{W/mK}$ $\quad$ $\nu = \mu/\rho = 3.794 \times 10^{-5}\ \text{m}^2/\text{s}$

Nusselt number can be determined from the given table:

D_i/D_o	Nu_i	Nu_o
0.00	–	3.66
0.05	17.46	4.06
0.010	11.56	4.11
0.25	7.37	4.23
0.50	5.74	4.43
1.00	4.86	4.86

Solution. The following assumptions can be made:

1. Thermal resistance of inner tube is not considered due to highly conductive tube material and its thickness is negligible.
2. Both fluid flows are fully developed.
3. Properties of both fluids are constant.

The overall heat transfer coefficient can be obtained from:

$$\frac{1}{U} = \frac{1}{h_i} + \frac{1}{h_o}$$

where h_i and h_o are the convection heat transfer coefficients inside and outside the tube, respectively.

The hydraulic diameter for a circular pipe is the diameter of the pipe itself, $D_h = D = 0.02$ m. The Reynolds number and average velocity of water are:

$$V = \frac{\dot{m}}{\rho A_c} = \frac{\dot{m}}{\rho\left(\frac{1}{4}\pi D^2\right)} = \frac{0.6}{990.1\left(\frac{1}{4}\pi 0.03^2\right)} = 0.86 \text{ m/s}$$

$$\text{Re} = \frac{VD}{\nu} = \frac{(0.86)(0.03)}{0.602 \times 10^{-6}} = 42,857 > 10,000$$

So, the water flow is turbulent.

For fully developed flow, Nusselt number can be estimated from:

$$\text{Nu} = \frac{hD}{k} = 0.023\text{Re}^{0.8}\text{Pr}^{0.4} = (0.023)(42857^{0.8})(3.91^{0.4}) = 201.5$$

$$h = h_i = \frac{k}{D}\text{Nu} = \frac{0.637}{0.03}(201.5) = 4278.5 \text{ W/m}^2\text{K}$$

Now it is required to repeat the above calculation for oil.

Hydraulic diameter of the annular space is obtained from:

$$D_h = D_o - D_i = 0.04 - 0.03 = 0.01 \text{ m}$$

$$V = \frac{\dot{m}}{\rho A_c} = \frac{\dot{m}}{\rho\left[\frac{1}{4}\pi\left(D_o^2 - D_i^2\right)\right]} = \frac{0.9}{852\left[\frac{1}{4}\pi\left(0.04^2 - 0.03^2\right)\right]} = 1.92 \text{ m/s}$$

$$\text{Re} = \frac{VD}{\nu} = \frac{(1.92)(0.01)}{3.794 \times 10^{-5}} = 506 < 2300$$

So, the flow of oil is laminar.

For fully developed flow, Nusselt number of the annular space is determined from table mentioned in the problem statement.

$$\frac{D_i}{D_o} = \frac{0.03}{0.04} = 0.75$$

$$\mathrm{Nu} = 5.30$$

$$h_o = \frac{k}{D_h}\mathrm{Nu} = \frac{0.138}{0.01}(5.30) = 73.14\ \mathrm{W/m^2K}$$

Therefore, the overall heat transfer coefficient is:

$$\frac{1}{U} = \frac{1}{h_i} + \frac{1}{h_o} = \frac{1}{4278.5} + \frac{1}{73.14}$$

$$U = 71.91\ \mathrm{W/m^2K}$$

Discussion. The overall heat transfer coefficient in a heat exchanger is dominated by the smaller heat transfer coefficient when the difference between the two values is large. To improve the overall heat transfer coefficient, finned surface can be used in the oil side.

Example 3.2. Water enters a heat exchanger that is counter flow at 20°C at a flow rate of 9 kg/s, where it is heated by a water stream that takes entry to the heat exchanger at 80°C at a flow rate of 3 kg/s. The specific heat of water is constant at 4.18 kJ/kg K. Calculate the maximum rate of heat transfer and the outlet temperatures of both fluids.

Solution. The maximum rate of heat transfer in the heat exchanger and the outlet temperatures are to be determined.

The following assumptions can be made:

1. Steady operating conditions
2. Well-insulated HE
3. Negligible changes in potential and kinetic energies
4. Constant properties of fluid

Heat capacity rates are:

$$C_h = \dot{m}_h c_{ph} = (3)(4.18) = 12.54\ \mathrm{KW/K}$$

$$C_c = \dot{m}_c c_{pc} = (9)(4.18) = 37.62\ \mathrm{KW/K}$$

$$C_{\min} = C_h = 12.54\ \mathrm{KW/K}$$

Maximum rate of heat transfer is calculated from:

$$\dot{Q}_{\max} = C_{\min}(T_{h,\,\mathrm{in}} - T_{c,\,\mathrm{in}}) = 12.54(80 - 20) = 752.4\ \mathrm{KW}$$

The temperatures of outlet of hot and cold streams can be estimated from:

$$\dot{Q} = C_c(T_{c,\mathrm{out}} - T_{c,\,\mathrm{in}})$$

$$T_{c,\ out} = T_{c,\ in} + \frac{\dot{Q}}{C_c} = 20 + \frac{752.4}{37.62} = 40^{o}\mathrm{C}$$

$$\dot{Q} = C_h(T_{h,\mathrm{in}} - T_{h,\ out})$$

$$T_{h,\ out} = T_{h,\ in} - \frac{\dot{Q}}{C_h} = 80 - \frac{752.4}{12.54} = 20^{o}\mathrm{C}$$

Discussion. The water is cooled to 20°C, where the cold water is heated up to 40°C. That occurs, as the mass flow rate of the hot water is one third of the cold water.

Example 3.3. A heat exchanger heats water from 30 to 90°C at a flow rate of 2.2 kg/s. The heating is done by geothermal water at 170°C at a flow rate of 3 kg/s. The inner tube has a diameter of 1.5 cm. The overall coefficient of heat transfer of the heat exchanger is 640 $\mathrm{W/m^2K}$. Implementing the effectiveness NTU method, calculate the length of the heat exchanger. The specific heat of hot fluid is 4.31 and the specific heat of cold fluid is 4.18 kJ/kg K, respectively.

Solution. The rates of heat capacity of both fluids are:

$$C_h = \dot{m}_h c_{ph} = (3)(4.31) = 12.93\ \mathrm{KW/K}$$

$$C_c = \dot{m}_c c_{pc} = (2.2)(4.18) = 9.20\ \mathrm{KW/K}$$

$$C_{\min} = C_c = 9.20\ \mathrm{KW/K}$$

$$c = \frac{C_{\min}}{C_{\max}} = \frac{9.20}{12.93} = 0.71$$

The maximum rate of heat transfer is calculated from:

$$\dot{Q}_{\max} = C_{\min}(T_{h,\ in} - T_{c,\ in}) = 9.20(170 - 30) = 1288\ \mathrm{KW}$$

The actual rate of heat transfer is:

$$\dot{Q} = [\dot{m}c_p(T_{\mathrm{out}} - T_{\mathrm{in}})]_{\mathrm{water}} = (2.2)(4.18)(90 - 30) = 551.76\ \mathrm{KW}$$

The effectiveness is calculated from:

$$\varepsilon = \frac{\dot{Q}}{\dot{Q}_{\max}} = \frac{551.76}{1288} = 0.428$$

The NTU of this counter flow heat exchanger can be determined from:

$$NTU = \frac{1}{c-1}\ln\left(\frac{\varepsilon - 1}{\varepsilon c - 1}\right) = \frac{1}{0.71 - 1}\ln\left(\frac{0.428 - 1}{0.428 \times 0.71 - 1}\right) = 0.68$$

The surface area of heat transfer is:

$$NTU = \frac{UA_s}{C_{\min}}$$

$$A_s = \frac{NTUC_{\min}}{U} = \frac{0.68 \times 9200}{640} = 9.78 \text{ m}^2$$

The tube length is:

$$A_s = \pi DL$$

$$L = \frac{A_s}{\pi D} = \frac{9.78}{\pi \times 0.015} = 208 \text{ m}$$

Discussion. Effectiveness NTU method works in a systematic and straight forward manner as well as it has better accuracy than other methods.

References

Afrand, M., Shahsavar, A., Sardari, P. T., Sopian, K., & Salehipour, H. (2019). Energy and exergy analysis of two novel hybrid solar photovoltaic geothermal energy systems incorporating a building integrated photovoltaic thermal system and an earth air heat exchanger system. *Solar Energy, 188*, 83–95. https://doi.org/10.1016/j.solener.2019.05.080

Alcaraz, A., Montalà, M., Valderrama, C., Cortina, J. L., Akbarzadeh, A., & Farran, A. (2018). Increasing the storage capacity of a solar pond by using solar thermal collectors: Heat extraction and heat supply processes using in-pond heat exchangers. *Solar Energy, 171*, 112–121. https://doi.org/10.1016/j.solener.2018.06.061

Ali, M., Ul-Hamid, A., Alhems, L. M., & Saeed, A. (2020). Review of common failures in heat exchangers – Part I: Mechanical and elevated temperature failures. *Engineering Failure Analysis, 109*, 104396. https://doi.org/10.1016/j.engfailanal.2020.104396

Badenhorst, H. (2016). A novel heat exchanger concept for latent heat thermal energy storage in solar power towers: Modelling and performance comparison. *Solar Energy, 137*, 90–100. https://doi.org/10.1016/j.solener.2016.08.006

Chen, S.-W., Chiu, W.-J., Lin, M.-S., Kuo, F.-J., Chai, M.-L., Lee, J.-D., Wang, J.-R., Lin, H.-T., Lin, W.-K., & Shih, C. (2017). 1D and Q2D thermal resistance analysis of micro channel structure and flat plate heat pipe. *Microelectronics Reliability, 72*, 103–114. https://doi.org/10.1016/j.microrel.2017.03.030

Farsi, A., & Dincer, I. (2019). Thermodynamic assessment of a hybrid particle-based concentrated solar power plant using fluidized bed heat exchanger. *Solar Energy, 179*, 236–248. https://doi.org/10.1016/j.solener.2018.12.046

Huang, C., Cai, W., Wang, Y., Liu, Y., Li, Q., & Li, B. (2019). Review on the characteristics of flow and heat transfer in printed circuit heat exchangers. *Applied Thermal Engineering, 153*, 190–205. https://doi.org/10.1016/j.applthermaleng.2019.02.131

Jakhar, S., Soni, M. S., & Gakkhar, N. (2017). An integrated photovoltaic thermal solar (IPVTS) system with earth water heat exchanger cooling: Energy and exergy analysis. *Solar Energy, 157*, 81–93. https://doi.org/10.1016/j.solener.2017.08.008

Joshi, P., & Tiwari, G. N. (2018). Energy matrices, exergo-economic and enviro-economic analysis of an active single slope solar still integrated with a heat exchanger: A comparative study. *Desalination, 443*, 85–98. https://doi.org/10.1016/j.desal.2018.05.012

Kim, W., Baik, Y.-J., Jeon, S., Jeon, D., & Byon, C. (2017). A mathematical correlation for predicting the thermal performance of cross, parallel, and counterflow PCHEs. *International Journal of Heat and Mass Transfer, 106*, 1294–1302. https://doi.org/10.1016/j.ijheatmasstransfer.2016.10.110

Liu, W., & Chow, T.-T. (2020). Study of water-filled double-glazing with submerged heat exchanger for solar absorption. *Applied Thermal Engineering, 171*, 115019. https://doi.org/10.1016/j.applthermaleng.2020.115019

Pandya, N. S., Shah, H., Molana, M., & Tiwari, A. K. (2020). Heat transfer enhancement with nanofluids in plate heat exchangers: A comprehensive review. *European Journal of Mechanics B: Fluids, 81*, 173–190. https://doi.org/10.1016/j.euromechflu.2020.02.004

Sarkar, D., Shah, K., Haji-Sheikh, A., & Jain, A. (2014). Analytical modeling of temperature distribution in an anisotropic cylinder with circumferentially-varying convective heat transfer. *International Journal of Heat and Mass Transfer, 79*, 1027–1033. https://doi.org/10.1016/j.ijheatmasstransfer.2014.08.060

Sarmiento, A. P. C., Soares, V. H. T., Carqueja, G. G., Batista, J. V. C., Milanese, F. H., & Mantelli, M. B. H. (2020). Thermal performance of diffusion-bonded compact heat exchangers. *International Journal of Thermal Sciences, 153*, 106384. https://doi.org/10.1016/j.ijthermalsci.2020.106384

Singh, M., Sharma, M. K., & Bhattacharya, J. (2020). Computational prediction of significant efficiency gain through multi-tank modular heat storage for solar thermal systems with variable-temperature input profile. *Thermal Science and Engineering Progress, 18*, 100551. https://doi.org/10.1016/j.tsep.2020.100551

Tyagi, S. K., Pandey, A. K., Pant, P. C., & Tyagi, V. V. (2012). Formation, potential and abatement of plume from wet cooling towers: A review. *Renewable and Sustainable Energy Reviews, 16*(5), 3409–3429. https://doi.org/10.1016/j.rser.2012.01.059

Vaivudh, S., Rakwichian, W., & Chindaruksa, S. (2008). Heat transfer of high thermal energy storage with heat exchanger for solar trough power plant. *Energy Conversion and Management, 49*(11), 3311–3317. https://doi.org/10.1016/j.enconman.2008.04.013

Wetter, M., Zuo, W., Nouidui, T. S., & Pang, X. (2014). Modelica Buildings library. *Journal of Building Performance Simulation, 7*(4), 253–270. https://doi.org/10.1080/19401493.2013.765506

Zaversky, F., Sánchez, M., & Astrain, D. (2014). Object-oriented modeling for the transient response simulation of multi-pass shell-and-tube heat exchangers as applied in active indirect thermal energy storage systems for concentrated solar power. *Energy, 65*, 647–664. https://doi.org/10.1016/j.energy.2013.11.070

Chapter 4

Solar thermal collector

Fayaz Hussain[1] and Md Hasanuzzaman[2]

[1]*Faculty of Electrical & Electronics Engineering, Ton Duc Thang University, Ho Chi Minh City, Viet Nam;* [2]*Higher Institution Centre of Excellence (HICoE), UM Power Energy Dedicated Advanced Centre (UMPEDAC), University of Malaya, Kuala Lumpur, Malaysia*

Chapter outline

4.1 Basic concept of solar thermal collectors

Among the other forms of renewable sources, solar, that is, energy from the sun has gained more focus due to its vast range of applications. This solar energy converts to thermal heat or electricity for use using solar collector technology. Its thermal application is dating back to thousands of years to present. However, with time the technological advancement has increased its types of applications and conversion efficiency. Solar thermal collector technology is being used for different applications to date (Barone et al., 2019; Qazi et al., 2021; Treichel & Cruickshank, 2021b). The basic concept of solar thermal collector involves when radiation from the sun hits a surface, some portion of it is absorbed, causing the surface temperature to increase. The increased temperature in form of heat is transported through a medium called

Technologies for Solar Thermal Energy. https://doi.org/10.1016/B978-0-12-823959-9.00008-8

fluid to the point of application (Fayaz et al., 2019). Solar collector efficiency is determined by absorption efficiency of the surface, minimized radiation losses back to the atmosphere, and the extraction of reasonable amount of heat energy in the collector (Fayaz et al., 2018). Basic concept of thermal collector is displayed in Fig. 4.1. Solar collectors involve two categories which are concentrating and nonconcentrating collectors (Evangelisti, De Lieto Vollaro, & Asdrubali, 2019) further discussed in next section.

4.2 Categorization of solar thermal collectors

Solar thermal collectors are categorized into two namely: nonconcentrating and concentrating. These categories are further classified into various collectors shown in Fig. 4.2.

The basic design and function, thermal analysis of each collector classified are discussed under each category.

4.3 Nonconcentrator collectors

This type of collectors comprises of flat plate, evacuated tube collectors and does not require a concentrating technique rather the amount of ray collected depends on the surface area. Large surface area in the case is important to increase efficiency of the collector (Treichel & Cruickshank, 2021a).

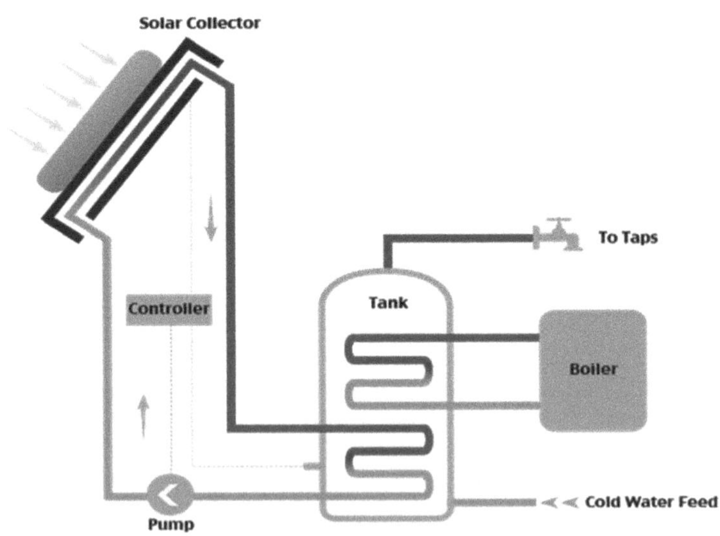

FIGURE 4.1 Basic concept of thermal collector (Mahdi & Bellel, 2014).

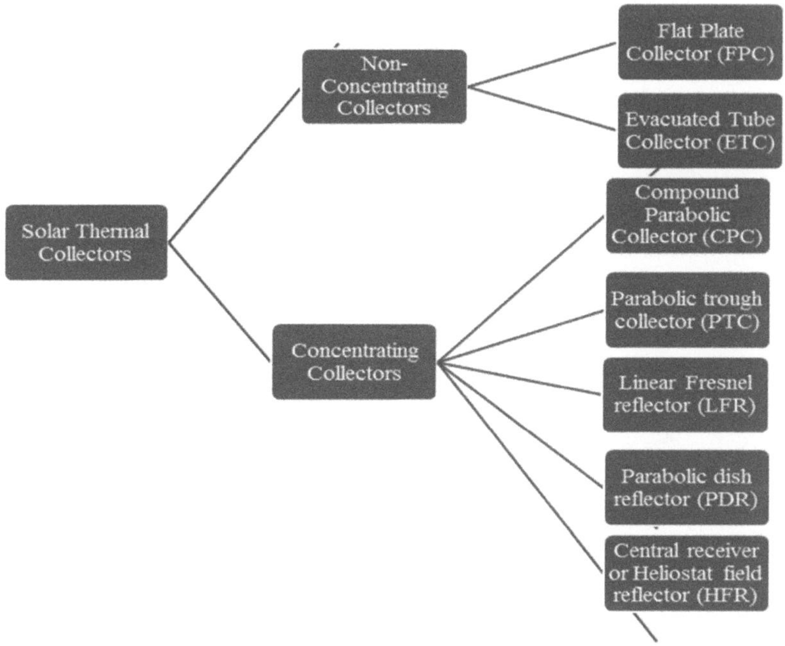

FIGURE 4.2 Categories and classification of solar collectors (Kumar, Hasanuzzaman, & Rahim, 2019).

4.3.1 Flat plate collector

Flat plate collectors which are nonconcentrating types are composed of flat thermal collector in design or shape. They are in two types called as glazed and unglazed. They can transfer heat through working fluid like water, refrigerants, air, etc. For flat plate collector, radiation hits transparent cover with surface of high absorptivity, the plate absorbed portions of the radiation energy which in turn converts into heat then transmit to the transport fluid through tubes for storage or use (Bezaatpour & Rostamzadeh, 2021).

Fig. 4.3 shows design of the flat plate collector. The following compartment are stated below:

- **Absorber plate**—Absorb radiation from the sun.
- **Glass cover**—Transfer energy to the absorber and minimizes radiative and convective heat loss in the collector.
- **Flow tube**—Consist of heating fluid.
- **Insulated metal box**—Reduce collector heat losses.

Thermal performance is important and for flat plate, it is analyzed in two ways: glazed and unglazed type.

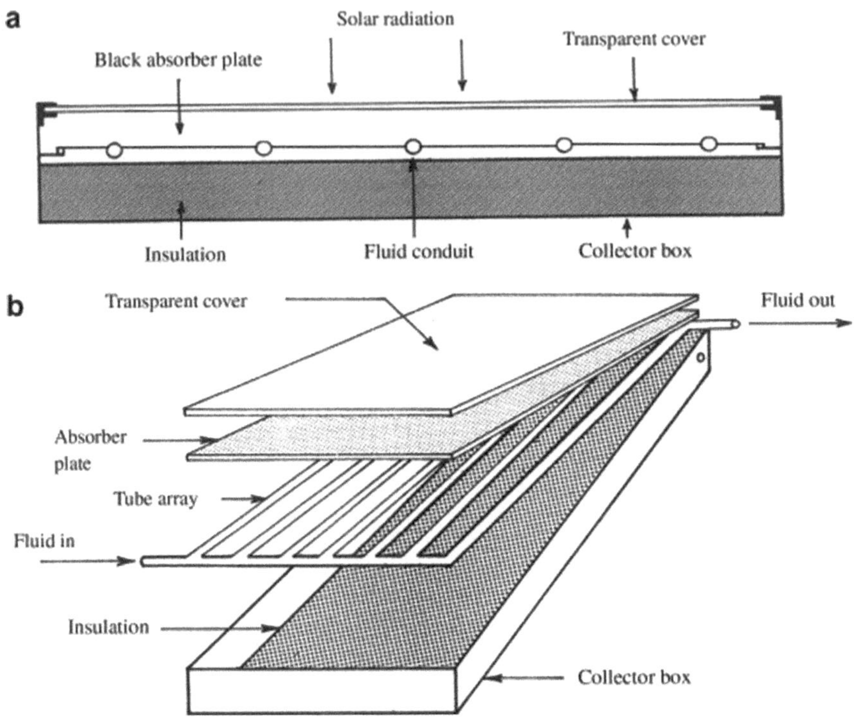

FIGURE 4.3 Basic design of flat plate collector (Sözen et al., 2008).

A. Glazed type FPC analysis

Fig. 4.4 shows a typical glazed type collector. If sunlight intensity *I*, in W/m^2, strike on the collector aperture with surface area *A*, in m^2,

1) The solar radiation amount received can be calculated

$$Q_i = I \cdot A \tag{4.1}$$

where

I is solar radiation intensity (W/m^2) and *A* is collector area (m^2).

Not all radiations are absorbed by aperture plate and part of the radiations are reflected back to surroundings. Therefore, the actual amount of solar radiations absorbed require the conversion factor which implies the % solar rays' transmission (i.e., penetrating the transparent cover) and the % absorbed. It is represented as product of transmission rate of the cover and the rate of the absorption plate. The new amount of radiations received:

$$Q_i = I(\tau\alpha) \cdot A \tag{4.2}$$

where τ is glazing transmission coefficient and α is plate absorption coefficient.

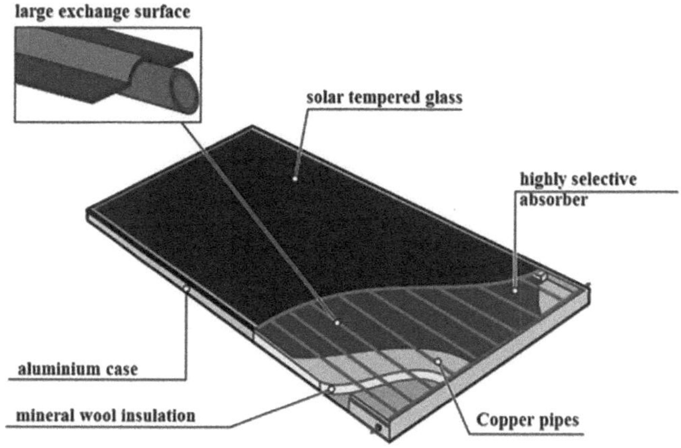

FIGURE 4.4 Typical glazed collector (Greco, Gundabattini, Gnanaraj, Masselli, 2020).

2) To determine the rate of heat loss Q_o

$$Q_o = U_L A(T_o - T_i) \tag{4.3}$$

where

U_L: Overall heat transfer coefficient
T_o: Temperature of collector, °C
T_i: Ambient temperature, °C

3) The rate of heat extraction from the collector

The rate of heat extraction at steady state is evaluated as the rate of useful energy extracted from the collector (Q_u), which is written as below (Islam & Morimoto, 2018):

$$Q_u = Q_i - Q_o = I\tau\alpha \cdot A - U_L A(T_o - T_i) \tag{4.4}$$

where

Q_u: Useful energy extraction, W
Q_o: Rate of heat loss, W
Q_a: Actual useful energy which can be written as follows:

$$Q_a = mc_p(T_o - T_i) \tag{4.5}$$

4) Heat removal factor derivation

Heat removal factor represents the ratio of the actual useful energy extracted to the useful energy (Islam & Morimoto, 2018), therefore:

$$F_R = \frac{mc_p(T_o - T_i)}{A[I\tau\alpha - U_L(T_i - T_a)]} \tag{4.6}$$

From Eq. (4.6), the actual useful energy gain becomes:

$$Q_u = F_R A[I\tau\alpha - U_L(T_i - T_a)] \tag{4.7}$$

5) Energy balance equation

Flat plate collector energy balance equation with area A is displayed:

$$I \cdot A\tau\alpha = Q_u + Q_o + \frac{de}{dt} \tag{4.8}$$

$$\frac{de}{dt} = \text{collector rate of internal energy storage}$$

In other words, collector efficiency η_c is the ratio of the useful energy extracted to the whole absorbed energy radiation.

$$\eta_c = \frac{Q_u}{I \cdot A} \tag{4.9}$$

Practically, efficiencies are gotten over finite period between 15 and 20 min. Therefore, average efficiency over a period time becomes:

$$\eta_c = \frac{\int_0^1 Q_u dt}{\int_0^1 I \cdot A dt} \tag{4.10}$$

where t is averaged performance over a period time.

6) The formulation of the optimization problem

The exergy efficiency η_{ex} is denoted below (Jafarkazemi & Ahmadifard, 2013):

$$\eta_{ex} = -U_1 A_p (T_p - T_a)\left(1 - \frac{T_a}{T_p}\right) \tag{4.11}$$

$$\text{Maximize } \eta_{\text{ex}} = \text{Eq. (4.11)}$$
$$\text{subject to}$$
$$\text{Eqs. (4.1)} - \text{(4.7)}$$
$$\text{and}$$
$$1 \le A_{\text{p}} \le 10,$$
$$0.001 \le \dot{m} \le 0.009, T_{out}, T_{\text{p}}, U_1, Q_{\text{u}}, \Delta P, C_p, \rho, F_{\text{R}}, F', \phi \ge 0$$

where A_{p} and $\dot{m}$ are independent parameters and Tout, T_p, U_1, Q_u, DP, C_p, r, F_R, F_0, and f are dependent parameters in the optimization procedure (Farahat, Sarhaddi, & Ajam, 2009).

7) Performance characteristic curve

Performance of FPC can be plotted through experiments by measuring these three parameters: solar irradiance-I, fluid inlet temperature-T_i, and ambient air temperature-T_a. The result plot denotes single line ($\Delta T/I$ − Curve) displayed in Fig. 4.5.

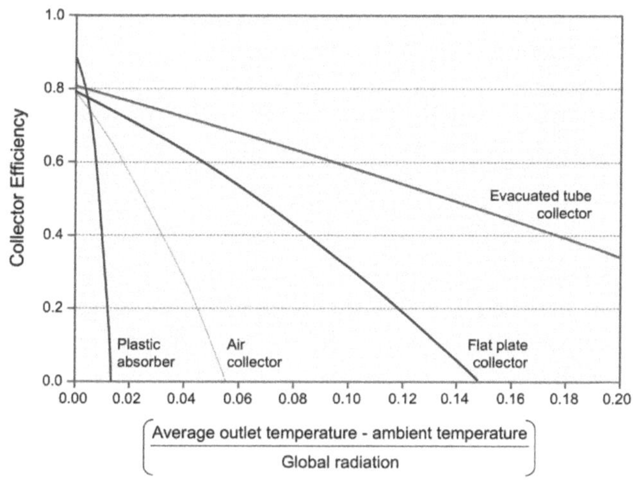

FIGURE 4.5 Performance characteristics curve of collector (Norton, 2006).

Efficiency η is determined and plotted against $(T_i - T_a)/I$. The slope represents the rate of heat loss of the collector. Experimentally, less slope occurs for collectors with cover sheets than without cover sheets. Therefore, cover sheeted collector have better performance.

B. Unglazed type FPC analysis

Schematic unglazed flat plate collector for its heat flow is shown in Fig. 4.6 below.

1) Solar-absorbing plate and the back wall energy balance equations are written as

$$q_{in} + q_{rad,W-C} = q_{conv,c-p} + q_{conv,loss} + q_{rad,loss} \tag{4.12}$$

$$q_{cond,w} = q_{conv,w-p} + q_{rad,w-c} \tag{4.13}$$

2) The absorbed energy is given by

$$q_{in} = \alpha_C I_T A_S \tag{4.14}$$

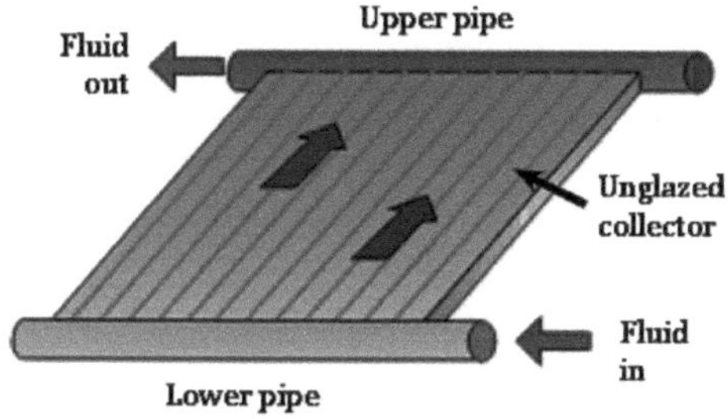

FIGURE 4.6 Unglazed type collector (Evangelisti et al., 2019).

where $A_s = (1-\sigma)A$—the frontal total area A minus the area hole. Porosity of the plate, σ, is given as $\sigma = 0.907(D/P)^2$, where α_c, I_T has been defined in Eq. (4.2).

3) The radiant exchange occurs between collector plate and back wall outer surface is (Motahar, 2010)

$$q_{rad,w-c} = \sigma_{sb}A\left(T_w^4 - T_c^4\right) \Big/ \left(\frac{1}{\varepsilon_w} + \frac{1}{\varepsilon_c} - 1\right) \tag{4.15}$$

where ε_w is infrared emittance of the back wall and ε_c is the infrared emittance of the collector plate. T_w is the back wall heat transfer temperature °C.

4) The incoming air from the collector to the plenum convects energy at the collector surface

$$q_{conv,c-p} = \dot{m}_c C_p\left(T_p - T_a\right) = \dot{m}_c C_p \varepsilon_{HX}(T_c - T_a) \tag{4.16}$$

where $T_p =$ plenum temperature °C$C_p =$ specific heat capacity of the plenum$\dot{m}_c = \rho V_s A$. The heat exchange effectiveness ε_{HX} becomes:

$$\varepsilon_{HX} = \frac{T_o - T_a}{T_c - T_a} = 1 - \exp\left(-\frac{KNu_D(1-\sigma)}{D\rho V_s C_p}\right) \tag{4.17}$$

For $0.1\% < \sigma < 5\%$ and $100 < Re_D < 2000$, from the correlation for Nusselt number predicts heat exchange effectiveness (Doğan, Sivrioğlu, & Başkaya, 2020; Saleh, 2012, p. 73):

$$Nu_D = 2.75\left[\left(\frac{P}{D}\right)^{-1.2} Re_D^{0.43} + 0.011\sigma Re_D\left(\frac{U_\infty}{V_s}\right)^{0.48}\right] \tag{4.18}$$

where $Re_D = V_h D/v$ is the Hole's Reynolds number with hole velocity of $V_h = V_s/\sigma$.

5) For unglazed, convective losses to the environment is not disregard therefore

$$q_{conv,loss} = \frac{Nu_{loss}\, k}{H} A(T_c - T_a) \tag{4.19}$$

$$Nu_{loss} = 0.82\frac{\rho C_p U_\infty v}{kV_s} C_f \tag{4.20}$$

6) The radiative losses to the environment are derived by

$$q_{rad,loss} = \varepsilon_c \sigma_{sb} A_s\left(T_c^4 - T_{sur}^4\right) \tag{4.21}$$

$$T_{sur}^4 = 0.5\left(T_{sky}^4 + T_{gr}^4\right) \tag{4.22}$$

$$T_{sky} = 0.0552\, T_a^{1.5},\ T_{gr} \cong T_a \tag{4.23}$$

where $T_{sky} =$ apparent sky temperature, $T_{gr} =$ ground temperature, $T_{sur} =$ surrounding temperature.

7) Energy is extracted can be calculated as

$$q_{cond,w} = U_{cond,w}A(T_{room} - T_w) \tag{4.24}$$

where T_{room} = room temperature °C

8) The convective heat transfer from outer wall to plenum air is

$$q_{conv,w-p} = \frac{KNu_H}{H}A(T_w - T_p) = \dot{m}_a C_p(T_o - T_p) \tag{4.25}$$

where T_o = outer temperature

$$Nu_H = \begin{cases} 0.0664\, Re_H^{0,5}\ Pr^{0.333} & Re_H < 5 \times 10^5 \\ (0.037\, R_e^{0.8} - 871)Pr^{0.333} & Re_H > 5 \times 10^5 \end{cases}$$

9) Pressure drop calculation

Considering the pressure drop across the collector as ΔP_c, the friction in the plenum as ΔP_{fric}, buoyancy force of the air as ΔP_{buoy}, and air acceleration of the system given as ΔP_{acc}. The required power related to the total pressure drop in the system ΔP_{tot} is:

$$\Delta P_{tot} = \Delta P_c + \Delta P_{fric} + \Delta P_{buoy} + \Delta P_{acc} \tag{4.26}$$

Pressure drop calculations include:

$$\Delta P_c = \frac{1}{2}\rho V_s \zeta \tag{4.27}$$

$$\zeta = 6.82\left(\frac{1-\sigma}{\sigma}\right)Re_D^{-0.236}$$

$$\Delta P_{fric} = f\frac{H\rho V_p^2}{2D_h} \tag{4.28}$$

$$V_p = \frac{1}{2}\frac{V_s H}{D_p},\ D_h = \frac{4(D_p \times W)}{2(D_p + W)}$$

$$\Delta P_{bouy} = \frac{1}{2}(\rho_o - \rho_a)gH \tag{4.29}$$

$$\Delta P_{acc} = \frac{1}{2}\rho(2V_p)^2 \tag{4.30}$$

The total fan power is determined with:

$$\dot{W}_{fan} = \frac{\dot{m}_c \Delta P_{tot}}{\rho} \tag{4.31}$$

11) The formulation of the optimization solution

To find a set of design parameters, $x = \{x_1, x_2, \ldots .x_n\}$ by applying optimization techniques. "x" might be minimization or maximization.

For example, maximize $f(x)$:

Subjected to:

$$h_j(x) = 0,\ j = 1, 2, \ldots, m$$

$$g_k(x) \leq 0,\ k = 1, 2, \ldots,\ p$$

$$x_i^L \leq x_i \leq x_i^U,\ i = 1, 2, \ldots ., n$$

where x_i^L and x_i^U are lower/upper band of x_i. Sequential quadratic programming method is applied. This approach searches a constrain maxima of a scalar function of various variables starting from an initial estimate. Objective function is the energy efficiency ψ of the unglazed collector below (Doğan et al. 2020).

$$\psi = 1 - \frac{\dot{E}x_{qloss} - \left[\dot{E}x_{qcond}\right]^{-} + \dot{I}}{\dot{E}x_{solar} + \dot{E}x_w + \left[\dot{E}x_{qcond}\right]^{+}} \tag{4.32}$$

where:

$$\dot{E}x_{solar} = I_T A_s \left(1 - \frac{4T_a}{3T_s} + \frac{1}{3}\frac{T_a}{T_s}\right)^4 \tag{4.33}$$

$$\dot{E}x_w = \dot{W}_{fan} \tag{4.34}$$

$$\dot{E}x_{qloss} = \left(q_{conv,loss} + q_{rad,loss}\right)\left(1 - \frac{T_a}{T_c}\right) \tag{4.35}$$

$$\dot{E}x_{qcond} = q_{cond}\left(1 - \frac{T_a}{T_w}\right) \tag{4.36}$$

The hole diameter and pitch are important design parameters relative to the pressure drop and heat transfer performance. It is required to find the optimum hole size and arrangement of the unglazed collector that maximize the exergy efficiency, ψ $(P,\ D)$ subject to the equality constrain functions, i.e., Eqs. (4.12)–(4.13).

Example 4.1. Determine solar radiation received and F_R with $6\ \text{W/m}^2\cdot\text{K}$ as overall heat loss coefficient. Heat transfer coefficient of the fluid tube is $1200\ \text{W/m}^2\cdot\text{K}$. The absorbed insolation during the first hour averages

90 W/m^2. The cover transmittance to solar radiation is 0.9. The solar plate absorptance 0.9, collector 1 m wide and 3 m long. Water flow rate is 0.02 kg/s. Temperature of water is 330 K = 27°C and ambient temperature 0°K (Take specific heat capacity of water to be 1 kJ/kg·K).

Solution. **i.** The solar radiation received Q_i:

$$Q_i = 90(0.9 \times 0.9) \cdot 3 = 218.7 \text{ W}$$

ii. The Heat removal factor F_R:

$$F_R = \frac{Q_a}{A.Q_u}, \ (300 \text{ K})$$

$$Q_a = 0.02 \times 1 \times 300 = 60 \text{ W}$$

$$Q_u = [(90 \times 0.9 \times 0.9) - 1200\ (0)] = 72.9 \text{ W}$$

$$F_R = \frac{60}{3 \times 72.9} = 0.274$$

Example 4.2. Calculate the top radiation losses to the surrounding with an unglazed collector, the specifications are given below:

Surface Area = 25 mm^2
Emittance = 0.95
Air and sky ambient temperature = 1°C
Heat transfer coefficient of the wind = 10 W/m^2 °C
Temperature mean plate = 100°C
Collector tilt = 45 degrees
Emittance (Glass) = 0.88
Plate porosity = 0.01

Solution. The following equation given below will be used to determine the radiation losses to the surrounding:

$$q_{rad,loss} = \varepsilon_c \sigma_{sb} A_s \left(T_c^4 - T_{sur}^4\right)$$

$$T_{sur}^4 = 0.5\left(T_{sky}^4 + T_{gr}^4\right)$$

$$T_{sky} = 0.0552\ T_a^{1.5},\ T_{gr} \cong T_a$$

i. $Tg \cong Ta \cong 1°C$
ii. $T_{sky} = 0.0552 \times 10^{1.5} = 1.7°C$
iii. $T^4_{sur} = 0.5\left(1.7^4 + 1^4\right) = 4.7°C$
iv. $q_{rad,\,loss} = 0.88 \times 0.01 \times 25\left(100^4 - 4.7^4\right) = 2.2 \times 10^7$ W

Applications

Flat plate collector is applied for residential/commercial water, pool, residential space heating, and industrial process heat. Solar cooling, air conditioning systems are also common applications. Flat plate collector efficiency for a liquid is (40–50) %.

Temperature range: Flat plate collector's temperature range are between 40 and 60°C.

4.3.2 Evacuated tube collectors

Evacuated tube collector is a tube-like collector consisting of series of evacuated tubes to raise water temperature for use when received sunlight. Vacuum, or evacuated spaces are used in the tubes to let in radiations and trap heat by reducing heat loss to the atmosphere. The tubes are made up of metal which act as absorber plate, fixed to a pipe (heat pipe) to dispatch the heat energy collected from the sun to the water for applications.

Evacuated tube collectors function with the basic principle of applying sunlight to heat liquids. Main constituents in the design of ETC in Fig. 4.7 are stated below:

Heat pipe—Transfer the heat to the fluid
Absorber plate—Absorbs the radiation entering the evacuated tube
Copper manifold—Heat exchanger
Thermal analysis of ETC is demonstrated below (Fig. 4.8):

1) Evacuated collector's energy efficiency is denoted as:

$$\eta_{th} = \frac{Q_u}{A_p S_{\text{eff}}} \tag{4.37}$$

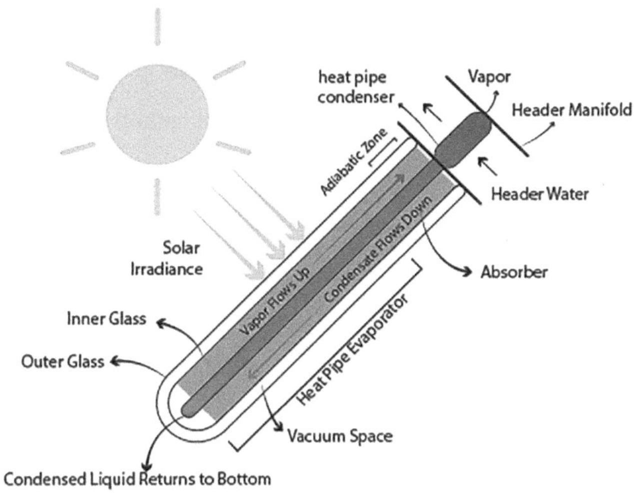

FIGURE 4.7 Solar evacuated tube collector (Elsheniti, Kotb, & Elsamni, 2019).

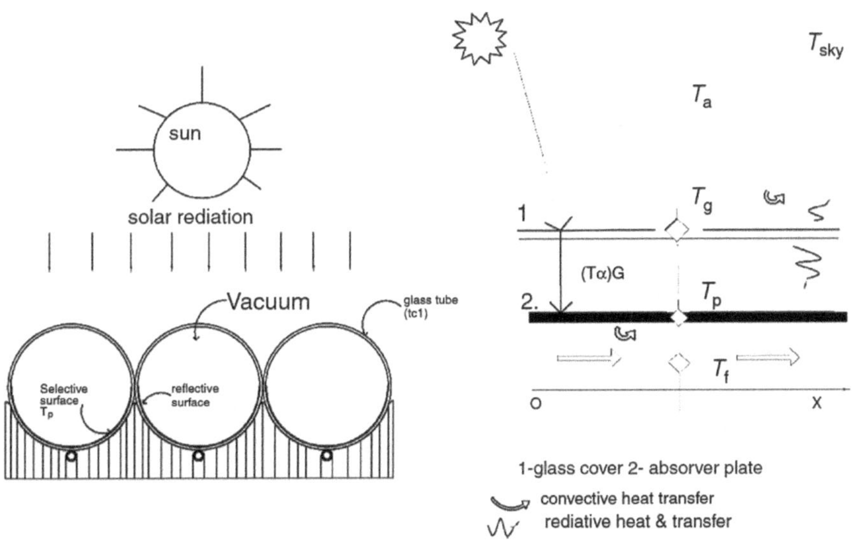

FIGURE 4.8 Typical evacuated tube collector (Arora, Chitkara, Udayakumar, & Ali, 2011).

S_{eff}—effective insolation coefficient on a singular tube. The useful energy extraction (Q_u) as follows:

$$Q_u = \dot{m}C_p(T_{out} - T_{in}) \tag{4.38}$$

T_{in}, T_{out}, Cp, $\dot{m}$ are the temperature of the inlet fluid, fluid outlet temperature, heat capacity, and mass flow rate of the agent fluid, respectively.

General energy balance equation is written as:

$$\sum \dot{E}_{in} - \sum \dot{E}_{out} = \sum \dot{E}_{dot} \tag{4.39}$$

where $\sum E_{in}$, $\sum E_{out}$, $\sum E_{dest}$ are the total exergy input, total exergy output, and total exergy destruction rates, respectively.

2) Exergy rate is denoted by:

$$\dot{E} = \dot{m}_f[(h - h_0) - T_0(S - S_0)] \tag{4.40}$$

3) The energy balance equation is written as follows:

$$\dot{E}_{in} + \dot{E}_{solar} + \dot{E}_{fin} = \dot{E}_{out} + IR \tag{4.41}$$

The solar radiation energy rate, according to Petela theorem, is calculated by (Petela, 2008, 2010; Saxena & Gaur, 2018):

$$\dot{E}_{\text{solar}} = S_{\text{eff}}A_p\left(1 - \frac{4T_a}{3T_{\text{sun}}} + \frac{1}{3}\left(\frac{T_a}{T_{\text{sun}}}\right)^4\right) \tag{4.42}$$

4) The energy efficiency is calculated based on second law of thermodynamics in (Saxena & Gaur, 2018):

$$\eta_E = \frac{\dot{E}_{useful.net}}{S_{\text{eff}}A_p\left(1 - \frac{4T_a}{3T_{\text{sun}}} + \frac{1}{3}\left(\frac{T_a}{T_{\text{sun}}}\right)^4\right)} \tag{4.43}$$

Example 4.3. Evacuated solar tube collector operating at 50% efficiency and the insolation coefficient is 80 W/m^2. Area of the plate is 3 m^2 and water flow rate is 0.02 kg/s (Take specific heat capacity of water C_p to be 1 kJ/kg·K). What is the operating temperature of the ETC?

Solution. To calculate for operating temperature, the efficiency equation is first used to find the useful energy gain:

$$\eta_{th} = \frac{Q_u}{A_p S_{\text{eff}}}$$

$$Q_u = \eta_{th} A_p S_{\text{eff}} = 0.5 \times 3 \times 80 = 120 \text{ W}$$

The operating temperature is further calculated with the following:

$$Q_u = \dot{m}C_p(T_{out} - T_{in})$$

$$T_o = \frac{Q_u}{mC_p} = \frac{120}{0.02} = 6000^\circ C \quad where,\ T_o = (T_{out} - T_{in})$$

Application

Residential/commercial water, space heating, cooling and industrial process heat. Efficiencies are typically (30–45)%. In general, ETCs will out-perform FPTs in cold weather because the efficiency of ETCs does not fall off as quickly when the outside air temperatures drop.

Temperature range: ETC has a moderate temperature range of 150–200°C.

4.4 Concentrator collectors

Concentrator consists of imaging and nonimaging collectors which are classified into various types as seen in Fig. 4.2 above. It is characterized by optical CR_o and geometrical CR concentration ratio. It traces the sun ray through some degree of freedom (DOF). Its advantage over nonconcentrator collector is its small area. Basic concept is shown in Fig. 4.9. Where:

$$CR_o = \frac{I_r}{I_o},\ CR = \frac{A_a}{A_r} \quad (4.44)$$

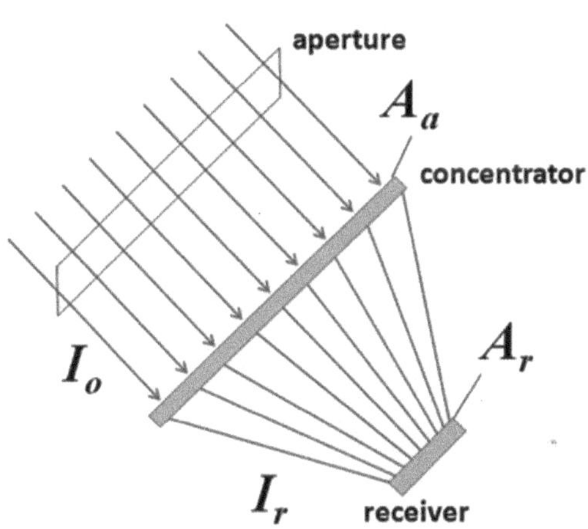

FIGURE 4.9 The basic design of concentrator collector (Nasrin, Hasanuzzaman, & Rahim, 2018a, 2018b; Thirugnanasambandam, Iniyan, & Goic, 2010).

But, the concentrator is perfect with regards to optical losses then, the optical and geometrical concentrator ratios are identical, that is, $CR_o = CR$.

The various solar concentrators will be discussed and analyzed by the characteristics.

4.4.1 Compound parabolic and Fresnel lens collectors

Compound parabolic collector (CPC) and linear Fresnel lens nonimaging collectors are ideal solar concentrating collectors. They have similarities but the slight difference is that, the sun focuses on flat linear Fresnel lens on the receiver, thus for CPC, the concentrator is not a flat lens, but a compound parabolic reflector (Figs. 4.10 and 4.11).

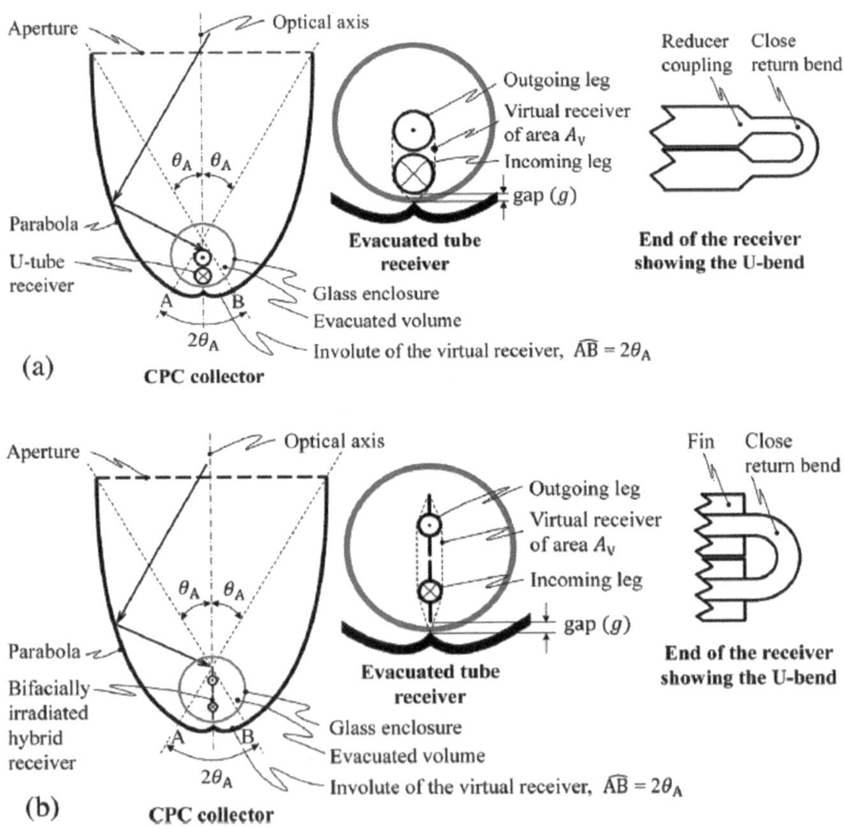

FIGURE 4.10 Cross-section of a CPC designed for (A) a U-tube receiver without fins, and (B) a hybrid of bifacially irradiated flat receiver and a U-tube receiver (Karwa, Jiang, Winston, & Rosengarten, 2015).

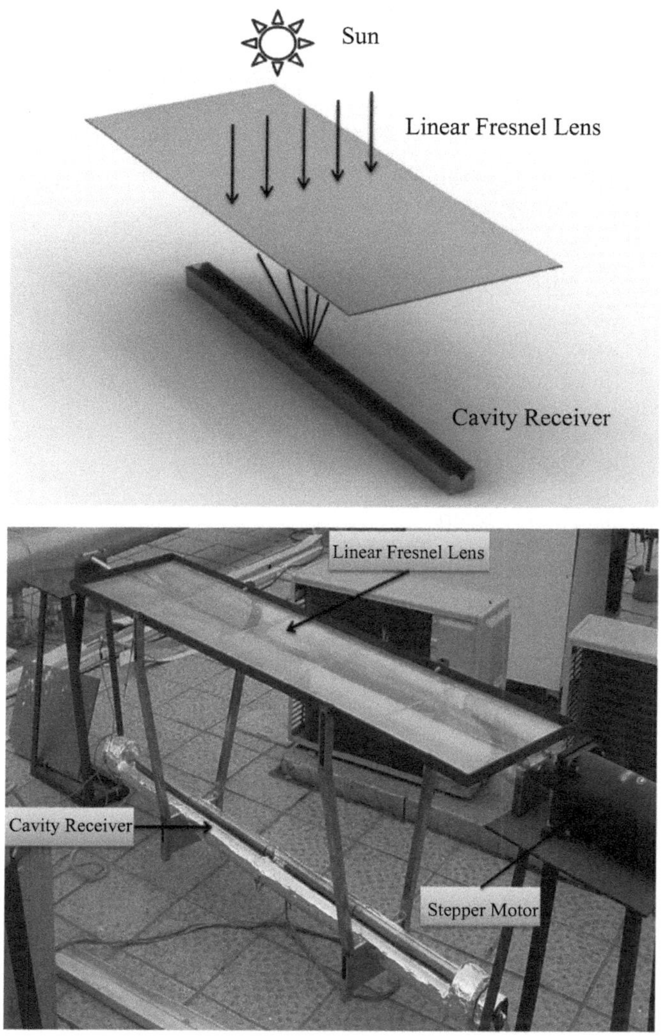

FIGURE 4.11 Flat linear Fresnel lens (Lin, Sumathy, Dai, & Zhao, 2014).

Thermal analysis can be evaluated, therefore, it's necessary to determine the collector efficiency factor, F'; coefficient loss, U_L and Heat removal factor, F_R of the collector (Marefati, Mehrpooya, & Shafii, 2018).

$$F_R = \frac{m_c \cdot C_p}{A_r \cdot U_L} \cdot \left[1 - \exp\left(\frac{-A_r \cdot U_L \cdot F'}{m_c \cdot C_p}\right)\right] \tag{4.45}$$

where A_r receiver area in (m^2) and collector efficiency F^I

$$F' = \frac{1/U_L}{1/U_L + \frac{D_{ro}}{h_{fi} D_{ri}} + \frac{D_{ro}}{2k_m} \cdot \left[\ln\left(\frac{D_m}{D_{ri}} \right) \right]} \tag{4.46}$$

D_{ri} and D_{ro} are inner and outer diameters. U_L is the thermal loss coefficient which can be determined in the following equation:

$$U_L = \left[\frac{A_r}{\left(h_{c,c-am} + h_{r,c-am}\right) \cdot A_c} + \frac{1}{h_{r,r-c}} \right]^{-1} \tag{4.47}$$

where $h_{r,r\text{-}c}$ is the coefficient of radiation from receiver to cover given below:

$$h_{r,r-c} = \frac{\sigma \cdot \left(T_r^2 + T_c^2\right) \cdot \left(T_r + T_c\right)}{\frac{1}{\varepsilon_r} + \left(\frac{1}{\varepsilon_c} - 1\right) \cdot \left(\frac{A_r}{A_c}\right)} \tag{4.48}$$

where T_c is glass cover temp.

$$T_c = \frac{A_r h_{r,r-c} T_r + A_c \left(h_{r,c-am} + h_{c,c-am}\right) T_{amb}}{A_r h_{r,r-c} + A_c \left(h_{r,c-am} + h_{c,c-am}\right)} \tag{4.49}$$

where T_r is the receiver temperature, calculated as:

$$T_r = T_{in} + \frac{Q_u}{A_r \cdot U_L \cdot F_R} (1 - F_R) \tag{4.50}$$

Coefficient of radiation between cover and the sky:

$$h_{r,c-am} = \varepsilon_c \cdot \sigma \cdot \left(T_c^2 + T_{sky}^2\right) \cdot \left(T_c + T_{sky}\right) \tag{4.51}$$

The Nusselt number is calculated with the following equation:

$$Nu = 0.023\, .\, Re^{0.8} \cdot Pr^{0.4} \tag{4.52}$$

Now, we can calculate the h_{fi} with the next equation:

$$Nu = \frac{h_{fi} \cdot D_{ri}}{k_{fi}} \tag{4.53}$$

Convection coefficient between the cover and air, $h_{c,c\text{-}am}$ is determined by equation:

$$h_{c,c-am} = \frac{8.6 \times V_{air}^{0.6}}{L^{0.4}} \tag{4.54}$$

Collector flow rate can be determined below:

$$m_c = m_a \cdot A_a \tag{4.55}$$

Useful energy delivered from the concentrator is derived:

$$Q_u = F_R\left(G_b \cdot \eta_{opt} \cdot A_a - A_r \cdot U_L \cdot (T_{in} - T_{am})\right) \tag{4.56}$$

Collector thermal efficiency is determined as follows:

$$\eta = \frac{Q_u}{G_b \cdot A_a} = F_R \cdot \left[\eta_{opt} - U_L \cdot \left(\frac{T_{in} - T_{am}}{G_b \cdot C}\right)\right] \tag{4.57}$$

A) Extra Equations for the Developed Model.

The thermodynamic limit of concentration for CPC as shown below:

$$CR_{CPC} = CR_{max,2D} = 1/sin\theta_{max} \tag{4.58}$$

B) Optimal analysis of CPC collector.

If we considered intercept factor γ, therefore:

$$\gamma = \left[I_{b,c} + \frac{I_{d,c}}{CR}\right] I_{tot,c} \tag{4.59}$$

where b is beam, d is diffuse, and tot is total flux incident on the collector aperture.

Optical efficiency denotes:

$$\eta_o = \rho_m^{\overline{n}} \tau_r \alpha_r \gamma \tag{4.60}$$

Thermal efficiency (with respect to total beam and diffuse plane flux) is written as (Marefati et al., 2018):

$$\eta_c = \rho_m^{\overline{n}} \tau_r \alpha_r \gamma - U_c\,(T_r - T_a)/(I_c CR) \tag{4.61}$$

where U_c is based on the aperture area (Marefati et al., 2018).

$$U_c = \varepsilon_e \sigma\,(T_r^2 + T_a^2)(T_r + T_a) + \left(\frac{A_k}{A_a}\right)\left(\frac{1}{R_k}\right)(CR) \tag{4.62}$$

where R_k is the conduction heat transfer resistance for entire conduction paths of total effective area A_k.

Application

Water desalination: CPCs are efficient and regarded as higher temperature collectors.

Temperature range: CPCs are used for the medium temperature range between 60°C and 150°C.

4.4.2 Parabolic trough collector

A parabolic trough collector (PTC) system has reflecting surface which resembles a parabolic shape. This reflecting surface is mostly made of reflecting mirrors or anodized aluminum sheets. PTC often traces the sun radiation with a DOF by applying each of the three situations: East to West, North to South, or Polar. Its absorbers are tubular that are covered in glass tube to minimize convective and radiative losses (Fig. 4.12).

The present model is done in five steps.

Step 1:

The thermal losses involve radiations and convection losses which is given by:

$$Q_{loss} = A_{co} \times \varepsilon_c \times \sigma \times \left(T_c^4 - T_{am}^4\right) + A_{co} \times h_{out} \times (T_c - T_{am}) \tag{4.63}$$

Cover temperature is assumed to be near to the ambient temperature because of ETC occurrence. Therefore, applying Taylor series:

$$T_c^4 - T_{am}^4 \approx 4 \times T_{am}^3 \times (T_c - T_{am}) \tag{4.64}$$

Eq. (4.64) validation is examined by comparing with other models. Eqs. (4.63) and (4.64) combination gives:

$$Q_{loss} = K_1 \times (T_c - T_{am}) \tag{4.65}$$

where $K_1 = A_{co} \times \varepsilon_c \times \sigma \times A \times T_{am}^3 + A_{co} \times h_{out}$

Step 2:

The thermal losses involve only radiation losses but not convection losses, therefore:

$$Q_{loss} = A_{ro} \times \varepsilon_r^* \times \sigma \times \left(T_r^4 - T_c^4\right) \tag{4.66}$$

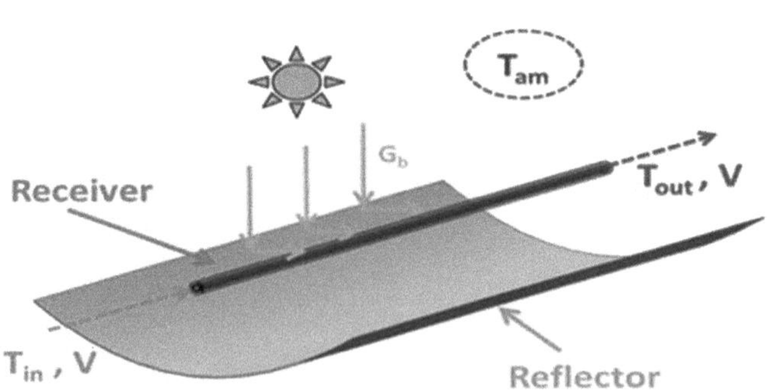

FIGURE 4.12 Typical parabolic trough collector (Bellos & Tzivanidis, 2018; Islam, Hasanuzzaman, & Abd Rahim, 2015).

With the following equation:

$$\varepsilon_r^* = \left[\frac{1}{\varepsilon_r} + \frac{1-\varepsilon_c}{\varepsilon_c} \times \frac{A_{ro}}{A_{ci}}\right]^{-1} \tag{4.67}$$

Eq. (4.66) can be written as below:

$$Q_{loss} = A_{co} \times \varepsilon_r^* \times \sigma \times \left(T_r^4 - T_{am}^4\right) - A_{co} \times \varepsilon_r^* \times \sigma \times \left(T_c^4 - T_{am}^4\right) \tag{4.68}$$

Eqs. (4.64) and (4.66) are combined as the following:

$$T_c^4 - T_{am}^4 = 4 \times T_{am}^3 / K_1 \times Q_{loss} \tag{4.69}$$

Eqs. (4.68) and (4.69), combined together:

$$Q_{loss} = k_2 \times \left(T_r^4 - T_{am}^4\right) \tag{4.70}$$

$$K_2 = A_{co} \times \varepsilon_r^* \times \sigma \times \left[1 + \frac{4 \times T_{am}^3 \times A_{ro} \times \varepsilon_r^* \times \sigma}{K_1}\right]^{-1}$$

Step 3:
Energy balance in the fluid volume is calculating the useful heat:

$$Q_u = m \times C_p \times (T_{out} - T_{in}) \tag{4.71}$$

And also derived from heat transfer to the fluid from the receiver:

$$Q_u = A_{ri} \times h \times \left(T_r - T_{fm}\right) \tag{4.72}$$

where T_{fm} is the fluid mean temperature:

$$T_{fm} = \frac{T_{in} + T_{out}}{2} \tag{4.73}$$

Using Eqs. (4.71)–(4.73), it can be written:

$$Q_u = K_3 \times (T_r - T_{in}), \tag{4.74}$$

$$\text{Where,} \quad K_3 = \left[\frac{1}{A_{ri} \times h} + \frac{1}{2 \times m \times c_p}\right]^{-1}$$

Step 4:
Simplifying Eq. (4.70) is written below:

$$Q_{loss} = K_2 \times \left(T_r^4 - T_{in}^4\right) + K_2 \times \left(T_{in}^4 - T_{am}^4\right) \tag{4.75}$$

Applying Taylor series, the temperature difference between receiver and fluid becomes:

$$T_r^4 - T_{in}^4 \approx 4 \times T_{in}^3 \times (T_r - T_{in}) \tag{4.76}$$

Using Eqs. (4.74) and (4.76), it can be written:

$$T_r^4 - T_{in}^4 \approx 4 \times T_{am}^3 / K_3 \times Q_u \quad (4.77)$$

Using Eqs. (4.75) and (4.77), it can be said:

$$Q_{loss} = \frac{4 \times T_{in}^3 \times K_2}{K_3} + K_2 \times \left(T_{in}^4 - T_{am}^4\right) \quad (4.78)$$

Step 5:

The absorbed energy is equal to the product of the optical efficiency and the direct beam irradiation:

$$\eta_{opt} \times Q_s = Q_u + Q_{loss} \quad (4.79)$$

Using Eqs. (4.78) and (4.79), it can be written:

$$Q_u = K_4 \times Q_s - K_5 \times \left(T_{in}^4 - T_{am}^4\right) \quad (4.80)$$

$$K_4 = \eta_{opt} \times \left[1 + \frac{4 \times T_{am}^3 \times K_2}{K_3}\right]^{-1} \quad (4.81)$$

$$K_5 = K_2 \times \left[1 + \frac{4 \times T_{am}^3 \times K_2}{K_3}\right]^{-1} \quad (4.82)$$

Parameter calculation

Solar collector thermal efficiency (η_{th}) is calculated as:

$$\eta_{th} = T_4 - K_5 \times \frac{\left(T_{in}^4 - T_{am}^4\right)}{A_a \times G_b} \quad (4.83)$$

Combining (4.79) and (4.80), thermal losses (Q_{loss}) is written as:

$$Q_{loss} = \left(\eta_{opt} - K_4\right) \times Q_s + K_5 \times \left(T_{in}^4 - T_{am}^4\right) \quad (4.84)$$

Combining (4.74) and (4.80), receiver temperature (T_r) is written as:

$$T_r = T_{in} + \frac{K_4}{K_3} \times Q_s - \frac{K_5}{K_3} \times \left(T_{in}^4 - T_{am}^4\right) \quad (4.85)$$

Combining (4.69) and (4.84), cover temperature (Tc) is written as:

$$T_c = T_{am} + \frac{\eta_{opt} - K_4}{K_1} \times Q_s - \frac{K_5}{K_1} \times \left(T_{in}^4 - T_{am}^4\right) \quad (4.86)$$

Combining (4.75) and (4.84), fluid outlet temperature ($Tout$) can be written as:

$$T_{out} = T_{in} + \frac{K_4}{m \times c_p} \times Q_s - \frac{K_5}{m \times c_p} \times \left(T_{in}^4 - T_{am}^4\right) \tag{4.87}$$

Combining (4.73) and (4.80), fluid mean temperature (T_{fm}) is written as:

$$T_{fm} = T_{in} + \frac{K_4}{2 \times m \times c_p} \times Q_s - \frac{K_5}{2 \times m \times c_p} \times \left(T_{in}^4 - T_{am}^4\right) \tag{4.88}$$

More equations for the model

1) Optical efficiency

The optical efficiency $\eta_{opt}(\theta)$ is calculated as the product of the incident angle modifier $K(\theta)$ to the maximum optical efficiency $\eta_{o,peak}$, which is obtained for zero incident angle:

$$\eta_{opt}(\theta) = K(\theta) \times \eta_{o,peak} \tag{4.89}$$

The peak optical efficiency, $\eta_{o,peak}$:

$$\eta_{o,peak} = \rho\tau\alpha\gamma \tag{4.90}$$

where

ρ = Reflectance of clean mirror (0.93–0.94).
τ = Transmittance of the absorber glass cover typically 0.93 but may increase above 0.96 when using antireflective coatings on each side of glass.
α = Absorptance (ranges from 0.94 to 0.95)
γ = Intercept factor

$$\gamma = \gamma_1\gamma_2\gamma_3 \tag{4.91}$$

γ_1—Geometrical errors in shape,
γ_2—Receiver supports and flexible bellows Shadowing,
γ_3—Deformation of mechanical support structure at the time of tracking

$$K(i) = \frac{\eta_o(i)}{\eta_o(i=0)} \tag{4.92}$$

$K(i)$ —incident angle modifier.

The end loss factor F_{el} is given:

$$F_{e1} = \frac{(L_c - Z)}{L_c} = 1 - \frac{r}{L_c}\tan i \tag{4.93}$$

If $r = f$ then,

$$F_{e1} \approx 1 - \frac{f}{L_c}\tan i \tag{4.94}$$

For r average value, therefore, aperture with width W:

$$F_{e1} = 1 - \frac{f}{L_c} 1 + \frac{W^2}{48 f^2} \tan i \tag{4.95}$$

Optical efficiency becomes:

$$\eta_o = \eta_{o \cdot peak} F_{el} F_s K(i) = \rho \tau \alpha \gamma F_{el} F_s K(i) \tag{4.96}$$

2) Thermal losses from PTC

The heat loss q_L may be written as:

$$q_L = A_k \frac{T_r - T_a}{R_k} + \varepsilon r \sigma \left(T_r^4 - T_a^4\right) A_r \tag{4.97}$$

3) Instantaneous efficiency η_c can be written as:

$$\begin{aligned} \eta_c &= \frac{q_u}{I_c A_a} = \frac{q_{abs} - q_l}{I_c A_a} \\ &= \eta_o - U_c \frac{(T_r - T_a) A_r}{I_c A_a} \\ &= \eta_o - U_c \frac{T_r - T_a}{I_c} \cdot \frac{A_r}{A_a} \end{aligned} \tag{4.98}$$

where T_r is temperature and A_r is area of the receiver. A_a is the aperture area. Variation in overall heat loss coefficient is expressed in second degree polynomial equation:

$$U_c = a + b(T_r - T_a) + c(T_r - T_a)^2 \frac{W}{m^2} \cdot K \tag{4.99}$$

a, b, and c are found experimentally. Typically, U_c is less than 5 W/m^2·K for absorber tubes with good vacuum.

Application

It can be used for domestic/industrial fluid heating, cooking and electricity production.

Temperature range: Between 150 and 400°C

4.4.3 Parabolic dish reflector

Fig. 4.13 shows the basic geometric design of parabolic dish reflector which uses a geometric concentrator instead of optical. The geometrical concentration ratio CR is calculated using Eq. (4.44) above.

For thermal analysis:

1) The optical efficiency is determined in the following:

$$\eta_o = \rho_c \tau_v \rho S \tag{4.100}$$

where $S = \dfrac{A_a - A_t}{A_a}$ and $A_t = A_a - Area\ base\ recibidor$

2) Heat absorbed by the dish can be calculated:

$$Q_{opt} = A_a \rho_c \tau_v \rho S I_b \tag{4.101}$$

3) At the receiver, the heat loss can be determined by:

$$Q_{loss} = A_r U_L (T_{rm} - T_{amb}) \tag{4.102}$$

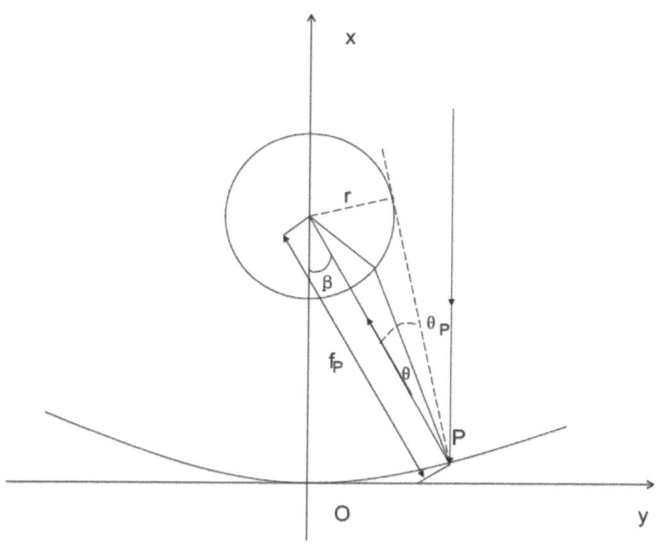

FIGURE 4.13 Geometric and dimension of parabolic dish reflector (Huang, Huang, Hu, & Chen, 2013).

Where these parameters have been defined above.

$U_L = \left[\frac{1}{h_w+h_r}\right]^{-1}$, $hr = 4\sigma\varepsilon_r Tair^3$, and $h_w = k_{air}/D_{out} * Nu$ Re $= V_{air} * D_{out}/Y_{air}$ and $Nu = 0.40 + 0.54Re^{0.52}$

Where $Y_{air} = \mu_{air}/\rho_{air}$

4) Useful energy delivered out is derived by:

$$Q_{out} = Q_{opt} - Q_{loss} \tag{4.103}$$

5) The instantaneous thermal efficiency becomes:

$$\eta_{inst} = \frac{Q_{out}}{A_a I_b} * 100 \tag{4.104}$$

Application

The useful energy from PDF can be used to generate clean electricity.

Temperature range: 80–900°C and can withstand oxidation during continuous and intermittent services.

4.4.4 Central receiver or heliostat field reflector

A central receiver collector tracks the solar and reflects the radiation into the receiver/absorber situated on a tower. CR involves a mirrors of large field called heliostats, receiver, and heliostats control tracking. Thermal energy absorbed by the receiver is transported to the medium fluid that will be kept for application.

A simple design of heliostat field is seen in Fig. 4.14. It is known as a central high temperature solar heat technologies [13]. It is important to analyze the thermal performance of this collector.

1) The rate of heat absorbed by irradiation is calculated by

$$Q_s = I \times A_{field} \tag{4.105}$$

2) The efficiency of the heliostat is denoted as

$$\eta_H = \frac{Q_{cen}}{Q_s} \tag{4.106}$$

where Q_{cen} is the rate of heat received by the central receiver and Q_s is the rate of heat received by irradiation.

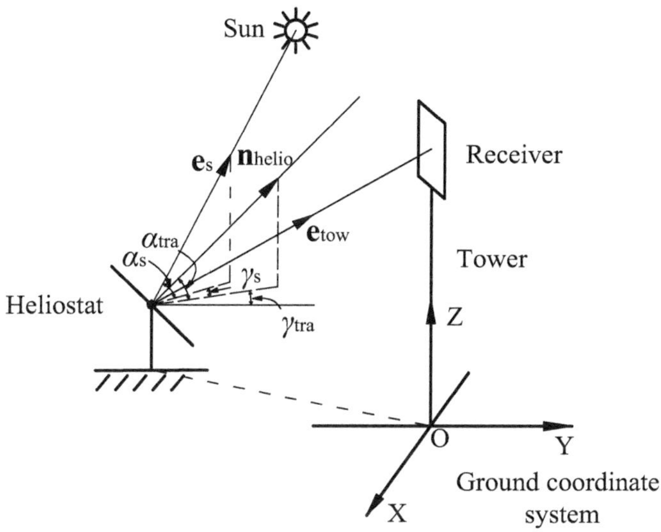

FIGURE 4.14 Heliostat solar collector (Yao, Hu, & Gao, 2015).

3) The rate of heat loss can be through emissivity, reflection, convection, and conduction

i. Rate of heat loss due to emissivity

$$Q_{cen,em} = \frac{\varepsilon_{avg} \times \sigma\left(T^4_{cen,surf} - T^4_o\right)A_{field}}{C} \tag{4.107}$$

where $\varepsilon_{avg} = \frac{\varepsilon_w}{\varepsilon_w + (1-\varepsilon_w)F_r}$. ε_w represents the wall emissivity, and F_r represents view factor. $T_{cen,\ surf}$ represents the temperature of the receiver's surface and T_0 represents the ambient temperature of the surroundings [13]. C is the concentration ratio.

ii. Due to reflection

$$Q_{cen,ref} = Q_{cen}\rho F_r \tag{4.108}$$

iii. Due to convection

$$Q_{cen,\ cov} = \frac{\frac{\left(h_{air,fc,insi}\left[T_{cen,surf} - T_o\right] + h_{air,nc,insi}\left[T_{cen,surf} - T_o\right]\right)A_{field}}{F_r}}{C} \tag{4.109}$$

where $h_{air,\ fc,\ insi}$ and $h_{air,\ nc,\ insi}$ are the force and natural convective heat transfer coefficients of the inside side of the receiver.

$h_{air,nc,insi} = 0.81(T_{cen,surf} - T_o)^{0.426}$ while $h_{air,fc,insi}$ can be determined by Nusselt Number: $Nu_{air,fc,insi} = 0{:}0287 * Re^{0.8}{}_{air,insi} * Pr^{1/3}{}_{air,insi}$

iv. Due to conduction

$$Q_{cen,cond} = \frac{\left(T_{cen,\ surf} - T_o\right)A_{field}}{\left[\frac{\delta_{insu}}{\lambda_{insu}} + \frac{1}{h_{air,o}}\right](C \times F_r)} \tag{4.110}$$

where δ_{insu} is the insulation thickness, λ_{insu} is the thermal conductivity of the insulation, and $h_{air,o}$ is the convective heat transfer coefficient of the outside air.

4) The rate of heat absorbed by the central receiver is given by

$$Q_{cen,abs} = m_s \times C_p(T_o - T_{in}) \tag{4.111}$$

5) The total heat received by the central receiver is given by

$$Q_{cen} = Q_{cen,em} + Q_{cen,ref} + Q_{cen,cov} + Q_{cen,cond} + Q_{cen,abs} \tag{4.112}$$

6) The central receiver thermal efficiency becomes

$$\eta_{cen} = \frac{Q_{cen,abs}}{Q_{cen}} \tag{4.113}$$

Applications

Central heliostat high temperature range is used for larger power production capacity and higher efficiency in comparison to other solar technologies.

Temperature range: Heliostat fields have an operating temperature range of 150–2000°C

References

Arora, S., Chitkara, S., Udayakumar, R., & Ali, M. (2011). Thermal analysis of evacuated solar tube collectors. *Journal of Petroleum and Gas Engineering, 4*(2), 74–82.

Barone, G., Buonomano, A., Forzano, C., Palombo, A., & Panagopoulos, O. (2019). Experimentation, modelling and applications of a novel low-cost air-based photovoltaic thermal collector prototype. *Energy Conversion and Management, 195*, 1079–1097. https://doi.org/10.1016/j.enconman.2019.04.082

Bellos, E., & Tzivanidis, C. (2018). Analytical expression of parabolic trough solar collector performance. *Designs, 2*(1), 9.

Bezaatpour, M., & Rostamzadeh, H. (2021). Design and evaluation of flat plate solar collector equipped with nanofluid, rotary tube, and magnetic field inducer in a cold region. *Renewable Energy, 170*, 574–586.

Doğan, A., Sivrioğlu, M., & Başkaya, S. (2020). Numerical investigation of heat transfer from heat sources placed in a horizontal rectangular channel. *Cumhuriyet Science Journal, 41*(3), 732–740.

Elsheniti, M. B., Kotb, A., & Elsamni, O. (2019). Thermal performance of a heat-pipe evacuated-tube solar collector at high inlet temperatures. *Applied Thermal Engineering, 154*, 315–325. https://doi.org/10.1016/j.applthermaleng.2019.03.106

Evangelisti, L., Vollaro, R. D. L., & Asdrubali, F. (2019). Latest advances on solar thermal collectors: A comprehensive review. *Renewable and Sustainable Energy Reviews, 114*, 109318. https://doi.org/10.1016/j.rser.2019.109318

Farahat, S., Sarhaddi, F., & Ajam, H. (2009). Exergetic optimization of flat plate solar collectors. *Renewable Energy, 34*(4), 1169–1174. https://doi.org/10.1016/j.renene.2008.06.014

Fayaz, H., Rahim, N. A., Hasanuzzaman, M., Nasrin, R., & Rivai, A. (2019). Numerical and experimental investigation of the effect of operating conditions on performance of PVT and PVT-PCM. *Renewable Energy, 143*, 827–841. https://doi.org/10.1016/j.renene.2019.05.041

Fayaz, H., Rahim, N., Saidur, R., & Hasanuzzaman, M. (2018). Techno-economic analysis of evacuated tube solar water heater using F-chart method. In *Paper presented at the IOP conference series: Materials science and engineering*.

Greco, A., Gundabattini, S., Gnanaraj, D. S., & Masselli, C. (2020). A comparative study on the performances of flat plate and evacuated tube collectors deployable in domestic solar water heating systems in different climate areas. *Climate, 8*(6), Article 78. https://doi.org/10.3390/cli8060078

Huang, W., Huang, F., Hu, P., & Chen, Z. (2013). Prediction and optimization of the performance of parabolic solar dish concentrator with sphere receiver using analytical function. *Renewable Energy, 53*, 18–26. https://doi.org/10.1016/j.renene.2012.10.046

Islam, M.k., Hasanuzzaman, M., & Abd Rahim, N. (2015). Modelling and analysis on the effect of different parameters on a parabolic-trough concentrating solar system. *RSC Advances, 5*(46), 36540–36546. https://doi.org/10.1039/C4RA12919A

Islam, M. P., & Morimoto, T. (2018). Advances in low to medium temperature non-concentrating solar thermal technology. *Renewable and Sustainable Energy Reviews, 82*, 2066–2093. https://doi.org/10.1016/j.rser.2017.08.030

Jafarkazemi, F., & Ahmadifard, E. (2013). Energetic and exergetic evaluation of flat plate solar collectors. *Renewable Energy, 56*, 55–63.

Karwa, N., Jiang, L., Winston, R., & Rosengarten, G. (2015). Receiver shape optimization for maximizing medium temperature CPC collector efficiency. *Solar Energy, 122*, 529–546. https://doi.org/10.1016/j.solener.2015.08.039

Kumar, L., Hasanuzzaman, M., & Rahim, N. A. (2019). Global advancement of solar thermal energy technologies for industrial process heat and its future prospects: A review. *Energy Conversion and Management, 195*, 885–908. https://doi.org/10.1016/j.enconman.2019.05.081

Lin, M., Sumathy, K., Dai, Y., & Zhao, X. (2014). Performance investigation on a linear Fresnel lens solar collector using cavity receiver. *Solar Energy, 107*, 50–62. https://doi.org/10.1016/j.solener.2014.05.026

Mahdi, K., & Bellel, N. (2014). Development of a spherical solar collector with a cylindrical receiver. *Energy Procedia, 52*, 438–448. https://doi.org/10.1016/j.egypro.2014.07.096

Marefati, M., Mehrpooya, M., & Shafii, M. B. (2018). Optical and thermal analysis of a parabolic trough solar collector for production of thermal energy in different climates in Iran with comparison between the conventional nanofluids. *Journal of Cleaner Production, 175*, 294–313.

Motahar, S. (2010). An analysis of unglazed transpired solar collectors based on exergetic performance criteria. *International Journal of Thermodynamics, 13*(4), 153–160.

Nasrin, R., Hasanuzzaman, M., & Rahim, N. A. (2018a). Effect of high irradiation and cooling on power, energy and performance of a PVT system. *Renewable Energy, 116*(Part A), 552–569. https://doi.org/10.1016/j.renene.2017.10.004

Nasrin, R., Hasanuzzaman, M., & Rahim, N. A. (2018b). Effect of high irradiation on photovoltaic power and energy. *International Journal of Energy Research, 42*(3), 1115–1131. https://doi.org/10.1002/er.3907

Norton, B. (2006). Anatomy of a solar collector: Developments in materials, components and efficiency improvements in solar thermal collector systems. *Refocus, 7*(3), 32–35. https://doi.org/10.1016/S1471-0846(06)70570-4

Petela, R. (2008). An approach to the exergy analysis of photosynthesis. *Solar Energy, 82*(4), 311–328. https://doi.org/10.1016/j.solener.2007.09.002

Petela, R. (2010). *Engineering thermodynamics of thermal radiation: For solar power utilization.* McGraw-Hill Education.

Qazi, A., Bhowmik, C., Hussain, F., Yang, S., Naseem, U., Adebayo, A. A., … Al-Rakhami, M. (2021). Analyzing the public opinion as a guide for renewable-energy status in Malaysia: A case study. *IEEE Transactions on Engineering Management*, 1–15. https://doi.org/10.1109/TEM.2020.3046749

Saleh, A. M. (2012). *Modeling of flat-plate solar collector operation in transient states.* Fort Wayne: Purdue University.

Saxena, G., & Gaur, M. K. (2018). Exergy analysis of evacuated tube solar collectors: A review. *International Journal of Exergy, 25*(1), 54–74.

Soriga, I., & Neaga, C. (2012). Thermal analysis of a linear Fresnel lens solar collector with black body cavity receiver. *International Journal of Low Carbon Technologies, 74*(4), 106–116.

Sözen, A., Menlik, T., & Ünvar, S. (2008). Determination of efficiency of flat-plate solar collectors using neural network approach. *Expert Systems with Applications, 35*(4), 1533–1539. https://doi.org/10.1016/j.eswa.2007.08.080

Thirugnanasambandam, M., Iniyan, S., & Goic, R. (2010). A review of solar thermal technologies. *Renewable and Sustainable Energy Reviews, 14*(1), 312–322.

Treichel, C., & Cruickshank, C. A. (2021a). Energy analysis of heat pump water heaters coupled with air-based solar thermal collectors in Canada and the United States. *Energy, 221*, 119801.

Treichel, C., & Cruickshank, C. A. (2021b). Greenhouse gas emissions analysis of heat pump water heaters coupled with air-based solar thermal collectors in Canada and the United States. *Energy and Buildings, 231*, 110594.

Yao, Y., Hu, Y., & Gao, S. (2015). Heliostat field layout methodology in central receiver systems based on efficiency-related distribution. *Solar Energy, 117*, 114–124. https://doi.org/10.1016/j.solener.2015.04.029

Chapter 5

Solar photovoltaic thermal systems

Khodadad Mostakim[1] and Md Hasanuzzaman[2]

[1]*Department of Mechanical Engineering, Rajshahi University of Engineering and Technology, Rajshahi, Bangladesh;* [2]*Higher Institution Centre of Excellence (HICoE), UM Power Energy Dedicated Advanced Centre (UMPEDAC), University of Malaya, Kuala Lumpur, Malaysia*

Chapter outline

5.1 Introduction

The sun is the source of all forms of energy on the planet earth. Sun's energy is generated through nuclear fusion reaction. Sun's energy, popularly known as

Technologies for Solar Thermal Energy. https://doi.org/10.1016/B978-0-12-823959-9.00005-2

solar energy, is a renewable energy source due to its inexhaustible supply. Furthermore, solar energy harvesting technologies are sufficiently ecofriendly with almost zero carbon emission. However, despite of many advantages, it has some practical drawbacks, the major problem being its dependency on availability of solar radiation. However, research is in motion to ensure effective use of solar energy at night. Solar energy can be harvested in two ways: firstly, as electricity through photovoltaic (PV) modules; secondly, heat through solar thermal collectors (STCs) (Kumar et al., 2019, 2021). Fig. 5.1 portrays different ways of utilizing solar energy.

5.2 Photovoltaic thermal technology

The combined photovoltaic thermal (PVT) system is an attractive solar technological invention. It is a hybrid system that combines STC and PV to produce heat and electricity from the same device (Hasanuzzaman et al., 2016; Jarimi et al., 2017; Naves et al., 2019; Obeidat, 2018; Rahman et al., 2015; Teo et al., 2012; Tonui & Tripanagnostopoulos, 2007). With the rise in PV module temperature, electrical efficiency drops substantially. The concept of PVT offers practicable solution to abate cell temperature recover the electrical

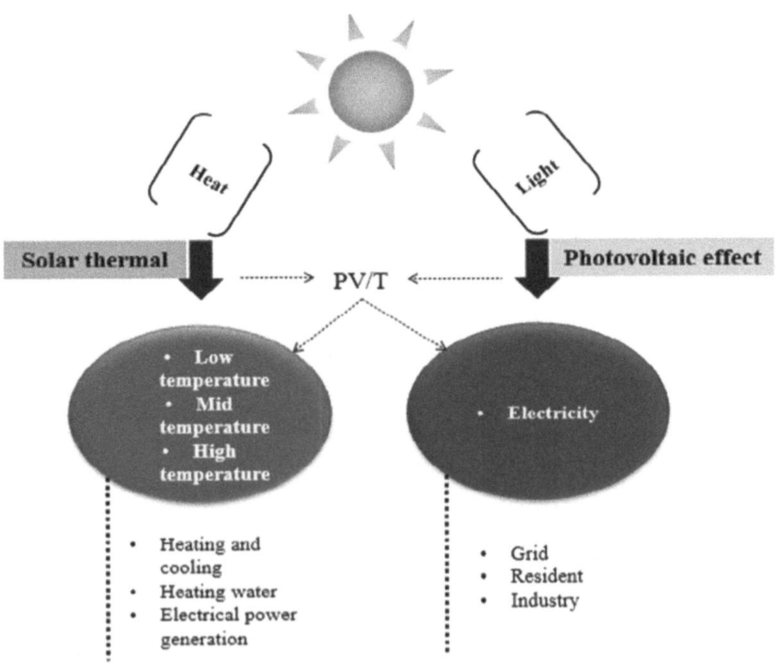

FIGURE 5.1 Different modes of utilization of solar energy.

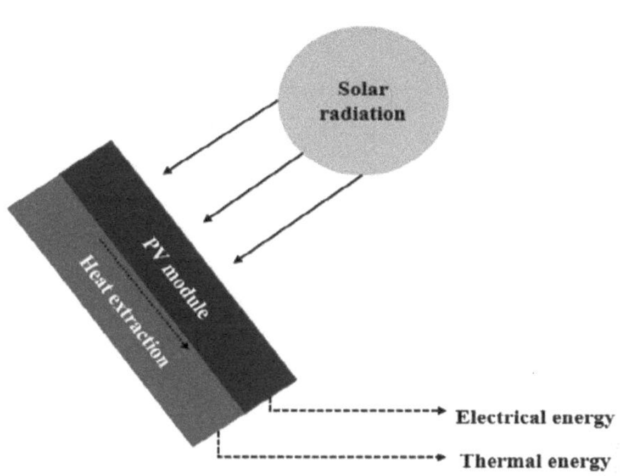

FIGURE 5.2 Basic concept of a PVT system.

efficiency. The concept of PVT was first introduced in the mid-70s (Wolf, 1976). But the technology has not yet been commercialized that much. Fig. 5.2 illustrates a basic concept of a PVT system.

Fig. 5.3 displays different components of a PVT system. PV module is the electrical component of this hybrid system. It can be polycrystalline, monocrystalline, amorphous silicon, etc. Another major element is the thermal absorber which is thermal component of this system. The whole system is enclosed in an enclosure and insulation is done so that there is less heat loss. Water and air are widely used as heat transfer fluid (HTF) to remove heat from the PV modules. Besides, different nanofluid-based HTFs have also been used in research level and found better results.

The PVT systems can be classified in different ways depending on medium or HTF for heat extraction, solar input, configuration, etc (Fig. 5.4). The heat extraction medium can be water, air, air-water, and nanofluid. Again, there are different systems to apply, such as forced and natural circulation. Furthermore, PVT can be categorized depending on the system configuration.

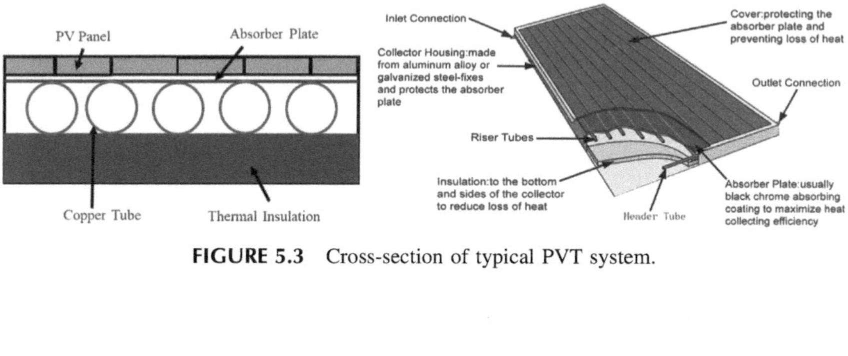

FIGURE 5.3 Cross-section of typical PVT system.

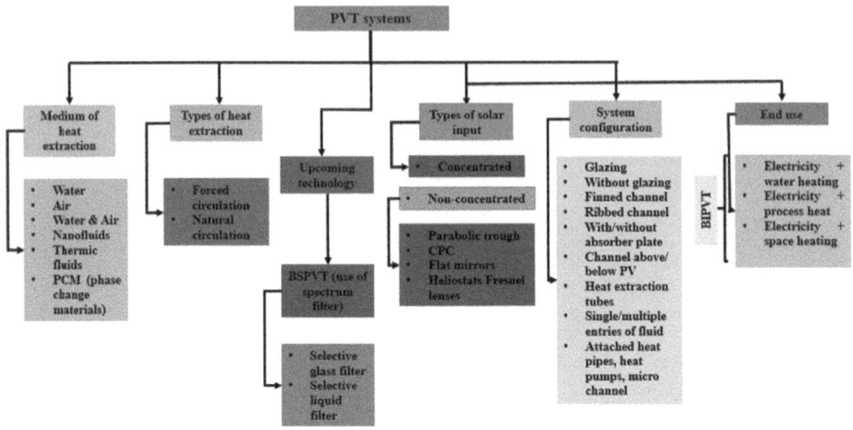

FIGURE 5.4 Broad classification of PVT system (Joshi & Dhoble, 2018).

Table 5.1 shows that the thermal efficiency of water-cooled PVT is higher than air-cooled PVT. Because the heat extraction capacity of water is higher than air. On the other hand, Othman et al. employed a combination of air and water as medium of heat extraction, which delivered a thermal efficiency of 76% (Othman et al., 2016). Moreover, thermal and electrical efficiency using phase change material is 55% and 20%, respectively (Gaur, Ménézo, & Giroux, 2017). Although the efficiency of air-based PVT is a bit less, the efficiency increases comparatively by integrating glazed and an unglazed system with it. Performance of BIPVT and BSPVT as new technology is quite good, but more research is required on these.

5.3 Solar cell or PV cell

Solar cell (also called PV cell) acts as the heart of the PV system. Fig. 5.5 shows that a PV cell is a semiconductor device that produces electric energy from sunlight through photoelectric effect. Different types of PV cells, their performance, and output utilization are discussed below.

5.3.1 Crystalline solar cell

The most common solar cells used in commercially available solar panels are crystalline silicon PV cells. Typically, solar cells are manufactured from single-crystalline silicon or multicrystalline silicon. Monocrystalline silicon cells are made from pseudosquare wafers of silicon, substrates are made from Czochralski float zone technology, and ribbon growth. Currently, multicrystalline silicon cells are made from square silicone substrates cut from ingots cast into quartz crystals. In order to decrease the amount of silicon light, ARC, TiO_2, and SiN are doped into the silicon surface.

TABLE 5.1 Research works on PVT technology and their findings.

PVT types	References	Country	Remarks
PVT/w	Long, Chow and Ji (2017)	China	• Experimental + simulation • Heat pipe assisted • Ee:10% • Te:35%
PVT/w	Singh and Tiwari (2017)	India	• Experimental • Used for active solar distillation system • Ee:7%–15% • Te:75%
PVT/a+w	Othman et al. (2016)	Malaysia	• Experimental • PV module: semitransparent • Double pass flow • Ee:17% • Te:76%
PVT/w-PCM	Gaur et al. (2017)	France	• Simulation • OM37 (PCM): store latent heat • Ee:16%–20% • Te:55%
PVT/w-BIPVT	Chen and Yin (2016)	USA	• Experimental • Roofing panel • Ee:14%–15% • Te:65%–75%
PVT/air	Fan, Kokogiannakis, Ma and Cooper (2017)	Australia	• Experimental + simulation • Dynamic model • More correct estimation than steady-state • Ee:8%–9% • Te:45%–63%

Continued

TABLE 5.1 Research works on PVT technology and their findings.—cont'd

PVT types	References	Country	Remarks
PVT/a-glazed + Unglazed	Tonui and Tripanagnostopoulos (2007a,b)	Greece	• Experimental + simulation • Fins + thin metal sheet + glazing • Improve performance • Ee:11%–12% • Te:50%–60%
PVT/a-BIPVT	Yang and Athienitis (2014)	Canada	• Experimental • Multiple air inlet • Higher performance
PVT/a-glazed	Shahsavar and Ameri (2010)	Iran	• Experimental • Thin metal sheet inside flow channel • Te:50%–60% • Ee:9%–10%
BSPVT/w	Joshi and Dhoble (2017)	–	• Liquid: Water • Transparent for response range of Si solar cell (750–1125 nm) • Absorption in IR region
BSPVT/d-w	Joshi, Dhoble and Jiwanapurkar (2016)	–	• Liquid: distilled water • Suitable for this system with Si solar cell
BSPVT/n-f	Crisostomo et al. (2017)	–	• Liquid: nanofluid • Ag + water • Nanofluids have shown strong plasmonic absorption in the visible spectrum with minimal scattering

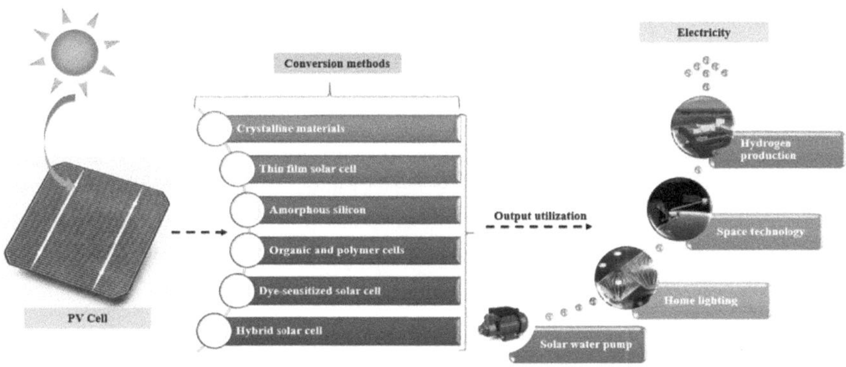

FIGURE 5.5 PV energy conversion pathway (Gorjian, 2020).

5.3.2 Thin-film solar cell

The new generation solar cell is thin-film solar cell and well known as thin-film PV cell, because it contains multiple thin-film layer of PV materials and film layers thickness is much less than typical P—N junction solar cells. Amorphous silicon, cadmium telluride, copper indium gallium deselenide materials are used in cell production. The functioning of thin-film solar cell is almost similar to the normal silicon wafer cells. The flexible arrangements of several layers in thin-film cell differentiates from silicon cells.

5.3.3 Amorphous solar cell

The oldest and mature type of thin film is amorphous silicon thin-film cells. Unlike traditional solar cell wafers, they are made of noncrystalline silicone. Due to amorphous it can absorb more sunlight comparative to c-Si and poly-Si cells. There are also positive coefficients in the output capacity of amorphous Si PV products (at high ambient temperature), for instance, in higher sunshine areas, the cell efficiency is higher. Further, the abundant, inexpensive, and nontoxic characteristics make is popular.

5.3.4 Organic and polymer solar cell

An organic photovoltaic (OPV) cell is a type of solar cell, which usually consists of either polymers or small molecules that are based on organic semiconductors (OSCs). A high level of conjugation is necessary for organic matter to be conductive or semiconductive. In additions, low dielectric constants are present in OSC's range of 0.3—0.5eV.4 for large E_b values. As a consequence, thermal energy alone in OSCs cannot achieve exciton dissociation. At least two different OSCs are necessary in an OPV in order to overcome this. However, the efficiency of polymer solar cell is above 10% (You et al., 2013).

5.3.5 Dye-sensitized solar cell

This type of solar cell has become more attractive in recent years due to its better PV performance under low light conditions, as well as for colors, appearance, and simple fabrication procedure. As a material, mesoporous TiO_2 electrode is used that is from colloidal TiO_2 nanoparticle. The recorded efficiency is just over 10%. However, the performance is 32% at 1000 lux in the indoor application because the absorption spectrum of the dye is equivalent to the emission spectrum of an indoor light source (Boschloo, 2019).

5.3.6 Hybrid solar cell

A hybrid solar cell is a PV device relying on the two semiconductors interface; one is organic, the other inorganic. A heterojunction of type II forms the interface between the two materials, at which the polymer usually serves as the material for electron donation, and electron acceptors are the inorganic semiconductor.

5.3.7 PV cell electrical parameters

In this section, fundamental expressions regarding solar cell have been discussed.

p-n junction

For holes like minority n-type carriers and the expression can be written as (Gray & engineering, 2011):

$$D_p \frac{P_n d^2}{dx^2} = \frac{\Delta P}{\tau_P} - G \tag{5.1}$$

and:

$$\frac{d^2(\Delta P)}{dx^2} = \frac{\Delta P}{L_{p^2}} - \frac{G}{D_p} \tag{5.2}$$

Generation and L_p are assumed to be constant, therefore the solution can be:

$$\Delta P = G\tau_P + A \exp\left(\frac{x}{L_p}\right) + B \exp\left(-\frac{x}{L_p}\right) \tag{5.3}$$

The hole current density with the boundary conditions of an p-n junction is to be found as:

$$I_p = \frac{qD_pP_{n0}}{L_p}\left(\exp\left(\frac{U}{U_T}\right) - 1\right)\exp\left(-\frac{x}{L_p}\right) - qGL_p \exp\left(-\frac{x}{L_p}\right) \tag{5.4}$$

Similarly, electron current density when the region is P:

$$I_n = \frac{qD_nP_{p0}}{L_n}\left(\exp\left(\frac{U}{U_T}\right) - 1\right)\exp\left(-\frac{x}{L_n}\right) - qGL_n\exp\left(-\frac{x}{L_n}\right) \tag{5.5}$$

Then, in solar cell the current density can be:

$$I = I_0\left(\exp\left[\frac{U}{U_T}\right] - 1\right) - qG\left(L_n + L_n + W\right) \tag{5.6}$$

Again, the current density generated by light, so that:

$$I_L = qG\left(L_n + L_p + W\right) \tag{5.7}$$

So, Eq. (5.6) can be written as:

$$I = I_0\left(\exp\left[\frac{U}{U_T}\right] - 1\right) - I_L \tag{5.8}$$

where

G: generation rate
L_n: electron diffusion length
L_p: hole diffusion length
I_0: saturation current
q: elementary charge
k: Boltzmann's constant
P: electrical power
U_T: temperature voltage
W: width of space charge region

From all equations it can be said that, charging carriers that are either generated in the space charging area or at a distance from the p-n junction contribute to the current and only this area is active in a solar cell.

Short-circuit current (I_{SC})

Short-circuit current means no voltage is available at the cell, i.e., the solar cell is short circuited and termed as I_{sc}. So, it is clear that, $I_{sc} = -I_L$ during short-circuit current. If the cell is running on a short circuit, $V = 0$ and the short-circuit current is defined as the current (I) flowing through the terminals. Now, the short-circuit current I_{sc} of a solar cell (low R_s and I_0, and high R_{sh}) can be demonstrated as $I_{sc} = -I_L$.

Open-circuit voltage (V_{OC})

The voltages available in the open circuit, V_{oc}, are the highest possible solar cell voltage and it's happening at zero current (Augusto et al., 2017;

Baruch, De Vos, Landsberg, & Parrott, 1995; Levy & Honsberg, 2006; Sinton & Cuevas, 1996).

$$V_{oc} = \frac{nkT}{q} \ln\left(\frac{I_L}{I_0} + 1\right) \tag{5.9}$$

The above equation indicates that V_{oc} increases with increase in temperature. Actually, this V_{oc} depends on saturation current. It also depends on light-generated current. But, in primary stage, I_0 goes up rapidly with temperature because of intrinsic carrier concentration (n_i). Again, from carrier concentration, the V_{oc} can be written as below (Green, 1981):

$$V_{oc} = \frac{kT}{q} \ln\left[\frac{(N_A + \Delta n)\Delta n}{n_i^2}\right] \tag{5.10}$$

where

T: temperature, K
N: ideality factor
I_0: dark saturation current
I_L: light generated current
n_i: intrinsic carrier density

Fill factor

To determine the maximum power from solar cell, the fill factor (FF) is essential because, it is related with V_{oc} and I_{sc}. Then the expression is (Jain & Kapoor, 2004):

$$FF = \frac{P_{MP}}{V_{oc} \times I_{sc}} = \frac{V_{MP} \times I_{MP}}{V_{oc} \times I_{sc}} \tag{5.11}$$

Now, the maximum theoretical FF can be determined after differentiating the solar cell power (w.r.t voltage). So that:

$$V_{MP} = V_{oc} - \frac{nkT}{q} \ln\left(\frac{qV_{MP}}{nkT} + 1\right) \tag{5.12}$$

To solve this equation, Lambert function is needed and the process is complex. The simplest way can be:

$$FF = \frac{v_{oc} - \ln(v_{oc} + 0.72)}{v_{oc} + 1} \tag{5.13}$$

where $v_{oc} = \frac{q}{nkT} \times V_{oc}$

Solar cell efficiency

The standard fundamental parameter is efficiency in comparison to the performance of one solar cell to another. The ratio of energy generated by the solar cell to input energy from the sun is represented as efficiency. Besides, the performance of the solar cell itself is reflected. Efficiency is dependent on the incident sunlight's spectrum and intensity and the solar cell temperature. Therefore, it is vital to measure the efficiency of the solar cell. The efficiency of a solar cell can be measured by determining the fraction of incidental electricity and is measured using below equations (Najafi & Woodbury, 2013).

$$P_{max} = V_{oc} I_{sc} FF \tag{5.14}$$

$$\eta = \frac{V_{oc} I_{sc} FF}{P_{in}} \tag{5.15}$$

where

η: efficiency
V_{oc}: open-circuit voltage
I_{sc}: short-circuit current

Example 5.1. Calculated solar cell power conversation efficiency at an intensity of 150 W/m^2. Use the following parameters to solve the problem:

Cell area = 7 cm^2
Open-circuit voltage = 0.35 V.
Short-circuit current = −6.9 mA.
Maximum output voltage, V_{max} = 0.21 V.
Maximum output current, I_{max} = − 5 mA.

Solution. Maximum power, $P_{max} = V_{max} \times I_{max} = 0.21 \times (-5) = -1.05$ mW

By rearranging Eq. (5.14):

$$FF = \frac{P_{max}}{V_{oc} \times I_{sc}} = \frac{(-1.05)}{0.35 \times (-6.9)} = 0.435$$

So, from Eq. (5.15):

$$\eta = \frac{V_{oc} I_{sc} FF}{P_{in}} = \frac{0.35 \times \left(6.9 \times 10^{-3}\right) \times 0.435}{150 \times \left(7 \times 10^{-4}\right)} = 1.0005\% \approx 1\%$$

Cell power conversion efficiency 1%.

Detailed balance

A detailed balance is used to determine the highest PV device's efficiency. However, the general expression is used to calculate the flux of absorption and the flux of emission of the solar cell and the expression is (Najafi & Woodbury, 2013; Tiedje, Yablonovitch, Cody, & Brooks, 1984):

$$\phi(E_a, E_b, \mu, T) = Cf \frac{2\pi}{h^3 C^2} \int_{E_a}^{E_b} \frac{E^2}{\exp\left(\frac{E-\mu}{kT}\right) - 1} dE \tag{5.16}$$

Boundary conditions

A common hypothesis is that an ideal ohmic contact may be treated in front contact, i.e.,

$$\Delta p(-W_N) = 0 \tag{5.17}$$

Further, the efficient recombination velocity model shows the combined effects of ohmic contact and the antireflective passivation layer. Therefore, the boundary condition at $x = -W_N$ is:

$$\frac{d\Delta p}{dx} = \frac{S_{F,\,eff}}{D_p}\Delta p(-W_N) \tag{5.18}$$

where $S_{F,\,eff}$ is the effective front surface recombination velocity.

The back contact is similar to as an ideal ohmic contact, that's why, $\Delta n(-W_N) = 0$

When solar cells are made from a back surface field (BSF), then the boundary condition will be (Khanna, 2019):

$$\left.\frac{d\,\Delta n}{dx}\right| = -\frac{S_{BSF}}{D_n}\Delta n\left(W_p\right) \tag{5.19}$$

The semiconductor is assumed to be nondegenerate under nonequilibrium condition, then:

$$p = n_i e^{\left(E_i - F_p\right)/kT} \text{and } n = n_i e^{(F_n - E_i)/kT} \tag{5.20}$$

It is obvious that $F_P = F_N = E_F$ under balanced conditions. So that the applied voltage is found to be:

$$qV = F_N(-W_N) - F_P(-W_P) \tag{5.21}$$

After assuming both quasi-Fermi energies remain constant within the region of depletion:

$$F_N(x) - F_p(x) = qV \tag{5.22}$$

The boundary condition at the edge of depletion area:

$$p_N(-x_N) = \frac{n^2}{N_D}e^{qV/kT} \text{ and } N_p\left(-x_p\right) = \frac{n^2}{N_A}e^{qV/kT} \tag{5.23}$$

When light hits at the front of the solar cell, then the generation rate can be written as:

$$G(x) = (1 - s)\int (1 - r\,(\lambda))\,f(\lambda)\alpha(\lambda)e^{-\alpha(x+W_N)}d\lambda \tag{5.24}$$

5.3.8 Performance of PV cell

The good or efficient solar cell have a high short-circuit current, high open-circuit voltage, a FF as close as possible to 1. The factor influencing of solar cell efficiency can be obtained from below equation (Yuan et al., 2019):

$$\eta = \frac{P_{max}}{P_{in}} = \eta_{ideal}\eta_{photon}FF\eta_V\eta_C^{int} \tag{5.25}$$

Maximum energy can obtain from an absorbed photon is E_G, the ideal efficiency can be written as below (Willoughby, 2014):

$$\eta_{ideal}(E_G) = \frac{I_{inc}E_G\frac{1}{q}}{P_{in}} = \frac{E_G}{P_{in}/A}\int_{\lambda<\lambda_G}^{0} f(\lambda)d\lambda \tag{5.26}$$

The photon efficiency is the result of photons that are reflected, transmitted, or absorbed into the solar cell and can be expressed as (Bozzola, Liscidini, & Andreani, 2012):

$$\eta_{photon} = \frac{I_{gen}}{I_{inc}} = \frac{\eta_C^{ext}}{\eta_C^{int}} \tag{5.27}$$

Additionally, the voltage efficiency ηV corresponds to the bandgap voltage of the open-circuit voltage:

$$\eta_V = \frac{V_{OC}}{\frac{1}{q}E_G} \tag{5.28}$$

Effect of temperature

As all other semiconductor device, solar can be affected with temperature. The band gap of semiconductor reduces with increasing temperature and the energy of electron is also increased. However, the mostly affected parameter is open-circuit voltage. Temperature depends on I_0, therefore, open-circuit voltage reduced with temperature (Singh & Singal, 1983).

$$I_0 = q \times Area\ (A) \times \frac{diffusivity\ (D) \times intrinsic\ carrier\ concentration\ (n_i^2)}{carrier\ diffusion\ length\ (L) \times doping\ (N_D)} \tag{5.29}$$

The intrinsic carrier concentration is (Najafi & Woodbury, 2013):

$$n_i^2 = 4\left(\frac{2\pi kT}{h^2}\right)^2 \left(m_e^* m_h^*\right)^{3/2} \exp\left(-\frac{E_{G0}}{kT}\right) = BT^3 \exp\left(-\frac{E_{G0}}{kT}\right) \tag{5.30}$$

where m_e^* is effective masses of electron, m_h^* is effective masses of holes, E_{G0} is the band gap, and B is constant.

Let the temperature-dependent parameters be neglected. Then substituting these equations back into the equation for I_0 and the expression of I_0 as follows (Najafi & Woodbury, 2013):

$$I_0 = qA\frac{D}{LN_D}BT^3 \exp\left(-\frac{E_{G0}}{kT}\right) \approx B'T^\gamma \exp\left(-\frac{E_{G0}}{kT}\right) \tag{5.31}$$

$$\begin{aligned} V_{OC} &= \frac{kT}{q}\ln\left(\frac{I_{SC}}{I_0}\right) = \frac{kT}{q}[\ln I_{SC} - \ln I_0] = \frac{kT}{q}\ln I_{SC} \\ &- \frac{kT}{q}\ln\left[B'T^\gamma \exp\left(-\frac{qV_{G0}}{kT}\right)\right] = \frac{kT}{q}\left(\ln I_{SC} - \ln B' - \gamma \ln T + \frac{qV_{G0}}{kT}\right) \end{aligned} \tag{5.32}$$

Rahman et al. showed that without cooling condition, as the solar cell temperature increased by 1°C, the electrical efficiency dropped by 0.06% (indoor operating condition) (Rahman, Hasanuzzaman, & Rahim, 2015). Chandrasekar, Suresh, & Senthilkumar (2013) investigated that electrical efficiency dropped significantly by 0.05% for any 1°C solar cell temperature increase without cooling. In another investigation, Rahman, Hasanuzzaman and Rahim (2017) exhibited that the electrical efficiency decreased by 0.22% as the solar cell temperature raised by 1°C (outdoor operating condition). This slight difference is possible due to different operating conditions.

Solar to electricity system

Solar cell helps to produce electricity under solar illumination by PV or regenerative photoelectrochemical cell. However, the maximum power can be obtained by multiplying the maximum current and maximum voltage ($P_{max} = I_{max}V_{max}$). The PV cell efficiency is the ratio of electric power output to input. The general expression for the cell is (Nielander et al., 2015):

$$\eta = \frac{P_{f,0} + P_{e,0}}{P_s + P_{e,i}} \tag{5.33}$$

The cell electrical efficiency can be obtained by setting the term related to chemical fuel output and input as zero, then it can will be:

$$\eta = \frac{P_{e,0}}{P_s} = \frac{I_{max} \times V_{max}}{P_s} \tag{5.34}$$

Solar to fuel system

In solar to fuel energy conversion, Gibbs free energy of the fuel can be used because it is the standard measurement of the fuel energy content, and the fuel forming reaction will be as water splitting, carbon dioxide reduction, etc. However, there must require a system that produces only fuel as output and solar power as input during the calculation. Also, by rearranging Eq. (5.33) and setting the term linked to the electric input and output as zero, the actual equation will be as below (Chen et al., 2010; Khaselev, Bansal, & Turner, 2001):

$$\eta_{SF} = \frac{P_{f,0} + P_{e,0}}{P_s + P_{e,i}} = \eta = \frac{P_{f,0} + 0}{P_s + 0} = \frac{A \times J_{op} \times E_{f,0} \times \varepsilon_{elec}}{P_s} \tag{5.35}$$

Moreover, the system that can produce H_2 and O_2 as output is called photo-driven water splitting system and the efficiency is termed as solar to hydrogen efficiency (η_{SH}). So, the expression is (Chen, Dinh, & Miller, 2013):

$$\eta_{SH} = \frac{A \times J_{op} \times 1.23 \times \varepsilon_{elec}}{P_s} \tag{5.36}$$

where

A: geometric area (cm^2).

J_{op}: operating current density ($A\ cm^{-2}$).

ε_{elec}: faradaic efficiency

Solar electricity to fuel system

Electrolysis includes the input of electricity as a chemical fuel for producing output power. Electrolysers work without power supply and no illumination power. That is, $P_{e,0} = 0$ and $P_s = 0$. The efficiency of electrolysis can be expressed as (Conibeer & Richards, 2007):

$$\eta_{EL} = \frac{P_{f,0} + P_{e,0}}{P_s + P_{e,i}} = \frac{P_{f,0}}{P_{e,i}} = \frac{Output}{Input} = \frac{E_{f,0}}{V_{e,i}} \tag{5.37}$$

The most exciting application is the water pumping system, which is driven by the PV module. Kaldellis, Meidanis and Zafirakis (2011) experimentally analyzed a water pumping system and investigated the performance of the system. During the analysis of the efficiency and total water quantity using solar potential, the authors used the equations, as shown in Table 5.2. The result showed that the system's total efficiency governed with solar energy under assessment is nearly 5%, and the average efficiency of the PV panel was 8%. A similar experiment is investigated by Mokeddem et al. (2011) where the system operated without any battery and control system. To calculate the full system performance, the author utilized some equations, as shown in Table 5.2. The author concluded that system efficiency never exceeds 9% with or without fuzzy operation. Another most interesting application is solar in

TABLE 5.2 Output utilization summary.

References	Output utilization	Utilized equations	Terms
Kaldellis et al. (2011)	Solar to water pumping	$E_{sol} = A_c \int_{t_0}^{t_0+\Delta t} G_T.dt$ (5.38) $E_{PV} = \int_{t_0}^{t_0+\Delta t} I_{PV}(G_T, \theta_c) U_{PV}(G_T, \theta_c).dt$ (5.39) $\eta_{PV} = \frac{E_{PV}}{E_{sol}}$ (5.40) $\Delta E_B^{\Delta t} = C_{b+\Delta t} U_{b+\Delta t} - C_b U_b$ (5.41) $E_w = \rho g h V_{tot}$ (5.42) $E_{PV} = E_w + \Delta E_b + \delta E_{loss}$ (5.43) $\eta_{tot} = \frac{E_w}{E_{sol}} = \eta_{PV}\eta_{BOS}\eta_p\eta_h$ (5.44)	Solar panel surface efficiency, E_{sol}; Specific time period, Δt; PV generator surface, A_c; Total irradiance, G_T; Panel energy output, E_{PV}; Efficiency of PV generator, η_{PV}; Input and output, I_{PV} and U_{PV}; Energy stored in system battery, ΔE_b; Battery capacity, C_b; Total water volume, V_{tot}; BOS: system components
Mokeddem et al. (2011)	Water pumping	$I = I_L + I_0\,[\exp[B(V + IR_s)] - 1]$ (5.45) $V_{OC}(G, 25) = \left(\frac{1}{B}\right) \ln\left[\frac{GI_{SC}(1,25)}{I_0}\right]$ (5.46) $I = I_{SC}(1 - \exp[-V_{OC} - V / C])$ (5.47)	Terminal current, I; Terminal voltage, V; Photocurrent, I_L; Diode reverse saturation current, I_0; Series resistance, R_s; $B = \frac{q}{nkT}$ *q: electronic charge,* *k: Boltzmann's constant,* *n: diode edeality factor* Curve fitting constant, C

Imani et al. (2008)	Solar at space environment	Electron-hole generation $$I_0 = I_{P0}\left(\frac{P}{P_0}\right)(1 + K_0(T - T_0)) - C\log\left(1 + \frac{\phi}{\phi_x}\right) \quad (5.48)$$ $$I_{P0} = RAP_0 \quad (5.49)$$ Equation stated for diodes $$J = \left(\frac{qD_P}{L_P}P_{n0} + \frac{qD_n}{L_n}n_{P0}\right)\left(e^{\frac{q(V+IR_s)}{kT}} - 1\right) + \frac{qn_iW}{2\tau_e}\left(e^{\frac{q(V+IR_s)}{2kT}} - 1\right) \quad (5.50)$$ Parallel resistance $$R_{sh} = R_{sh0}\exp(K_1(T - T_0)) \quad (5.51)$$ Series resistance $$R_s = R_{s0}\exp(1 + K_2(T - T_0)) \quad (5.52)$$	Responsivity of cell, R; Irradiance, P; Constant, K_0; Reference temperature T_0; Radiation fluence, ϕ; Value of parallel resistant for reference temperature, R_{sh0};

space environment. Imani et al. (Imani, Rezaie, & Taherbaneh, 2008) modeled and evaluated the condition of terrestrial solar panel in space environment. For modeling the system, author used equations and analyzed the space characterization parameters such as radiation, irradiation, and temperature. Simulations have shown that the individual solar panel in LEO satellites can be used for two months. However, a five-year mission with 20 mm cover glass may also be helpful.

5.4 Energy conversion in different types of PVT systems

PVT system can convert energy in two different ways as thermal and electrical energy. There are different types of PVT system. The energy conversion systems of different types of PV/T system, their performance, and output utilization are discussed in the subsequent sections.

5.4.1 Energy conversion in PVT/water system

In water type PVT system, water is used as a medium to extract heat. Glazed, unglazed, fully covered, partially covered collector by PV module, transparent, semitransparent, and PCM-based systems are included in this type of PVT (Islam et al., 2021; Nahar, Hasanuzzaman, & Rahim, 2017b). Lastly, the output will utilize by water heating, feeding water for solar still.

He, Zhang and Ji (2011) carried out an experimental analysis regarding PVT system using natural circulation of water included flat plate collector, PVT collector, PV module. Author calculated electrical and thermal energy conversion by using some formulas that are mentioned below (He et al., 2011):

Evaluation criterion of the PV/T system

As the evaluation indicator of the comprehensive performance of the PV/T system, primary energy saving efficiency (n_f) is proposed (Dubey & Tiwari, 2008):

$$\eta_f = \frac{\eta_e}{\eta_{power}} + \eta_{th} \tag{5.53}$$

Here, thermal efficiency can be obtained by experimentally using Eq. (5.54) (He et al., 2011):

$$n_{th} = MC_pK_{th} \tag{5.54}$$

Here,

$$K_{th} = \frac{T_f - T_i}{HA_c} \tag{5.55}$$

For natural circulation of water, the thermal efficiency can be written as (Huang & Du, 1991):

$$\eta_{th} = \alpha - U\frac{(T_f - T_a)}{H} \tag{5.56}$$

where

η_{th}: thermal efficiency
η_e: electrical efficiency
η_{power}: electric power generation
m: mass of water
C_p: specific heat of water
H: solar radiation
α: daily average thermal efficiency
U: coefficient of heat loss
T_a: ambient temperature

Example 5.2. Let, Kareem investigated on a PVT system under natural circulation of water. The result showed that the daily efficiency of that system was higher than other traditional thermosiphon system and the daily electrical (average) efficiency was $\eta_e = 10\%$. In order to calculate the daily efficiency, Kareem used some parameters which are mentioned below:

Solar radiation on the surface of collector, $H = 16.14\ \mathrm{MJ/m^2}$
Initial water temperature $= 34.9°\mathrm{C}$
Final water temperature $= 59.1°\mathrm{C}$
Ratio, $\frac{mass\ of\ water}{Area\ of\ collector} = \frac{M}{A_c} = 66.1\ \ \mathrm{kg/m^2}$
Determine the daily efficiency (η_{th}) of this PVT system.

Solution. Useful energy of the collector per area:

$$= \frac{MC_p\,(T_f - T_i)}{Area} = 66.1 \times 4200 \times 24.2 = 6.72 \times 10^6\ \mathrm{J/m^2}$$

From Eq. (5.55), the daily efficiency $\eta_{th} = \frac{6.72 \times 10^6}{H} = \frac{6.72 \times 10^6}{16.14 \times 10^6} = 41.64 \approx 42\%$

5.4.2 Energy conversion in glazed PVT/water system

Rommel et al. analyzed about the single-glazed flat plate PVT system and investigated to reduce thermal resistance from solar cell to selective absorber (Dupeyrat, Ménézo, Rommel, & Henning, 2011; Fayaz, Rahim, Hasanuzzaman, Nasrin, & Rivai, 2019, Fayaz, Rahim, Hasanuzzaman, Rivai, &Nasrin, 2019; Nahar, Hasanuzzaman, Rahim, & Parvin, 2019). In order to measure the thermal efficiency of a standard thermal collector, the authors used Eq. (5.57) (Dupeyrat et al., 2011):

$$\eta_{th} = (\tau\alpha)_{eff} - U\frac{\Delta T}{G} \tag{5.57}$$

Some of the incoming radiation is converted into electricity and therefore cannot be transferred to heat, thus altering the thermal efficiency of PVT.

Thermal efficiency (He et al., 2011):

$$\eta^*_{th} = \eta_{th}(thermal\ efficiency\ without\ extraction\ efficiency) - \eta_{elec}\ (electrical\ efficiency) \tag{5.58}$$

where

G: incoming solar radiation

$(\tau\alpha)_{eff}$: effective transmission

Bhattarai et al. (2012) investigated the performance of a water type PVT with single glazing, an absorber plate, and copper tubes. Flowing water in the attached tubes extracts heat from the absorber plate, as shown in Fig. 5.6. Here, the total efficiency is equal to the sum of the electrical efficiency and thermal efficiency which is denoted by Eq. (5.59):

$$\text{Overall efficiency, } \eta_o = \eta_{el} + \eta_{th} \tag{5.59}$$

Arefin, Islam, Zunaed and Mostakim (2020) carried out a novel integrated PVT system where flat plate collector was located outside the PV panel and water was passed by a pump through a piping network. In that experiment, the

FIGURE 5.6 Real experimental setup of a PVT system Bhattarai, et al. (2012).

useful energy of collector (Q_u) and instantaneous efficiency of collector (η_{iec}) was measured using Eqs. (5.60) and (5.61), respectively.

$$Q_u = mC_p\ (T_{out} - T_{in}) \tag{5.60}$$

$$\eta_{iec} = \frac{Q_u}{A_p I_s} \tag{5.61}$$

where

A_p: plate area (m^2).

I_s: solar-radiation intensity (W/m^2).

Example 5.3. Natural forced circulation of water is passing through a PVT collector at an intensity of 1000 W/m^2. The water inlet and outlet temperatures are 25°C and 49°C, respectively. The area of the PVT collector is 3.1 m^2 and mass of passing water is 1 kg in 60 s. Calculate the instantaneous efficiency of collector.

Solution. Given that:

Solar-radiation intensity, $I_s = 1000\ W/m^2$

Plate area, $A_p = 3.1\ m^2$

Mass of water = 1 kg

Time = 60 s

Useful energy of collector Q_u per second = $mC_p\ (T_{out} - T_{in}) = \frac{1}{60} \times 4200 \times 24 = 1680$ W

So, the instantaneous efficiency of collector, $\eta_{iec} = \frac{Q_u}{A_p I_s} = \frac{1680}{3.1 \times 1000} = 54\%$

5.4.3 Energy conversion in unglazed PVT/water and PVT-PCM systems

Wu, Chen and Xaio (2018) established a method to evaluate the heat transfer characteristics and performance of a PVT system. They implemented a cooling channel above a PV module on their water-cooled PVT system and considered mass flow rate, cooling channel height, the intensity of solar, inlet water temperature during an investigation. However, the authors deemed average Nusselt number (top and bottom surface) for various mass flow rate (m), where G was 1000 W/m^2, H was 5 mm, and inlet temperature T was 298.15K. From Fig. 5.7A, during water flow through the cooling channel, the Nu_{bottom} fell gradually, but at a final stage, it reached a particular value. Simultaneously, the heat transfer performance improved along with increasing of m. Moreover, the temperature difference was decreased, resulting in the improvement of Nu_{bottom}. Also, the results indicated that higher cooling channel height leads to higher temperature difference among PV panel and PVT as shown in Fig. 5.7B.

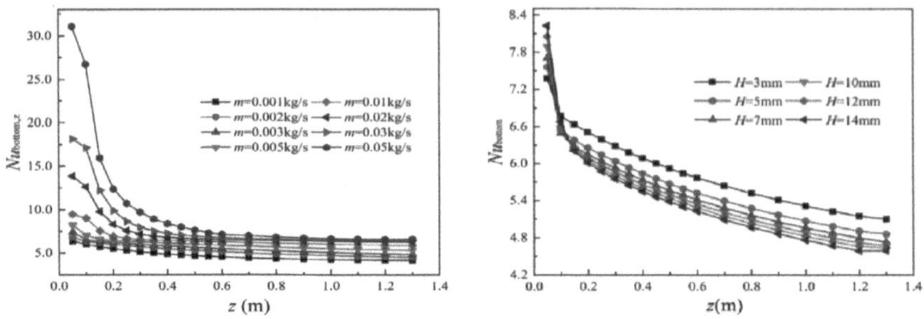

(a) (b)

FIGURE 5.7 Effect of mass flow rate (A) (Wu et al., 2018) and channel height (B) (Wu et al., 2018).

Moreover, the output performance of the water type PVT system can be measured by below-mentioned equations (Nahar, Hasanuzzaman, & Rahim, 2017a; Nasrin, Rahim, Fayaz, & Hasanuzzaman, 2018):

$$\text{Solar energy incident on the PVT, } E_{in} = P_c \tau_g \alpha_c IA \tag{5.62}$$

$$\text{Electrical power output, } E_{el} = \eta_{sc} P_{sc} \tau_g \alpha_{sc} IA \left[1 - \mu_{sc}(T_{sc} - T_r)\right] \tag{5.63}$$

$$\text{Thermal output, } E_{th} = mC_p \left[T_{out} - T_{in}\right] = \rho AVC_p [T_{out} - T_{in}] \tag{5.64}$$

$$\text{Electrical efficiency, } \eta_{el} = \frac{E_{el}}{E_{in}} \tag{5.65}$$

$$\text{Thermal efficiency, } \eta_{th} = \frac{E_{th}}{E_{in}} \tag{5.66}$$

where

μ_{sc}: thermal coefficient of solar cell efficiency
P_{sc}: packing factor
τ_g: transmissivity of glass
α_{sc}: absorptivity of solar cell

Example 5.4. The cell temperature of a 1.5 m^2 PVT collector is limited to 55°C by means flowing water at a velocity of 0.05 m/s through 10 mm copper pipe of the thermal collector. The water inlet temperature is the same as ambient temperature and equal to 25°C, while the outlet temperature reaches as high as 42°C under an irradiation of 1000 W/m^2. Calculate the electrical and thermal output of the PVT collector. Also, determine the electrical, thermal, and overall efficiencies. Given: thermal coefficient of solar cell efficiency = 0.00045/K, packing factor = 0.86 for mono and 0.95 for poly-crystalline cell, glass transmittivity = 0.9, absorptivity = 0.9.

Solution. Solar energy incident in the PVT:

$$E_{in} = P_c \tau_g \alpha_c IA = 0.95 \times 0.9 \times 0.9 \times 1000 \times 1.5 = 1154.25 \text{ W}$$

Electrical power output:

$$E_{el} = \eta_{sc} P_{sc} \tau_g \alpha_{sc} IA\, [1 - \mu_{sc}(T_{sc} - T_r)] = 0.16 \times 0.95 \times 0.9 \times 0.9 \times 1000 \times 1.5 \times [1 - (0.00045)\{(55+273) - (25+273)\}] = 182.2\ \text{W}$$

Thermal output, $E_{th} = mC_p\, [T_{out} - T_{in}] = \rho AVC_p [T_{out} - T_{in}] = 1000 \times \left(\frac{\pi}{4} \times 0.01 \times 0.01\right) \times 0.05 \times 4185 \times (42 - 25) = 279.24\ \text{W}$

Electrical efficiency, $\eta_{el} = \frac{E_{el}}{E_{in}} = 15.8\%$

Thermal efficiency, $\eta_{th} = \frac{E_{th}}{E_{in}} = 24.2\%$

So, Overall efficiency $= (15.8 + 24.2)\ \% = 40\%$

5.4.4 Energy conversion in PVT/air system

Bambrook and Sproul (2012) worked on the way of maximizing the output of PVT/air system. As illustrated in Fig. 5.8, this system was single glaze, single pass, and open loop. To improve the efficiency, authors have increased the air flow rates and reduced the temperature of PV module. In order to, reduce temperature, the excess fan is required, and air mass flow rate was ranged between 0.03 and 0.05 kg/m^2. The below-mentioned equation is used to measure the total energy of PVT (Bambrook & Sproul, 2012):

$$\text{Total value} = E_{Th} \times \$0.10 + \left(E_{pv} + E_{fan}\right) \times \$0.19 \tag{5.67}$$

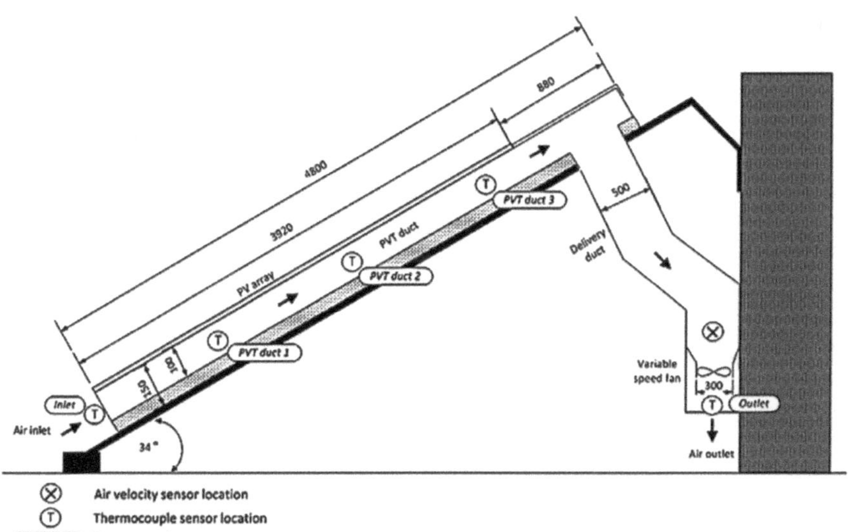

FIGURE 5.8 Open loop PVT/air system (Bambrook & Sproul, 2012).

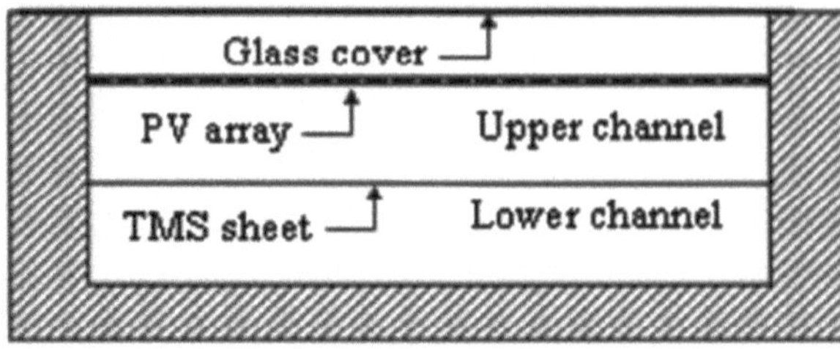

FIGURE 5.9 Cross-sectional view of the PV/T system used by Shahsavar and Ameri (2010).

Shahsavar and Ameri (2010) exhibited a direct-coupled PV/T air collector model using the thin metal aluminum sheet that suspended between upper air channel and lower air channel. This investigation was done with both natural and forced convection air flows over the bottom face of the PV array, as shown in Fig. 5.9. About 25 mm thick fiberglass was used to insulate the channel wall for reducing heat loss to the environment. The forced convection acquired using DC fans (12V, 0.55A, 6.6W). For the mass flow of cooling air of 0.1 kg/s, the highest possible efficiency of 70% had been measured.

Tonui and Tripanagnostopoulos (2008) worked on-air type PV/T system with air cooling and applied two methods to improve the heat transfer rate. In the center of the panel, a thin sheet of metal was suspended, and the fins linked to the PV wall for enhancing heat transfer from the PV module. The performance resulted better with a glazed and unglazed PV module that represents in Fig. 5.10.

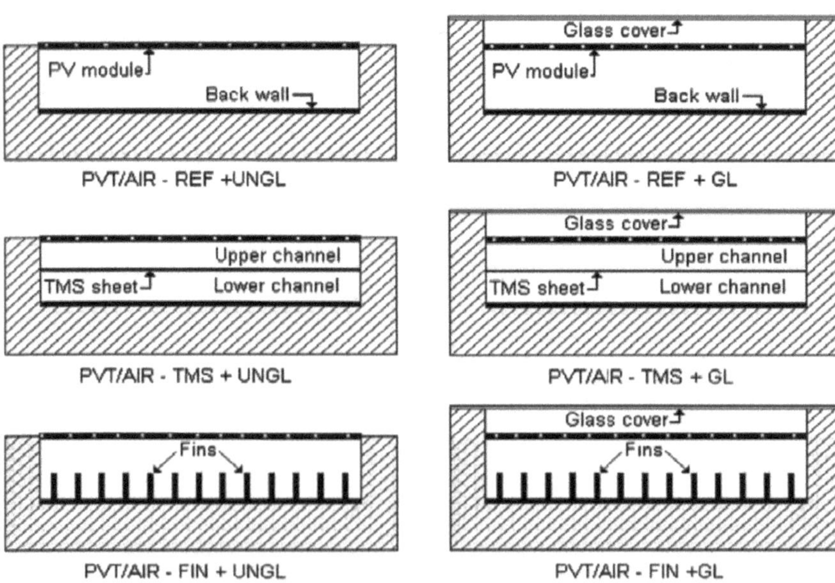

FIGURE 5.10 Cross-sectional view of PV/T air collector. *Adapted from Tonui, J. K., & Tripanagnostopoulos, Y. (2008). Performance improvement of PV/T solar collectors with natural air flow operation.* Solar Energy, 82*(1), 1–12.*

References

Arefin, M. A., Islam, M. T., Zunaed, M., & Mostakim, K. J. C. E. (2020). *Performance analysis of a novel integrated photovoltaic–thermal system by top-surface forced circulation of water.*

Augusto, A., Herasimenka, S. Y., King, R. R., Bowden, S. G., & Honsberg, C. (2017). Analysis of the recombination mechanisms of a silicon solar cell with low bandgap-voltage offset. *Journal of Applied Physics, 121*(20), 205704.

Bambrook, S., & Sproul, A. (2012). Maximising the energy output of a PVT air system. *Solar Energy, 86*(6), 1857–1871.

Baruch, P., De Vos, A., Landsberg, P., & Parrott, J. (1995). On some thermodynamic aspects of photovoltaic solar energy conversion. *Solar Energy Materials and Solar Cells, 36*(2), 201–222.

Bhattarai, S., et al. (2012). Simulation and model validation of sheet and tube type photovoltaic thermal solar system and conventional solar collecting system in transient states. *Solar Energy Materials and Solar Cells, 103*, 184–193. https://doi.org/10.1016/j.solmat.2012.04.017. In this issue.

Boschloo, G. (2019). Improving the performance of dye-sensitized solar cells. *Front. Chem., 7*, 77. https://doi.org/10.3389/fchem.2019.00077

Bozzola, A., Liscidini, M., & Andreani, L. C. J. O. E. (2012). Photonic light-trapping versus Lambertian limits in thin film silicon solar cells with 1D and 2D periodic patterns. *Optics Express, 20*(102), A224–A244.

Chandrasekar, M., Suresh, S., Senthilkumar, T. J. E. C., & Management. (2013). Passive cooling of standalone flat PV module with cotton wick structures. *Energy Conversion and Management, 71*, 43–50.

Chen, Z., Dinh, H. N., & Miller, E. (2013). *Photoelectrochemical water splitting* (Vol. 344). Springer.

Chen, Z., Jaramillo, T. F., Deutsch, T. G., Kleiman-Shwarsctein, A., Forman, A. J., Gaillard, N., ... Sunkara, M. J. J.o. M. R. (2010). Accelerating materials development for photoelectrochemical hydrogen production: Standards for methods, definitions, and reporting protocols. *Journal of Materials Research, 25*(1), 3–16.

Chen, F., & Yin, H. J. A. E. (2016). Fabrication and laboratory-based performance testing of a building-integrated photovoltaic-thermal roofing panel. *Applied Energy, 177*, 271–284.

Conibeer, G., & Richards, B. J. I. J.o. H. E. (2007). A comparison of PV/electrolyser and photoelectrolytic technologies for use in solar to hydrogen energy storage systems. *International Journal of Hydrogen Energy, 32*(14), 2703–2711.

Crisostomo, F., Hjerrild, N., Mesgari, S., Li, Q., & Taylor, R. A. J. A.e. (2017). A hybrid PV/T collector using spectrally selective absorbing nanofluids. *Applied Energy, 193*, 1–14.

Dubey, S., & Tiwari, G. (2008). Thermal modeling of a combined system of photovoltaic thermal (PV/T) solar water heater. *Solar Energy, 82*(7), 602–612.

Dupeyrat, P., Ménézo, C., Rommel, M., & Henning, H.-M. J. S. E. (2011). Efficient single glazed flat plate photovoltaic–thermal hybrid collector for domestic hot water system. *Solar Energy, 85*(7), 1457–1468.

Fan, W., Kokogiannakis, G., Ma, Z., & Cooper, P. J. R. E. (2017). Development of a dynamic model for a hybrid photovoltaic thermal collector–solar air heater with fins. *Renewable Energy, 101*, 816–834.

Fayaz, H., Rahim, N. A., Hasanuzzaman, M., Nasrin, R., & Rivai, A. (2019). Numerical and experimental investigation of the effect of operating conditions on performance of PVT and PVT-PCM. *Renewable Energy, 143*, 827–841. https://doi.org/10.1016/j.renene.2019.05.041

Fayaz, H., Rahim, N. A., Hasanuzzaman, M., Rivai, A., & Nasrin, R. (2019). Numerical and outdoor real time experimental investigation of performance of PCM based PVT system. *Solar Energy*, 135–150. https://doi.org/10.1016/j.solener.2018.12.057

Gaur, A., Ménézo, C., & Giroux, S. J. R. E. (2017). Numerical studies on thermal and electrical performance of a fully wetted absorber PVT collector with PCM as a storage medium. *Renewable Energy, 109*, 168–187.

Gray, J. L. (2011). The physics of the solar cell. In *Handbook of photovoltaic science and engineering* (2nd ed., pp. 82–128).

Green, M. A. J. S. S. E. (1981). Solar cell fill factors: General graph and empirical expressions. *Solid State Electronics, 24*(8), 788–789.

Hasanuzzaman, M., Malek, A. B. M. A., Islam, M. M., Pandey, A. K., & Rahim, N. A. (2016). Global advancement of cooling technologies for PV systems: A review. *Solar Energy, 137*, 25–45. https://doi.org/10.1016/j.solener.2016.07.010

He, W., Zhang, Y., & Ji, J. (2011). Comparative experiment study on photovoltaic and thermal solar system under natural circulation of water. *Applied Thermal Engineering, 31*(16), 3369–3376. https://doi.org/10.1016/j.applthermaleng.2011.06.021

Huang, B., & Du, S. (1991). *A performance test method of solar thermosyphon systems.*

Imani, S., Rezaie, A. H., & Taherbaneh, M. (2008). Modeling and evaluation of a terrestrial solar panel in space environment. In *Paper presented at the 2008 IEEE international conference on sustainable energy technologies.*

Islam, M. M., Hasanuzzaman, M., Rahim, N. A., Pandey, A. K., Rawa, M., & Kumar, L. (2021). Real time experimental performance investigation of a NePCM based photovoltaic thermal system: An energetic and exergetic approach. *Renewable Energy, 172*, 71–87. https://doi.org/10.1016/j.renene.2021.02.169

Jain, A., & Kapoor, A. (2004). Exact analytical solutions of the parameters of real solar cells using Lambert W-function. *Solar Energy Materials and Solar Cells, 81*(2), 269–277.

Jarimi, H., Bakar, M. N. A., Othman, M., & Din, M. (2017). Bi-fluid photovoltaic/thermal PV/T solar collector with three modes of operation: Experimental validation of a theoretical model. In *Mediterranean green buildings & renewable energy* (pp. 445–464). Springer.

Joshi, S. S., & Dhoble, A. S. (2017). Experimental investigation of solar photovoltaic thermal system using water, coconut oil and silicone oil as spectrum filters. *Journal of the Brazilian Society of Mechanical Sciences and Engineering, 39*(8), 3227–3236.

Joshi, S. S., & Dhoble, A. S. (2018). Photovoltaic -thermal systems (PVT): Technology review and future trends. *Renewable and Sustainable Energy Reviews, 92*, 848–882. https://doi.org/10.1016/j.rser.2018.04.067

Joshi, S., Dhoble, A., & Jiwanapurkar, P. (2016). Investigations of different liquid based spectrum beam splitters for combined solar photovoltaic thermal systems. *Journal of Solar Energy Engineering, 138*(2), 021003.

Kaldellis, J., Meidanis, E., & Zafirakis, D. J. A. E. (2011). Experimental energy analysis of a standalone photovoltaic-based water pumping installation. *Applied Energy, 88*(12), 4556–4562.

Khanna, V. K. (2019). *Flexible electronics* (Vol. 3). IOP Publishing Limited.

Khaselev, O., Bansal, A., & Turner, J. J. (2001). High-efficiency integrated multijunction photovoltaic/electrolysis systems for hydrogen production. *International Journal of Hydrogen Energy, 26*(2), 127–132.

Kumar, L., Hasanuzzaman, M., & Rahim, N. A. (2019). Global advancement of solar thermal energy technologies for industrial process heat and its future prospects: A review. *Energy Conversion and Management, 195*, 885–908. https://doi.org/10.1016/j.enconman.2019.05.081

Kumar, L., Hasanuzzaman, M., Rahim, N. A., & Islam, M. M. (2021). Modeling, simulation and outdoor experimental performance analysis of a solar-assisted process heating system for industrial process heat. *Renewable Energy, 164*, 656–673. https://doi.org/10.1016/j.renene.2020.09.062

Levy, M. Y., & Honsberg, C. (2006). Rapid and precise calculations of energy and particle flux for detailed-balance photovoltaic applications. *Solid-State Electronics, 50*(7–8), 1400–1405.

Long, H., Chow, T.-T., & Ji, J. (2017). Building-integrated heat pipe photovoltaic/thermal system for use in Hong Kong. *Solar Energy, 155*, 1084–1091.

Mokeddem, A., Midoun, A., Kadri, D., Hiadsi, S., & Raja, I. A. (2011). Performance of a directly-coupled PV water pumping system. *Energy Conversion and Management, 52*(10), 3089–3095.

Nahar, A., Hasanuzzaman, M., & Rahim, N. A. (2017a). Numerical and experimental investigation on the performance of a photovoltaic thermal collector with parallel plate flow channel under different operating conditions in Malaysia. *Solar Energy, 144*, 517–528. https://doi.org/10.1016/j.solener.2017.01.041

Nahar, A., Hasanuzzaman, M., & Rahim, N. A. (2017b). A three-dimensional comprehensive numerical investigation of different operating parameters on the performance of a photovoltaic thermal system with pancake collector. *Journal of Solar Energy Engineering, 139*(3). https://doi.org/10.1115/1.4035818

Nahar, A., Hasanuzzaman, M., Rahim, N. A., & Parvin, S. (2019). Numerical investigation on the effect of different parameters in enhancing heat transfer performance of photovoltaic thermal systems. *Renewable Energy, 132*, 284–295. https://doi.org/10.1016/j.renene.2018.08.008

Najafi, H., & Woodbury, K. A. (2013). Optimization of a cooling system based on Peltier effect for photovoltaic cells. *Solar Energy, 91*, 152–160. https://doi.org/10.1016/j.solener.2013.01.026

Nasrin, R., Rahim, N. A., Fayaz, H., & Hasanuzzaman, M. (2018). Water/MWCNT nanofluid based cooling system of PVT: Experimental and numerical research. *Renewable Energy, 121*, 286–300. https://doi.org/10.1016/j.renene.2018.01.014

Naves, A. X., Barreneche, C., Fernández, A. I., Cabeza, L. F., Haddad, A. N., & Boer, D. (2019). Life cycle costing as a bottom line for the life cycle sustainability assessment in the solar energy sector: A review. *Solar Energy, 192*, 238–262. https://doi.org/10.1016/j.solener.2018.04.011

Nielander, A., Shaner, M., Papadantonakis, K. J. E. E. S., Francis, S.a., & Lewis, N. S. (2015). *Energy & Environmental Science, 8*, 16–25.

Obeidat, F. (2018). A comprehensive review of future photovoltaic systems. *Solar Energy, 163*, 545–551. https://doi.org/10.1016/j.solener.2018.01.050

Othman, M. Y., Hamid, S., Tabook, M., Sopian, K., Roslan, M., & Ibarahim, Z. (2016). Performance analysis of PV/T Combi with water and air heating system: An experimental study. *Renewable Energy, 86*, 716–722.

Rahman, M. M., Hasanuzzaman, M., & Rahim, N. A. (2015). Effects of various parameters on PV-module power and efficiency. *Energy Conversion and Management, 103*, 348–358. https://doi.org/10.1016/j.enconman.2015.06.067

Rahman, M. M., Hasanuzzaman, M., & Rahim, N. A. (2017). Effects of operational conditions on the energy efficiency of photovoltaic modules operating in Malaysia. *Journal of Cleaner Production, 143*, 912–924. https://doi.org/10.1016/j.jclepro.2016.12.029

Shahsavar, A., & Ameri, M. (2010). Experimental investigation and modeling of a direct-coupled PV/T air collector. *Solar Energy, 84*(11), 1938–1958. https://doi.org/10.1016/j.solener.2010.07.010

Shiva Gorjian, A. S. (2020). *Photovoltaic solar energy conversion technologies, applications and environmental impacts*. Academic Press, Elsevier.

Singh, R., & Singal, C. (1983). Temperature dependence of the open-circuit voltage of an N+-P-P+ silicon solar cell under high illumination levels. *Solar Cells, 10*(2), 155–175.

Singh, D., & Tiwari, G. N. (2017). Performance analysis of basin type solar stills integrated with N identical photovoltaic thermal (PVT) compound parabolic concentrator (CPC) collectors: A comparative study. *Solar Energy, 142*, 144–158.

Sinton, R. A., & Cuevas, A. (1996). Contactless determination of current–voltage characteristics and minority-carrier lifetimes in semiconductors from quasi-steady-state photoconductance data. *Applied Physics Letters, 69*(17), 2510–2512.

Teo, H. G., Lee, P. S., & Hawlader, M. N. A. (2012). An active cooling system for photovoltaic modules. *Applied Energy, 90*(1), 309–315. Retrieved from http://www.sciencedirect.com/science/article/pii/S0306261911000201.

Tiedje, T., Yablonovitch, E., Cody, G. D., & Brooks, B. G. J. (1984). Limiting efficiency of silicon solar cells. *IEEE Transactions on Electron Devices, 31*(5), 711–716.

Tonui, J., & Tripanagnostopoulos, Y. J. (2007a). Air-cooled PV/T solar collectors with low cost performance improvements. *Solar Energy, 81*(4), 498–511.

Tonui, J. K., & Tripanagnostopoulos, Y. (2007b). Improved PV/T solar collectors with heat extraction by forced or natural air circulation. *Renewable Energy, 32*(4), 623–637. https://doi.org/10.1016/j.renene.2006.03.006

Tonui, J. K., & Tripanagnostopoulos, Y. (2008). Performance improvement of PV/T solar collectors with natural air flow operation. *Solar Energy, 82*(1), 1–12.

Willoughby, A. (2014). *Solar cell materials: Developing technologies*. John Wiley & Sons.

Wolf, M. (1976). Performance analyses of combined heating and photovoltaic power systems for residences. *Energy Conversion, 16*(1–2), 79–90.

Wu, S.-Y., Chen, C., & Xiao, L. (2018). Heat transfer characteristics and performance evaluation of water-cooled PV/T system with cooling channel above PV panel. *Renewable Energy, 125*, 936–946.

Yang, T., & Athienitis, A. K. (2014). A study of design options for a building integrated photovoltaic/thermal (BIPV/T) system with glazed air collector and multiple inlets. *Solar Energy, 104*, 82–92.

You, J., Dou, L., Yoshimura, K., Kato, T., Ohya, K., Moriarty, T., … Li, G. (2013). A polymer tandem solar cell with 10.6% power conversion efficiency. *Nature Communications, 4*(1), 1–10.

Yuan, J., Huang, T., Cheng, P., Zou, Y., Zhang, H., Yang, J. L., … Wang, R. J. N.c. (2019). Enabling low voltage losses and high photocurrent in fullerene-free organic photovoltaics. *Nature Communications, 10*(1), 1–8.

Further reading

Green, M. A. (2014). Crystalline silicon solar cells. In *Clean electricity from photovoltaics* (pp. 87–137).

Jaiganesh, K., & Duraiswamy, K. (2013). Experimental study of enhancing the performance of pv panel integrated with solar thermal system. *International Journal of Engineering and Technology, 5*(4), 3419–3426.

Smets, A. H., Jäger, K., Isabella, O., Swaaij, R. A., & Zeman, M. (2015). *Solar Energy: The physics and engineering of photovoltaic conversion, technologies and systems*. Cambridge: UIT.

Chapter 6

Solar thermal power plant

Moham Ed Abdur Razzaq[1] and Md Hasanuzzaman[2]
[1]Department of Mechanical Engineering, Chittagong University of Engineering & Technology, Chattogram, Bangladesh; [2]Higher Institution Centre of Excellence (HICoE), UM Power Energy Dedicated Advanced Centre (UMPEDAC), University of Malaya, Kuala Lumpur, Malaysia

Chapter outline

Technologies for Solar Thermal Energy. https://doi.org/10.1016/B978-0-12-823959-9.00001-5

6.1 Introduction

The utilization of renewable energy sources needs to be extended owing to the quick depletion of conventional fuel reserves and their significant environmental impact. It is therefore important to identify appropriate alternatives and prepare a transition to other sources of energy that have a minimal impact on the environment and are available in abundance to meet demand and ensure the sustainable energy security. As a green, abundant, and nondepleting resource, solar energy is continuously receiving more and more attention from research community. The sun is a nearly limitless radiant heat source at around 5800 K. Solar radiation can be used to operate energy-intensive engineering processes as a source of energy. However, compared to traditional power generation schemes, the implementation of solar energy for power generation purposes is expensive, so a new method is required to address this difficulty.

Approximately 18% of the world's end-use energy was delivered in the form of electricity, while 66% were supplied as hydrocarbon fossil fuel (coal, oil, and natural gas) powerplants including steam cycle thermal power plants, combined cycle thermal power plants, and gas turbine power plants (Key World Energy Statistics 2020, 2020); 40% of the energy consumed as fossil fuels is used for transport and 27% for industrial uses (Key World Energy Statistics 2020, 2020). In the electricity sector, to reduce the lion share of the global greenhouse gas emissions, a replacement of fossil fuels with renewable sources of energy is needed. The most mature alternative is to use concentrated solar power (CSP) technology for electricity, counting in 2016 with a market size of approximately 5 GW of installed capacity. There has been growing concern about the substantial benefits and production potential of CSP, also referred to as solar thermal power (STP) or CSP generation technology. Right now, the biggest obstacle it encounters is to cut the price of producing power so that it can compete with fossil fuels. It is projected that reliable and economic CSP generation technology will eventually mature and become highly competitive commercially over the next two decades. Stable and consistent power production, low costs, and excellent technological and economic advantages are characteristics of CSP generation technology; the strategic improvement of this technology is of great concern for carbon neutral economy and sustainable development (Wang, 2019).

Only CSP has traditionally been termed and referred instead of solar thermal energy. The term STE has only become common in recent years and some institutions have shifted the CSP concept to a higher level to cover both STE and concentrated photovoltaics (CPV). Some specialists, however, do use CSP to refer and replace STE, and CSP does not provide CPV in these instances. The underlying meaning of CSP therefore differs from organizations to organizations; it is an ambiguous concept, without a specific rationale, and is not used in this chapter. To apply internationally to the scientific knowledge established to concentrate and transform solar radiation into value-added thermal energy (i.e., CST technology), the term concentrated solar thermal (CST) is used. Concentrated solar thermal power (CSTP) specifies to power produced from CST mechanisms in this chapter.

6.2 Basic concept of solar thermal power plant

The concept of "solar thermal power plant" involves power plants that first transform solar radiation into heat energy. The generated heat energy is consequently indoctrinated by a thermal engine into useful rotational (mechanical) energy and then transformed into electricity. To achieve the utmost efficiency, high temperatures are needed for thermodynamic reasons. Improving the energy flux density of the incident solar radiation on a collector achieves certain high temperatures. Practically, under the following steps, the STP generation process is implemented:

- Concentrating solar radiation into a collector system;
- Enhancing radiation flux density (i.e., solar radiation concentration on a receiver), if appropriate;
- Solar radiation absorption (i.e., alteration of the radiation energy within the receiver into heat energy);
- Thermal energy transfer to an energy transformation medium;
- Using a heat engine (e.g., a steam turbine) to transform thermal energy into mechanical output;
- Installing a generator to transform mechanical energy into electrical energy.

The simplified graphic of a standard CSTP is shown in Fig. 6.1. The first portion, beginning at the prime source of energy (i.e., solar), the concentrator, which raises the flux density of absolute solar radiation such that the density of the solar flux on the receiver's surface can be up to the multiplication of several thousand greater than the flux density of solar energy hitting the surface of the globe. The latter of these is typically of the order of 1000 W/m^2 or less. Usually, this concentration is consummated by projecting solar radiation obtained onto a receiver with a surface shorter than that of the concentrator. Concentrated solar radiation striking the receiver's outer surface is converted

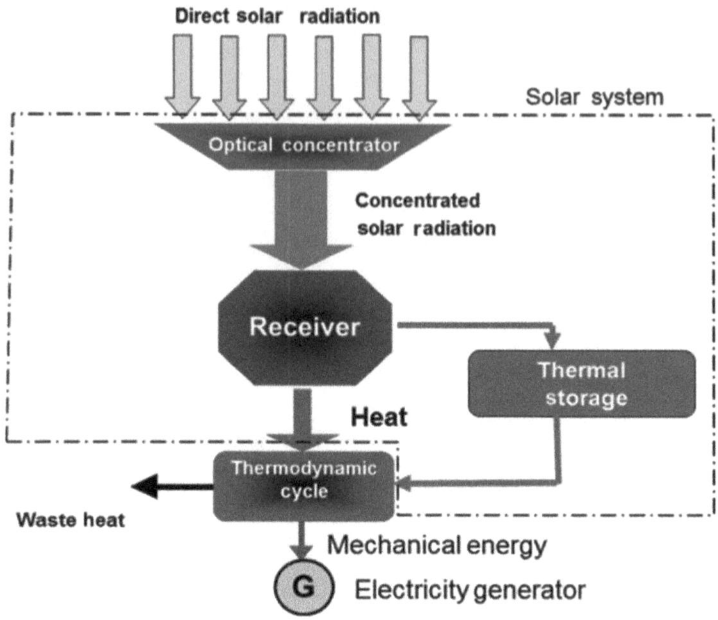

FIGURE 6.1 Simplified schematic of a concentrating solar thermal power plant (Zarza-Moya, 2018).

to heat by raising the working fluid's enthalpy that circulates within the receiver. Many working fluids (e.g., water, molten sodium chloride, or air) are used nowadays.

Sunlight encroaching on a concentrator must be directed appropriately to ensure that it reaches the receiver after being mirrored, per the principle of optics. Therefore, a very significant disadvantage of CST is that solar radiation can only be used; diffuse light is not useful since it foments in all directions and cannot be effectively focused (Zarza-Moya, 2018). In the receiver, the heat absorbing fluid gain the thermal energy as latent or sensible heat, which is subsequently transformed into mechanical energy utilizing thermodynamic cycle to produce electricity or transferred to an energy storage system to be used afterward when solar radiation is not accessible. The power block where heat is transformed to electrical energy includes a thermal cycle, the accessory for the cycle, and a generator.

6.3 Solar thermal power generation technologies

CSTPs are currently available in a variety of technologies. The following section will discuss briefly about each type of CSTP generation technologies available.

6.3.1 Solar tower power plant

Solar tower power plant (STPP) (Fig. 6.2A) has already reached commercial application and utilizes an array of sun light following heliostats which focus solar radiations upon a centralized receiver system fixed at the height of a tower to generate electrical energy. In a two-axis tracking system, heliostats are made up of many flat or slightly concave mirrors which track the sun. Heat is transported to an HTF in a central receiver, which is then moved to heat exchangers to operate a steam boiler. The steam boiler thus transforms mechanical energy into electrical energy, which is then supplied to the electricity grid.

6.3.2 Parabolic through solar power plant

A parabolic through solar power plant (PTSPP) (Fig. 6.2B) contains a group of reflectors bent in a parabolic shape in one dimension to concentrate solar energy on an absorber tube installed in the line of foci. From sunrise to sunset, the reflectors and absorber tubes spin in time with the solar. The absorber is made of a metal and a glass tube covering to minimize the loss of convective heat, with either air or vacuum between the two. A special material with considerably larger solar irradiance absorption and low heat emission is used to cover the metal tube (Behar, 2018).

6.3.3 Parabolic dish solar power plant

Parabolic stirling engine power plants have been designed to produce energy in kW for industrial applications and are more appropriate for power delivery in small settlements (Behar, 2018). Two axes of the parabolic dish track the sun. The incident rays of the sun are then concentrated on a spot. The receiver receives solar radiation to the Stirling engine and transfers the thermal energy (Fig. 6.2C).

6.3.4 Linear Fresnel reflector solar power plant

Rows and rows of flat or moderately slanted mirrors capture sunlight on a downward pointing linear receiver in linear Fresnel reflectors. The receiver is a permanent structure that is placed over a tower and is mounted on a single or multiple axis tracker, following the sun's linear reflectors. The key advantage of LFR technologies is that compact curved mirrors and fixed receivers are easy to build, requiring lower capital costs. Linear Fresnel reflector solar power plants (LFRSPPs) are less effective at producing electricity than other technologies. Modern micro linear Fresnel reflectors employ two adjacent receivers for each mirror row and therefore require lower resources to produce a given output than PTSPP (Fig. 6.2D) (Behar, 2018).

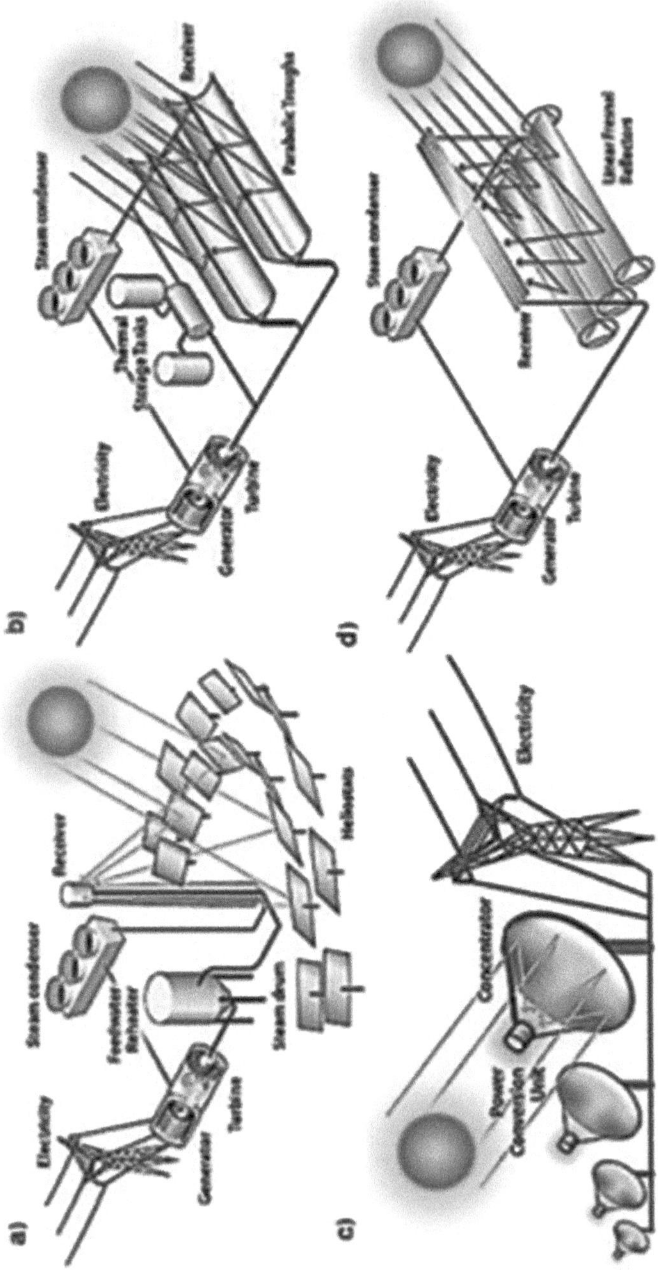

FIGURE 6.2 Concentrated solar thermal power plant technologies; (A) STPP, (B) PTSPP, (C) PDSPP, (D) LFRSPP (Gorjian & Ghobadian, 2015).

6.3.5 Solar chimney power plant

The three key components of a solar chimney power plant (SCPP) (Fig. 6.3) are an energy conversion unit with one or more turbine generators, a collector, and a chimney. The air underneath the transparent collector roof gets heated by the solar radiation, producing a temperature gradient. The density gradient would therefore be formed by the temperature variation. As the collector drops from its entrance to its middle, buoyancy forces will cause air to direct from the collector inlet to the chimney entrance, where a turbine will rotate the flowing air to generate electricity (Behar, 2018). The parameters of several levels of solar thermal systems are shown in Table 6.1.

6.4 Solar position modeling

During solar thermal applications, solar radiation must be accepted as a black body emitter that contain 6000°K temperature and 0.3−3 μm of wavelength. Solar power that comes to the globe surface is the primary determinant of astronomical and geographical factors including latitude and longitude, height, angle of solar declination, angle of solar hour, quality of air, and meteorological conditions. It is possible to categorize solar radiation either as direct or distributed. Concentration of solar specifically uses direct irradiation that hasn't been fragmented in the stratosphere, namely solar radiation. Many concentrators follow an uni-axis or dual-axis rotating approach of monitoring

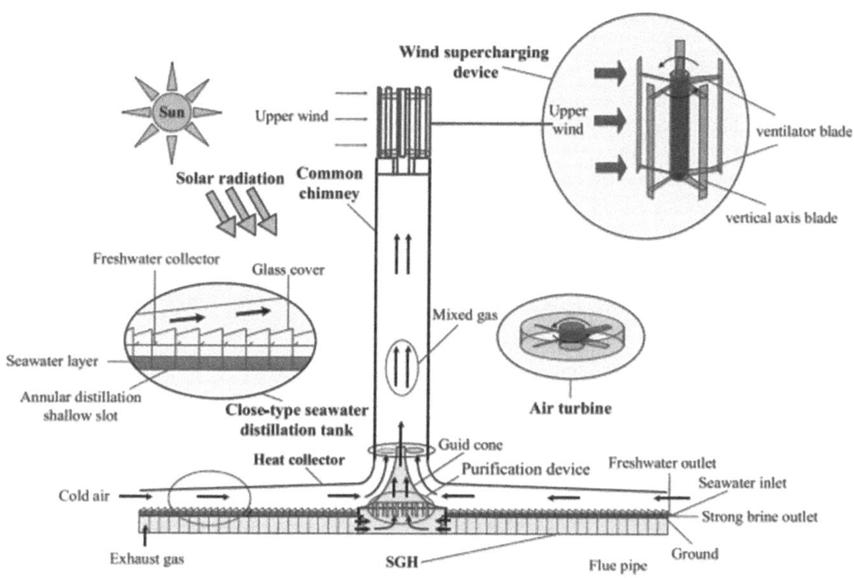

FIGURE 6.3 Solar chimney power plant (Zuo et al., 2020).

TABLE 6.1 Specifications of different solar thermal plant (Krishna, Faizal, Saidur, Ng, & Aslfattahi, 2020).

	Solar tower	Parabolic through	Parabolic dish	Linear Fresnel
Generation capacity (MW)	10–100	10–100	0.01–1	5–250
Solar concentration ratio	600–1000	50–90	<3000	25–170
Solar to electricity conversion (%)	10–22	10–16	16–29	8–12
Working temperature (°C)	300–1200	150–800 (with gaseous HTF, can reach >700)	300–1500	150–1400
Power cycle	Steam Rankine cycle; Brayton cycle	Steam Rankine cycle	Steam Rankine cycle; Sterling cycle; Brayton cycle	Organic Rankine cycle; steam Rankine cycle
Advantages	Highly efficient; Easy to integrate with oil and gas based powerplant	Highly reliable and durable; Highly compatible with oil and gas-fired dual power plants.	Appropriate for high-volume, high-efficiency production; No need for cooling with water	Easy to construct; Suitable with oil and gas-fired combined power stations
Associated cost	High	Low	Very high	Low

the movement of the solar to increase the overall efficiency of the solar radiation. These concentrators are referred to as concentrators of monitoring. This following section describes the mathematical equations required for solar field modeling for various collectors.

6.4.1 Solar angle

a) *Declination angle, δ:* It is the angle included in the line linking the center of the Planet to that of the earth's sun and equatorial plane. It's value varies on an annual basis and changes daily. The declination angle changes in the range of -23.45 degrees $= \delta \leq 23.45$ degrees (Fig. 6.4). Using the following equation, the average declination angle of a given day can be found (Türk Togrul & Onat, 1999):

$$\delta = 23.45^\circ \sin\left(360^\circ \times \frac{284 + n}{365}\right) \tag{6.1}$$

The nth day of the year is indicated by the number n, which is the date serial number; for instance, $n = 1$ indicates 1st January of the year. It is easy to obtain the date serial number "n" by calculation from Table 6.2.

b) *Duration of sunshine, H_{dl}:* It is formed by the angle of localized longitude and daily solar declination angle and describes the disparity between the hours of daily dawn and sunset. The following expression can be used to quantify it (Yilmaz & Söylemez, 2014):

$$H_{dl} = 2\cos^{-1}(-\tan \phi \tan \delta)/15 \tag{6.2}$$

where ϕ means local latitude.

c) *Solar hour angle, ω:* It corresponds to the earth's rotation-induced local meridian line and is the sun's angular divergence. At the midday, $\omega = 0^\circ$; ω having negative quantity in the early hour of the day and a positive quantity in the evening. The following numeric is applied to enunciate solar hour angle (Yilmaz & Söylemez, 2014):

$$\omega = 0.25^\circ(AST - 720) \tag{6.3}$$

d) *Solar zenith angle, θ_z* (Yilmaz & Söylemez, 2014):

$$\cos \theta_z = \cos \delta \cos \phi \cos \omega + \sin \delta \sin \phi \tag{6.4}$$

e) Solar altitude, α_s:

$$\alpha_s = 90^\circ - \theta_z \tag{6.5}$$

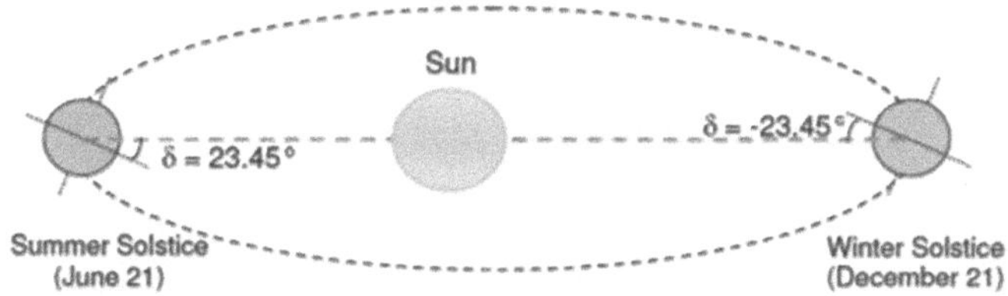

FIGURE 6.4 Declination angle (Mohammad et al., 2018).

TABLE 6.2 Date and date serial number (Türk Togrul & Onat, 1999).

Month	i	ii	iii	iv	v	vi	vii	viii	ix	x	xi	xii
n (tth day of the month)	T	t+31	t+59	t+90	t+120	t+151	t+181	t+212	t+243	t+273	t+304	t+334

f) Solar azimuth angle, γ_s:

$$\gamma_s = arc\cos\left(\frac{\sin\delta\cos\phi - \cos\delta\cos\omega\sin\phi}{\cos\alpha_s}\right) - 180^\circ \tag{6.6}$$

$$\gamma_s = -\gamma_s if \sin\omega > 0 \tag{6.7}$$

6.4.2 Solar tracking angle

a) *Slope of light-receiving surface, β:* It's the angular position between horizontal surface and inclined light absorbing face, $0° \leq \beta \leq 180$ degrees, when $\beta > 90$ degrees, the surface is angled down.

b) *Incidence angle, θ* (Yilmaz & Söylemez, 2014)*:*

$$\begin{aligned}\cos\theta &= \sin\delta\sin\phi\cos\beta - \sin\delta\cos\phi\sin\beta\cos\gamma \\ &\quad + \cos\delta\cos\phi\cos\beta\cos\omega + \cos\delta\sin\phi\sin\beta\cos\gamma\cos\omega \\ &\quad + \cos\delta\sin\beta\sin\gamma\sin\omega \\ &= \cos\theta_z\cos\beta + \sin\theta_z\sin\beta\cos(\gamma_s - \gamma)\end{aligned} \tag{6.8}$$

In Eq. (6.10), the angle of azimuth of the sun light absorbing face perpendicular against the longitudinal surface is defined as γ, and the forward vector is defined as south by west.

6.4.3 Direct normal beam insolation

The direct normal beam reflects the amount of solar power that enters the earth's atmosphere that is not dispersed and has been consumed by the atmosphere. The word normal here refers to the direct radiation that is registered in its normal direction on a plane. The design of the direct beam normal insolation (I_b) is assumed to be the average of the highest I_b and the average I_b after 2 h of sunshine and 2 h of sunset for a particular location. In addition, global insolation (I_g), is computed by summing diffused insolation (I_d) and I_b (Yilmaz & Söylemez, 2014).

$$I_g = I_d + I_b \tag{6.9}$$

The direct normal irradiance (DNI) falling normally on the plane is computed as (Yilmaz & Söylemez, 2014):

$$DNI = I_b \cdot \cos(\theta_z) \tag{6.10}$$

The following procedure is used to calculate the effective DNI falling perpendicular to the plane (Yilmaz & Söylemez, 2014):

$$I = DNI \cdot \cos\theta \tag{6.11}$$

6.5 Design of a parabolic through solar thermal power plant

Parabolic through solar thermal power plant (PTSTPPs) are the most efficient STP generation technology currently available. A PTC collects the energy from the sun and transforms it to useable thermal energy through HTF that flows across the solar field. Usually, PTCs are worked at a maximum temperature of 400°C and as HTF synthetic oil is widely evolved. To evaluate heat and sizing loss of the power plant during feasibility study, the heat transfer assessment of such collectors is relevant and also allows the implications of collector loss and HTF flow control techniques on total plant efficiency to be assessed. This section will provide a detailed overview of PTSPP including design of *parabolic trough collector* and *power plant design*. Fig. 6.5 demonstrates the various components and stages involve in the PTSTPP.

6.5.1 Parabolic through collector design

The solar field is comprised of a succession of parallel rows of multiple collectors. A linear imaginary concentrator with an optimal concentration ratio is the main element in the PTC field. It comprises (Fig. 6.6) a reflector (mirror), heat receiver, support structure, and tracking mechanisms. Fig. 6.7 shows the schematic of a PTC system.

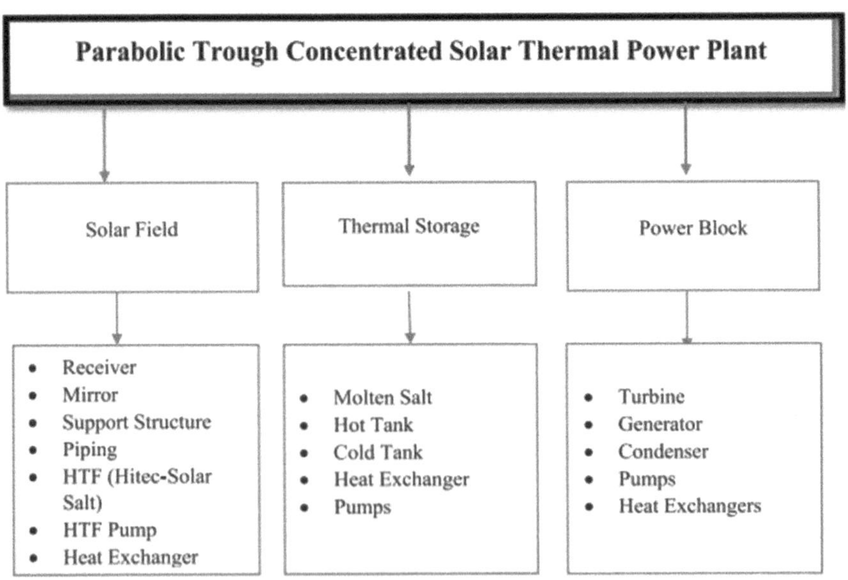

FIGURE 6.5 Various components and design parameters of a PTSPP (Bishoyi & Sudhakar, 2017).

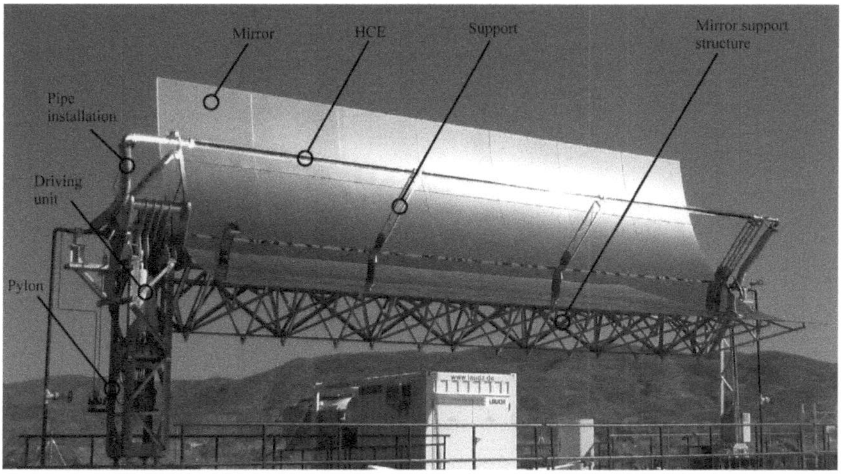

FIGURE 6.6 PTSC structure and components (Yılmaz & Mwesigye, 2018).

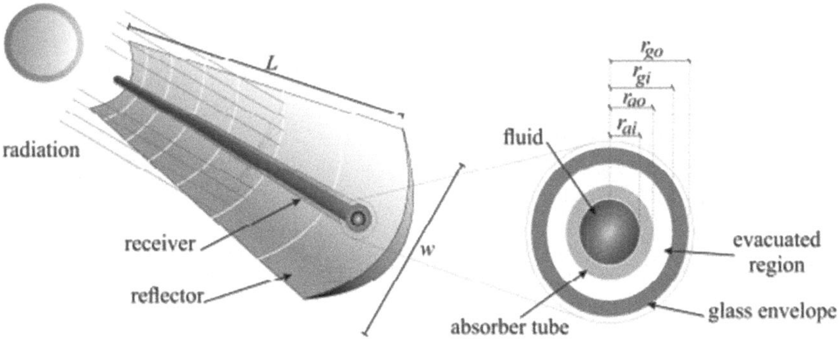

FIGURE 6.7 Schematic diagram of a PTC system (Quezada-García, Sánchez-Mora, Polo-Labarrios, & Cázares-Ramírez, 2019).

Geometric configuration

The geometry of a parabola is depicted in Fig. 6.8, where the fixed-point F is referred to as the focus and the V is referred to as the vertex. If the center is taken at *V* and x-axis directed toward the axis of the parabola, the formula of the parabola is (Shawky & Khalil, 2020):

$$y = \frac{x^2}{4f} \tag{6.12}$$

where f is the length of foci.

If the center is shifted to the F:

$$y = \frac{x^2}{4f} - f \tag{6.13}$$

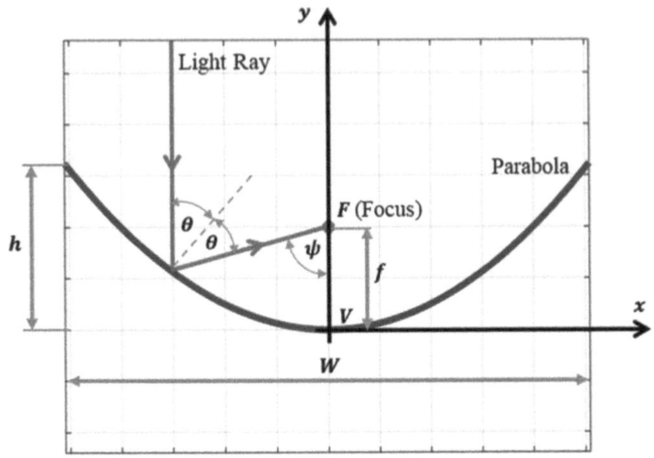

FIGURE 6.8 Parabola geometry (Shawky & Khalil, 2020).

The height of the curve h for a certain segment of the parabolic curve (W) is (Shawky & Khalil, 2020):

$$h = \frac{W^2}{16f} \tag{6.14}$$

The rim angle ψ_{rim} can be evaluated using the parabola dimensions as (Shawky & Khalil, 2020):

$$\tan(\psi_{rim}) = \frac{1}{\left[\frac{W}{8h} - \frac{2h}{W}\right]} \tag{6.15}$$

The arc length S can be calculated as (Shawky & Khalil, 2020):

$$S = \left[\frac{W}{2}\sqrt{\left(\frac{4h}{W}\right)^2 + 1}\right] + 2f \ln\left[\frac{4h}{W} + \sqrt{\left(\frac{4h}{W}\right)^2 + 1}\right] \tag{6.16}$$

When the collector is in proper shape and alignment, the cylindrical absorber outer diameter required for all solar images to be intercepted is calculated to be (Shawky & Khalil, 2020):

$$D = 2r_{rim} \sin 0.267 = \frac{W \sin 0.267}{\sin \psi_{rim}} \tag{6.17}$$

Here r_{rim} is the rim radius.

$$r_{rim} = \frac{2f}{1 + \cos \psi_{rim}} \tag{6.18}$$

In the construction and development of PTC that are precise and have fewer profile mistakes, geometrical relations provided here are helpful. Design and analysis of PTC systems, specifically optical analysis utilizing ray tracing approaches, require accurate system geometry specification.

Solar energy absorption modeling

In basic context, the thermal model of PTC is carried out by measuring the useful power obtained by the fluid, Q_u (Behar et al., 2015),with an energy balance between the system's absorbed solar power and environmental thermal losses, Q_{loss}.

$$Q_u = DNI \cdot \eta_{opt} \cdot \eta_{shading} \cdot K(\theta) \cdot \zeta(\theta) \cdot \cos\theta A_c - Q_{loss} \tag{6.19}$$

Here, θ is the angle of incidence, η_{clean} is the cleanliness of the collector surface, and A_c is the collector aperture area. η_{opt} is the maximum optical efficiency when the collector aperture area is normal to the direct beam radiation. The optical efficiency (Behar et al., 2015) is the multiplication of the collector reflectance (ρ_c), the intercept factor γ, the transmittance of glass envelope τ, and the absorbance of the absorber or receiver α.

$$\eta_{opt} = \rho_c \gamma \tau \alpha \tag{6.20}$$

The optical performance losses due to the incidence angle are defined by the incidence angle modifier (IAM) $K(\theta)$. As the angle of incidence rises, these losses arise because of increased reflection and absorption by the glass envelope. $K(\theta)$ can be derived from the following numerics developed by Dudley et al. (1994):

$$K(\theta) = \cos\theta + 0.000884(\theta) - 0.00005369(\theta)^2 \tag{6.21}$$

The geometrical end losses ($\zeta(\theta)$) are called the implications of the angle of incidence on the effective collector aperture area. This factor, often described as the end effect, accounted for the fraction of the length of the receiver that is not illuminated by the incident of sun rays on the aperture area. Gaul and Rabl (1980) have suggested the following equation to predict the end losses.

$$\zeta(\theta) = 1 - \left\{ \left(\frac{f}{l_c}\right) \cdot \left[1 + \left(\frac{A_c^2}{48 \cdot f^2}\right)\right] \cdot \tan(\theta) \right\} \tag{6.22}$$

Lippke (1995) has also proposed a basic model, which can be represented as follows:

$$\zeta(\theta) = 1 - \frac{f}{l_c} \tan\theta \tag{6.23}$$

Here l_c is the length of the collector (m).

Solar shading is another element that influences the absorption of solar radiation and adjacent PTCs generate shading. This happens often at low angles of the sun (at dawn or at sunset) and depends solely on the spacing of the lines. The shading efficiency ($\eta_{shading}$) is the ratio of the available area of the mirror aperture to the total area of the aperture, i.e., the mirror illumination area.

$$\eta_{shading} = \frac{W_{eff}}{W} = \frac{L_{space}}{W} \cdot \frac{\cos \theta_z}{\cos \theta} \tag{6.24}$$

where W_{eff} is the effective width of the mirror unshaded area (m), W is the collector aperture actual width, and L_{space} is the length of row spacing between throughs.

Solar receiver design

The essential component of a PTC is the heat collection element (HCE), which is just an absorber tube encased by a glass envelope (Fig. 6.9). Usually, the absorber is a stainless steel tube with a specific absorber surface that ensures the optical and radiative characteristics needed. In the region in which the surface emits radiation, specific surfaces combine substantial solar radiation absorption with low temperature emittance (Bishoyi & Sudhakar, 2017; Islam et al., 2015). The absorber is protected from corrosion and thermal loss by the glass envelope, which is a sealed antireflective glass tube. The vacuum environment is mainly used for reducing heat losses at larger heating rates and shielding the surface of the selective absorber from being oxidized. In order to reach the desired vacuum enclosure and thermal expansion gap between the steel tubing and the glass envelope, the HCE employs traditional glass for metal seals and metal bellows at either end. Fig. 6.10 shows the heat loss interactions among components and receiver.

The heat transfer modeling of the receiver needs some assumptions that include the following:

➢ The flow is fully developed

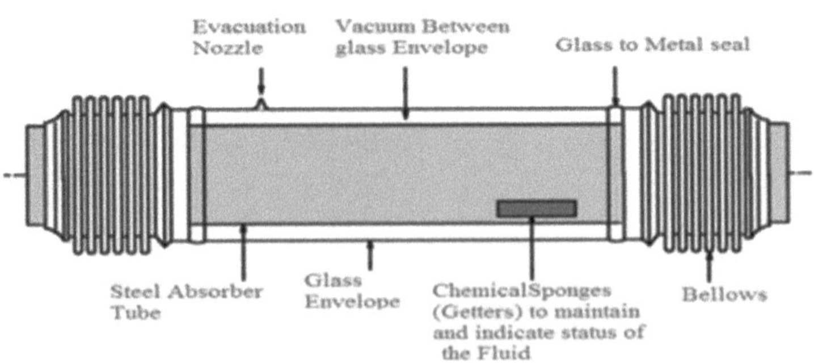

FIGURE 6.9 Sectional view of a parabolic through receiver (Bishoyi & Sudhakar, 2017).

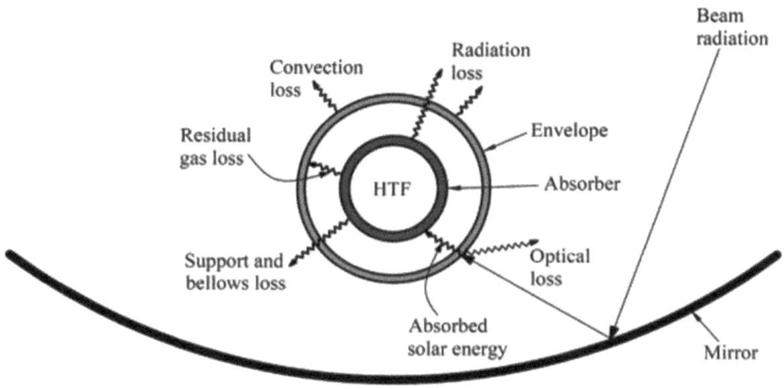

FIGURE 6.10 Typical losses from a PTC (Yılmaz & Mwesigye, 2018).

- The surface of the concentrator is reflected secularly.
- There is no difference in collector lengths, such as constant diameters.
- Inside the absorber, no free surface exists.
- The fluid is incompressible.
- The sky is considered to be a black body at a temperature similar to long-wave radiation.

The heat transfer model is developed by the conservation of energy approach between the HTF and the atmosphere. Fig. 6.11 depicts the heat transmission in the cross section of the solar collector as well as the thermal resistance model utilized in the heat transfer investigation.

The reflected solar energy from concentrator (mirror) is consumed by the glass envelope $\dot{Q}_{e\text{-}abs}$, and the surface of the absorber $\dot{Q}_{a\text{-}abs}$. Portion of absorbed heat in the absorber is transferred by forced convection to the HTF $\dot{Q}_{a\text{-}f,conv}$, rest of the energy radiated back to the glass enclosure by radiation $\dot{Q}_{a\text{-}e,rad}$, natural convection $\dot{Q}_{a\text{-}e,conv}$ and dissipated along the brackets by

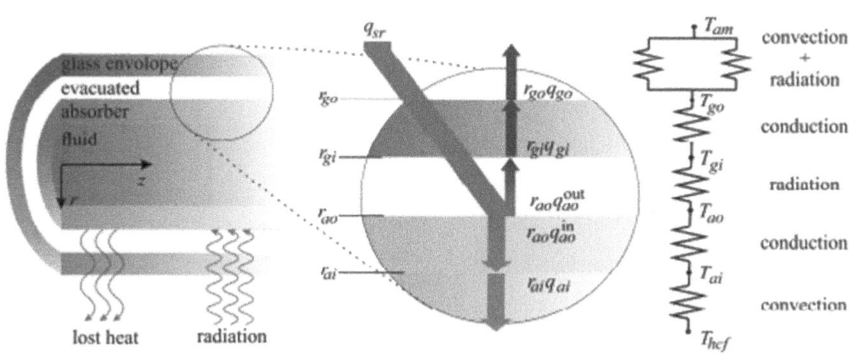

FIGURE 6.11 Heat transfer modeling of the receiver (Quezada-García et al., 2019).

conduction $\dot{Q}_{brac,\ cond}$. Via conduction, the heat dissipates from the absorber, transfers along the glass enclosure, and together with the heat consumed in the glass enclosure, $\dot{Q}_{e\text{-}abs}$ is released to the atmosphere through convection $\dot{Q}_{e\text{-}sa,conv}$ and radiation $\dot{Q}_{e\text{-}sa,rad}$ to the sky.

Transfer of heat from the absorber to the heat transfer fluid

Applying energy transfer principle in the control volume assuming $h \approx C_pT$, the following partial differential equation is found:

$$A_{i,a}\rho_f C_{p,f}\frac{\partial T_f}{\partial t} = -\dot{m}_f\frac{\partial}{\partial z}\left(C_{p,f}T_f + \frac{V_f^2}{2}\right) + \dot{Q}'_{a-f,conv} \tag{6.25}$$

$$V_f = \frac{\dot{m}_f}{\rho_f A_{i,a}} \tag{6.26}$$

Here, $A_{i,a}$ is the absorber inside cross-sectional area for internal diameter $D_{i,a}$. $\dot{m}_f$ is the mass flow rate of HTF, and ρ_f is the density of the HTF.

$$A_{i,a} = \frac{\pi}{4}{D_{i,a}}^2 \tag{6.27}$$

The convective heat transfer from the absorber to HTF is given by:

$$\dot{Q}'_{a-f,conv} = \pi Nu_f k_f (T_a - T_f) \tag{6.28}$$

Here, Nu_f is the Nusselt number, k_f is the HTF thermal conductivity, and wall temperature of the absorber is T_a.

Gnielinski correlation (Gnielinski, 1976) is established for fully developed turbulent flow (Re_D>2300) in a circular tube.

$$Nu_f = \frac{\left(\frac{C_f}{2}\right)(Re_D - 1000)Pr}{1 + 12.7(C_f/2)^{\frac{1}{2}}(Pr^{\frac{2}{3}} - 1)}\left(\frac{Pr}{Pr_w}\right)^{0.11} \tag{6.29}$$

$$Re_D = \frac{V_f D_{i,a}}{\upsilon_f} \tag{6.30}$$

Here, C_f denotes the Fanning Friction, Re_D defines the Reynolds number, Pr denotes the Prandtl number, and υ_f denotes the kinematic viscosity of the HTF. The friction coefficient C_f is calculated using Filonenko correlation (Sadik, Shah, & Aung, 1987):

$$C_f = (1.58 \ln Re_D - 3.28)^2 \tag{6.31}$$

For fully developed laminar flow $C_f = \frac{16}{Re_D}$ (6.32).

The coefficient of convection heat transfer for rough tubes (e.g., solar field piping) can be measured using the coefficient of friction defined by the Colebrook and White equation (Sadik et al., 1987):

$$\frac{1}{\sqrt{C_f}} = 3.48 - 1.7372 \ln\left(\frac{2\varepsilon}{D_{i,a}} + \frac{9.35}{Re_D\sqrt{C_f}}\right) \quad 5 \le \text{Re}\varepsilon \le 70 \tag{6.33}$$

where $Re_\varepsilon = \frac{V_f \varepsilon}{\upsilon_f}$ and ε is the roughness of the tube.

Sadik et al. (1987) provides the best empirical correlation for computation of practical friction factor in a rough tubes. This equation is valid for $4000 \le Re_D \le 10^8$ and $2 \times 10^{-8} \le \varepsilon \le 0.1$:

$$\frac{1}{\sqrt{C_f}} = 3.48 - 1.7372 \ln\left(\frac{2\varepsilon}{D_{i,a}} - \frac{16.2426}{Re_D} \ln B_2\right) \tag{6.34}$$

$$B_2 = \frac{\left(2\varepsilon/D_{i,a}\right)^{1.1098}}{6.0983} + \left(\frac{7.149}{Re_D}\right)^{0.8981} \tag{6.35}$$

C_f for transition flow ($2100 \le Re_D \le 4000$) can be computed using the empirical correlation proposed by Sadik et al. (1987):

$$C_f = 0.0054 + \frac{2.3 \times 10^{-8}}{Re_D^{\frac{-3}{2}}} \tag{6.36}$$

Because the entrance length for a long tube (10z/D60) is small, the approximate Nusselt number for the whole tube is considered to be equal to the value related with a thermally completely formed turbulent flow. For short tubes, the Al-Arabi correlation (Sadik et al., 1987) can be used to obtain the mean Nusselt number for thermally developing flow.$\frac{Nu_{f,m}}{Nu_{f,\infty}} = 1 + \frac{C_{tr}}{z/D}$, where $Nu_{f,\infty}$ is the Nusselt number for fully developed flow and C_{tr} is the transition coefficient, D (Sadik et al., 1987) is the diameter of the tube, and z is the length.

$$C_{tr} = \frac{\left(\frac{z}{D}\right)^{0.1}}{Pr^{\frac{1}{6}}}\left(0.68 + \frac{3000}{Re_D{}^{0.81}}\right) \tag{6.37}$$

For concentric annulus flow, turbulent correlation mentioned above can used by using a hydraulic diameter D_h.

Conduction heat transfer through absorber wall

Heat is transmitted to the inner surface of the absorber through conduction from the outside surface. Fourier's law of conduction can be used to evaluate conduction heat flux per unit length.

$$\dot{Q}'_{a,o-a,i,cond} = \frac{2\pi k_a (T_{i,a} - T_{o,a})}{\ln\left(D_{o,a}/D_{i,a}\right)} \tag{6.38}$$

Here, $T_{i,a}$ and $T_{o,a}$ is the absorber inner and outer surface temperature, respectively, k_a is the thermal conductivity of the absorber material, and $D_{i,a}$ and $D_{o,a}$ is the inside and outside diameter of the absorber tube, respectively.

Transfer of heat from the absorber to the glass envelope

Convection heat transfer and radiation heat transfer are used to transmit heat between the absorber and the glass enclosure. Applying energy balance equation, the following PDE is obtained:

$$\begin{aligned} A_a \rho_a C_{p,a} \frac{\partial T_a}{\partial t} = {} & A_a \frac{\partial}{\partial z}\left(k_a \frac{\partial T_a}{\partial z}\right) + \dot{Q}'_{a,abs} - \dot{Q}'_{a-f,conv} \\ & - \dot{Q}'_{a-e,conv} - \dot{Q}'_{a-e,rad} - \dot{Q}'_{cond,bracket} \end{aligned} \tag{6.39}$$

Here, A_a denotes the area of cross section of the absorber, $A_a = \left(\frac{\pi}{4}\right)\left(D_{o,a}^2 - D_{i,a}^2\right)$, $\dot{Q}'_{a,abs}$ defines the solar absorption in the absorber per receiver length of the receiver, $\dot{Q}'_{a-f,conv}$ is the per unit length heat transfer by convection from absorber to HTF, $\dot{Q}'_{a-e,conv}$ denotes per unit length convection heat transfer from the absorber to envelope, $\dot{Q}'_{a-e,rad}$ defines the per unit length radiation heat transfer from absorber to glass envelope, and $\dot{Q}'_{cond,bracket}$ denotes the per unit length heat conduction through brackets. The pressure of the annulus affects heat transfer from the absorber tube's outer surface to the glass envelope's inner surface. The radiation is combined with molecular conduction at low pressure, less than 1 torr. Instead of the conduction process, free convection takes place at higher pressures (>1 torr) (Behar et al., 2015).

Vacuum in annulus (p < 1 Torr) At lower pressures, the thermal conductivity of the gas in the annulus is a function of temperature, and the number of collisions between molecules is comparatively low which is known as free molecule regime, and it employs the concept of conduction heat transmission (Kuehn & Goldstein, 1976). The efficient heat transfer coefficient, based on this theory, is given by Sparrow et al. (2004):

$$h_{a,o-e} = \frac{k_{gas}}{\frac{D_{o,a}}{2}\ln\left(D_{i,e}/D_{o,a}\right) + \beta_b \lambda \left(D_{i,e}/D_{o,a} + 1\right)} \tag{6.40}$$

$$\dot{Q}'_{a-e,conv} = h_{a,o-e} \pi D_{o,a} (T_a - T_e) \tag{6.41}$$

Here, k_{gas} is the thermal conductivity is the gas in the annulus space at the average annulus gas temperature $\overline{T}_{a-e} = (T_a + T_e)/2$, $D_{i,e}$ denotes the glass enclosure diameter, T_a is the receiver tube temperature, T_e is the glass envelope temperature, λ is a molecule's mean free path between collisions, and β_b is the interaction coefficient. The β_b is determined using the following formula: $\beta_b = \frac{(2-\alpha)(9\gamma-5)}{2\alpha(\gamma+1)}$ here α is the accomodation coefficient, and γ defines specific heats ratio. The true number of the accommodation coefficient, which is directly tied to the molecular weight of the gas and the roughness of the surfaces, remains a major mystery (Kuehn & Goldstein, 1976). Values ranging from 0.01 to a unit have been recorded by experimental findings of different materials and gases. The mean free path can be calculated using the following formula:$\lambda = 2.331 \times 10^{-20} \frac{\overline{T}_{a-e}}{P_{annulus}\delta^2}$ where $P_{annulus}$ is the gas pressure at annulus, and δ is the annulus gas molecular diameter.

Annulus pressure higher than 1 Torr It has been shown that the conduction layer model correctly predicts the heat transport for horizontal tubes. This correlation follows an adaptive approach to estimate the average bulk temperature and is established on comprehensive experimental and numerical heat transport data (Sadik et al., 1987). The convective heat transfer is given by Sadik et al. (1987):

$$\dot{Q}'_{a-e,conv} = \pi k_{gas} Nu_{a-e}(T_a - T_e) \tag{6.42}$$

The correlation for convection part is given by Kuehn and Goldstein (1976):

$$Nu_{D_i conv} = \frac{hD_i}{k_{gas}} = \frac{2}{\ln\left(\dfrac{1+\frac{2}{Nu_i}}{1-\frac{2}{Nu_o}}\right)}$$

where $Nu_i = \left\{\left[0.518 Ra_{Di}^{\frac{1}{4}} f_2(Pr)\right]^{15} + \left(0.1 Ra_{Di}^{\frac{1}{3}}\right)^{15}\right\}^{1/15}$ and.

$Ra_{Di} = \frac{g\beta(T_a - \overline{T})_b D_i^3}{\upsilon^2} Pr$ where g known as the acceleration due to gravity, $\beta = \frac{1}{\overline{T}_{a-e}}$ stands for the volumetric thermal coefficient, and υ known as the kinematic viscosity. Properties of the HTF are determined at average temperature of $\overline{T} = \frac{T_a - T_e}{2}$, $D_i = D_{o,a}$ and $D_o = D_{i,e}$. The following equations have been collected from the (Sparrow et al., 2004).

$$Nu_o = \left(\left\{\left[\left(\frac{1}{1-e^{-\frac{1}{4}}}\right)^{\frac{5}{3}} + \left[0.587 f3(Pr) Ra_{Do}^{\frac{1}{4}}\right]^{\frac{5}{3}}\right]^{\frac{3}{5}}\right\}^{15} + \left(0.1 Ra_{Do}^{\frac{1}{3}}\right)^{15}\right)^{1/15} \tag{6.43}$$

$$Ra_{Do} = \frac{g\beta(T_b - T_e)D_o{}^3}{\upsilon^2} Pr \tag{6.44}$$

$$f2(Pr) = \left[1 + \left(\frac{0.559}{Pr}\right)^{\frac{3}{5}}\right]^{-\frac{5}{12}} \tag{6.45}$$

$$f3(Pr) = \left[\left(1 + \frac{0.6}{Pr^{0.7}}\right)^{-5} + \left(0.4 + 2.6Pr^{0.7}\right)^{-5}\right]^{-\frac{1}{5}} \tag{6.46}$$

For the conduction part, heat transfer by conduction when the Rayleigh number approches to zero is given by Sparrow et al. (2004):

$$Nu_{D_i cond} = \frac{2}{\cosh\left\{\frac{\left[D_i{}^2 + D_o{}^2\right]}{2D_iD_o}\right\}} \tag{6.47}$$

For any Rayleigh number, the overall Nusselt number is valid and given by Sparrow et al. (2004):

$$Nu_{a-e} = \left[(Nu_{D_i conv})^{15} + (Nu_{D_i cond})^{15}\right]^{1/15} \tag{6.48}$$

The average bulk temperature is determined by Bellos and Tzivanidis (2018):

$$\frac{\overline{T}_b - T_e}{T_a - \overline{T}_b} = \frac{\overline{Nu}_{D_i conv}}{\overline{Nu}_{D_o conv}} \tag{6.49}$$

$$\overline{Nu}_{D_i conv} = \frac{2}{\ln\left[1 + \frac{2}{Nu_i}\right]} \text{ and } \overline{Nu}_{D_o conv} = \frac{2}{-\ln\left[1 - \frac{2}{Nu_o}\right]}$$

Radiation heat transfer from the absorber to the envelope

The preceding equation is used to calculate the heat transfer via radiation between the absorber tube and the glass envelope (Forristall, 2003).

$$\dot{Q}'_{a-e,rad} = \frac{\sigma\pi D_{o,a}\left(T_{o,a}{}^4 - T_{i,e}{}^4\right)}{\left[\frac{1}{\varepsilon_{o,a}} + \frac{\left(1 - \varepsilon_{i,e}\right)D_{o,a}}{\varepsilon_e D_{i,e}}\right]} \tag{6.50}$$

where σ is the Stefan–Boltzmann constant, $T_{i,e}$ is the surface temperature of the inner glass envelope, $\varepsilon_{o,a}$ emissivity of the absorber outside surface, ε_e is the emissivity of the glass envelope.

Conductive heat transport through the glass envelope

By the following equation, the conductive heat transfer through the envelope can be derived (Bellos & Tzivanidis, 2018):

$$Q_{i,e-o,e,cond} = \frac{2\pi k_e (T_{i,e} - T_{o,e})}{\ln\left(\frac{D_{o,e}}{D_{i,e}}\right)} \tag{6.51}$$

where k_e is the thermal conductivity of the glass envelope, $T_{i,e}$ and $T_{o,e}$ are the inner and outer temperature of the glass envelope, and $D_{o,e}$ and $D_{o,i}$ are the outer and inner diameter of the glass envelope.

Heat transfer from HCE support bracket

The solar receiver is positioned by support brackets mounted on the collector construction at the collector focal line and mounted at every end of solar receiver. The heat losses through the bracket were ignored by some studies. The heat loss from a bracket can be estimated using the given equation (Yılmaz & Mwesigye, 2018):

$$Q_{bracket-amb} = \left(\frac{Nu_{bracket} P_{bracket} k_{bracket}{}^2 A_{bracket}}{D_{bracket} L_{HCE}{}^2}\right)^{0.5} (T_{bracket} - T_{amb}) \tag{6.52}$$

Here, $Nu_{bracket}$ is the Nusselt number at the bracket base temperature, $P_{bracket}$ is the perimeter of bracket, $k_{bracket}$ is the thermal conductivity of the bracket at $T_{bracket}$, $A_{bracket}$ is the minimum area of cross section of the bracket, $T_{bracket}$ is the base temperature of the bracket, $D_{bracket}$ is the nominal diameter of the bracket, and L_{HCE} is the length of the heat collector element.

Heat loss from the glass envelope to the atmosphere

From the outer surface of the glass envelope, heat is being transferred to the ambient by convection and radiation. Convection heat transfer has two modes: with wind (forced) and without wind (natural). The radiation heat transfer is normally between the glass envelope and either the sky or the collector's surface, but when the solar receiver is supposed to be surrounded simply by the sky's surface, the highest heat loss of radiation happens. Applying energy balance equation:

$$\begin{aligned} A_e \rho_e C_{p,e} \frac{\partial T_e}{\partial t} = & A_e \frac{\partial}{\partial z}\left(k_e \frac{\partial T_e}{\partial z}\right) + \dot{Q}'_{e,abs} + \dot{Q}'_{a-e,conv} \\ & + \dot{Q}'_{a-e,rad'} - \dot{Q}'_{e-sa,conv} - \dot{Q}'_{e-s,rad'} \end{aligned} \tag{6.53}$$

Here $\dot{Q}'_{e,abs}$ stands for the absorbed solar energy in the envelope per receiver length, $\dot{Q}'_{e-sa,conv}$ denotes the per unit length convection heat transfer between the glass envelope and the surroundings per unit length, and $\dot{Q}'_{e-s,rad'}$ denotes the radiation heat transfer from the glass envelope to the sky for per unit length. Thermal conductivity k_e and heat capacity ρC_p are evaluated using following polynomial equations (Assael, Gialou, Kakosimos, & Metaxa, 2004):

$$k_e = k_{e,o} \sum_j a_j \left(\frac{T}{298.15} \right)^j \quad (6.54)$$

$$(\rho C_p) = (\rho C_p)_{e,o} \sum_j b_j \left(\frac{T}{298.15} \right)^j \quad (6.55)$$

The following equation is used to estimate convective heat transfer between the envelope and atmosphere (Bellos & Tzivanidis, 2018):

$$\dot{Q}'_{e-sa,conv} = \pi h_e D_{o,e} (T_e - T_{amb}) \quad (6.56)$$

where $h_e = \frac{Nu_e k_e}{D_{o,e}}$, $D_{o,e}$ is the envelope outer diameter, T_e is the temperature of envelope, and T_{amb} is the atmosphere temperature. For no wind case, Churchill and Chi (Sparrow, Abraham, & Tong, 2004) suggested the following relation for horizontal cylinder under natural convection:

$$Nu_e = \left[0.60 + 0.387 \left\{ \frac{Ra_{D_e}}{\left[1 + (o.559/Pr)^{\frac{9}{16}}\right]^{16/9}} \right\}^{1/6} \right]^2 \quad (6.57)$$

where the Rayleigh number (Ra_{De}) for ambient air is dependent on the outer diameter of the glass envelope.

For wind case:

$$Nu_e = 0.193 (Re_{D_e})^{0.618} \, Pr^{0.33} \quad (6.58)$$

Properties of fluid are estimated at average temperature $\overline{T} = \frac{T_e + T_{amb}}{2}$.

Heat transfer loss by radiation causes due to the difference in temperature between envelope and the sky. This amount can be evaluated according to the given equation (Leon & Kumar, 2007):

$$\dot{Q}'_{e-s,rad'} = \sigma \pi D_{o,e} \varepsilon_e \left(T_e^{\,4} - T_{sky}^{\,4} \right) \quad (6.59)$$

Here ε_e is the envelope emissivity, T_{sky} is the sky temperature, T_e is the envelope outer surface temperature. Sky temperature is developed by Martin and Berdahl (Yılmaz & Mwesigye, 2018):

$$T_{sky} = 0.0553 T_{amb}^{\,1.5} \quad (6.60)$$

Pressure drops from the piping system

To measure the power needed for pumping the HTF in the solar field and from the solar field to the power block, the pressure loss is evaluated. The loss of pressure relies on the properties of the HTF, as well as the pipes and the fittings inner diameter. The drop of pressure in a pipe section i is given by:

$$\Delta P_{Li} = \rho_i g h_i \tag{6.61}$$

Using the Darcy–Weisbach equation, the friction head in the pipe segment i is determined:

$$h_i = 2C_{fi}\frac{L_i}{D_i}\frac{V_i^2}{g} \tag{6.62}$$

where C_{fi} denotes the coefficient of friction, L_i known as the pipe length i, V_i stands for the velocity of flow, D_i denotes the pipe inner diameter, and g is known as the acceleration due to gravity.

The coefficient of friction can be estimated from the following correlation developed by Chen (Sadik et al., 1987):

$$\frac{1}{\sqrt{C_{fi}}} = 3.48 - 1.7372\ \ln\left[2\frac{\varepsilon_{pipe}}{D_i} - \frac{16.2426}{Re_D}\ln\left(\frac{\left(2\varepsilon_{pipe}/D_i\right)^{1.1098}}{6.0983} + \left(\frac{7.149}{Re_D}\right)^{0.8981}\right)\right] \tag{6.63}$$

where Re_D is known as the Reynolds number, ε_{pipe} stands for roughness of the pipe. This empirical formula is true for $4000 \le Re_D \le 10^8$ and $2 \times 10^{-8} \le \varepsilon/a \le 0.1$.

Pressure drops in the fittings (elbow, expansion valves, joint, cross, etc) can be evaluated by Behar (2018):

$$L_{f_i} = K_{ft}\left(\frac{D_i}{2C_{f_i}}\right) \tag{6.64}$$

$$h_{fi} = 2C_{fi}\frac{L_{f_i}}{D_i}\frac{V_i^2}{g} \tag{6.65}$$

$$\Delta P_{L_{f_i}} = \rho_i g h_{fi} \tag{6.66}$$

where K_{ft} is the coefficient for different fittings. Table 6.3 conveys the information of K_{ft} for various types of solar field fittings.

The total pressure lost in the power block is the summation of pressure loss in the heat exchanger and the header of the piping that connects the power block with the solar field.

$$\Delta P_{PB} = \Delta P_{PB_{piping}} + \Delta P_{HX} \tag{6.67}$$

Here, ΔP_{HX} defines the drop of pressure in the heat exchanger, and $\Delta P_{PBpiping}$ called as the drop of pressure in the piping of power block. There is a minimal operating pressure that exists for different HTFs. The overall pressure drop in

TABLE 6.3 Various pipe fittings have different *K* values (Behar, 2018).

Fitting	K_{ft}
Gate valve	0.19
Check valve	2.5
Globe valve	10.00
Standard elbow	0.9
Medium elbow	0.75
Long elbow	0.6
Weldolet	1.8
Ball joint	4.73
Flexible hose	20.60

the piping of solar field should be applied to that pressure. The minimum pressure for the selected HTFs is given in Table 6.4.

HTF pump

A variable speed pump is adopted because the HTF mass flow rate is related to the solar intensity. The minimal flow rate of the pump selected is the flow needed at the solar field. The theoretical pump power necessary for the solar field is given by Behar (2018):

$$\dot{W}_{P,n} = \rho_{HTF} g \Delta P_{SHGT} \tag{6.68}$$

TABLE 6.4 Minimum and Maximum pressure and temperature for different HTFs.

HTFs	T_{min} (°C)	T_{max} (°C)	P_{min} (bar)
VP-1	37.8	400	15
Dowtherm Q	37.8	330	6
Dowtherm RP	37.8	350	6
Solar salt	260	621	6
Hitec	149	538	6
Hitec XL	150	5001	6

At variable speeds, the law of pump is used to evaluate the pump work required (Behar, 2018):

$$\dot{W}_P = \dot{W}_{P,n}\left(\frac{\dot{m}_{HTF}}{\dot{m}_{HTF,n}}\right)^3 \tag{6.69}$$

Here m_{HTF} and $m_{\mathrm{HTF,n}}$ are the actual and nominal mass flow rate, respectively.

HTF pump efficiency can be estimated by Behar (2018):

$$\eta_P = \eta_{P,n}\left(\alpha_P + 2(1-\alpha_P)\left(\frac{\dot{m}_{HTF}}{\dot{m}_{HTF,n}}\right) - (1-\alpha_P)\left(\frac{\dot{m}_{HTF}}{\dot{m}_{HTF,n}}\right)^2\right) \tag{6.70}$$

With $\alpha_P = -0.4$ is proposed by Lippke (Behar, 2018).

The total electric power needed by the pump of the solar field is (Behar, 2018):

$$\dot{W}_{P,net} = \frac{\dot{W}_P}{\eta_P} \tag{6.71}$$

6.6 Design of a solar tower power plant

A solar tower aided power plant mainly consists of the "solar part" and the "power block." The solar part composed of a field of heliostats, a solar tower, a receiver, and HTF flow system. The receiver mounted at the top of a tower absorbs solar energy from the DNI reflection of heliostats and converts in to useful thermal energy (heat). A molten salt is typically used as HTF to carry the converted heat to the thermal energy storage (TES) and then to the power block. Fig. 6.12 shows a typical solar tower powerplant with energy storage facility.

6.6.1 Heliostat field modeling

Fig. 6.13 shows the variables to define a heliostat field layout. The heliostat field includes plenty of heliostats that can reflect solar radiations to the receiver. The heliostat's thermal energy reflected to the receiver can be computed as (Li, Yang, et al., 2018):

$$Q_{hel} = Q_s \cdot \eta_{hel} \tag{6.72}$$

where Q_{hel} denotes the solar energy reflected by the heliostats, Q_s is the solar radiation energy falling on the heliostats, and η_{hel} is the heliostat efficiency.

$$Q_s = N_{hel} \cdot A_{hel} \cdot DNI/10^6 \tag{6.73}$$

where N_{hel} is the number of heliostats in the field and A_{hel} is the heliostat area.

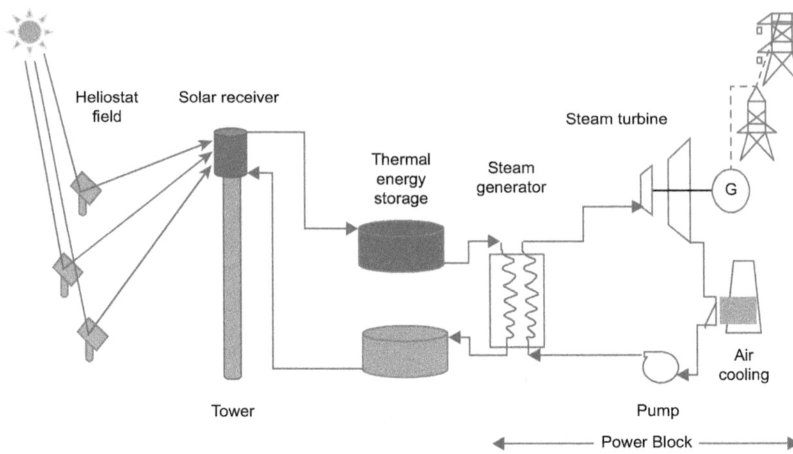

FIGURE 6.12 Schematic diagram of a typical STPP (Behar, Sbarbaro, & Morán, 2020).

In Fig. 6.13, DM shows the length between the two neighboring heliostats center and can be computed as (Collado & Guallar, 2012):

$$DM = DH + desp = \left[\sqrt{1 + \left(LW/_{LH}\right)^2} + x\right]LH \qquad (6.74)$$

The azimuthal spacing between two contiguous heliostats can be controlled by the optimizing parameter x.

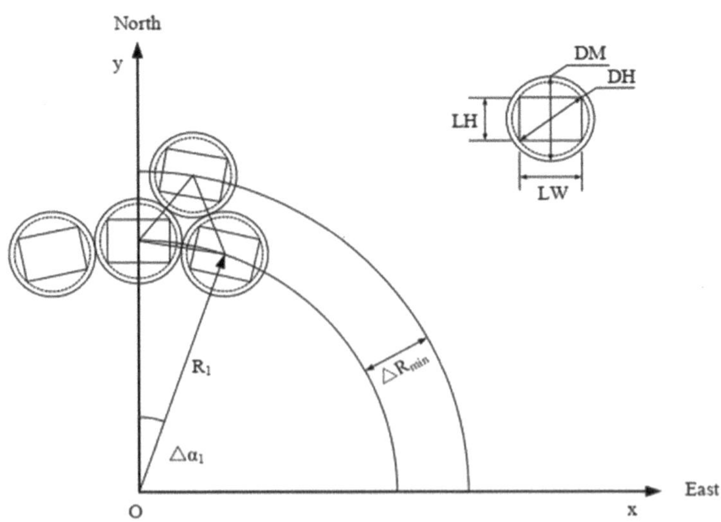

FIGURE 6.13 Fundamental variables for heliostat field (Li, Zhai, et al., 2018).

The minimal radial increment can be calculated by (Collado & Guallar, 2012):

$$\Delta R_{min} \cong DM \cdot \cos 30^{\circ} \tag{6.75}$$

To modify the radial distance in different areas, an optimization factor (y_i) is presented. The updated expression for radial distance, therefore, is as follows (Collado & Guallar, 2012):

$$\Delta R_i = y_i \cdot DM \cdot \cos 30^{\circ} \ (i = 1, 2, 3\ldots) \tag{6.76}$$

The amount of rows in different sections can be estimated as (Collado & Guallar, 2012):

$$N_{rows_1} = round\left[\frac{R_2 - R_1}{\Delta R_1}\right] \tag{6.77}$$

$$N_{rows_2} = round\left[\frac{R_3 - R_2}{\Delta R_2}\right] \tag{6.78}$$

The amount of heliostats in a single row can be evaluated as (Collado & Guallar, 2012):

$$Nhel_1 = \frac{2\pi}{\Delta \alpha_1} = \frac{2\pi R_1}{DM} \tag{6.79}$$

$$Nhel_2 = \frac{4\pi R_1}{DM} \tag{6.80}$$

$$Nhel_3 = \frac{8\pi R_1}{DM} \tag{6.81}$$

The total amount of heliostats in the solar field can be estimated by (Collado & Guallar, 2012):

$$Num = \sum N_{rows_i} \cdot Nhel_i \ (i = 1, 2, 3) \tag{6.82}$$

The required heliostat field land area can be determined by (Collado & Guallar, 2012):

$$field\_land_{area} = \pi (R_{last} + 0.5DM)^2 \tag{6.83}$$

The efficiency of the heliostat η_{hel} ($\eta(x,y,t)$), which can be presented by the mirror reflectivity, atmospheric attenuation factor, shading and blocking factor, cosine factor, and interception factor (Li, Yang, et al., 2018):

$$\eta(x,y,t) = \rho \cos \omega(x,y,t) f_{at}(x,y,t) f_{sb}(x,y,t) f_{int}(x,y,t) \tag{6.84}$$

Here ρ denotes the actual reflectivity of the mirror, cos(x,y,t) stands for the cosine factor, f_{at} (x,y,t) is the attenuation factor for atmosphere, f_{sb} (x,y,t) is the

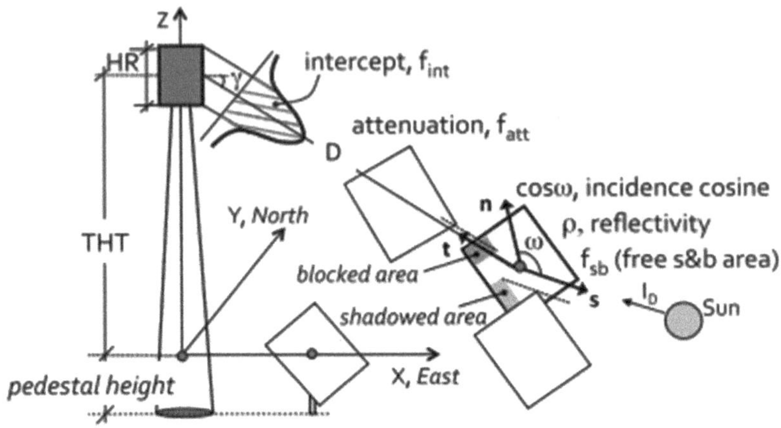

FIGURE 6.14 Nomenclature of the heliostat optical efficiency (Collado & Guallar, 2013).

shadowing and blocking factor, f_{int} (x,y,t) is the intercept factor. Fig. 6.14 shows the nomenclature of the optical efficiency of heliostat.

$$\cos(x, y, t) = \cos\left(arc\ \cos\left(\vec{S} \cdot \vec{R}\right) / 2\right) \tag{6.85}$$

where $\vec{S}$ is the incident vector directing to the sun and can be evaluated according to the following equation (Li, Zhai, & Yang, 2017):

$$\vec{S} = (-\cos\alpha \sin\beta, -\cos\alpha\cos\beta, \sin\alpha) \tag{6.86}$$

The reflector vector ($\vec{R}$) can be determined by the coordinates of the heliostat centers and the receiver, which is directing to the receiver.

The atmospheric attenuation factor is developed to measure the beam losses of the reflected light, which can be viewed as a relationship of distance between the heliostat and the receiver (D_{hr}) and can be calculated as (Collado & Guallar, 2013):

$$f_{at} = 0.99321 - 0.000176D_{hr} + 1.97 \times 10E - 8{D_{hr}}^2 \quad (D_{hr} \leq 1000\ \text{m}) \tag{6.87}$$

$$f_{at} = e^{-0.0001106D_{hr}} \quad (D_{hr} > 1000\ \text{m}) \tag{6.88}$$

The interceptor factor is evaluated according to the HFLCAL model (Schmitz, Schwarzböl, Buck, & Pitz-Paal, 2006):

$$f_{int} = \frac{1}{2\pi{\sigma_{tot}}^2}\int_{x'}\int_{y'} \exp\left(-\frac{x'^2 + y'^2}{2{\sigma_{tot}}^2}\right) dy'dx' \tag{6.89}$$

where σ_{tot} is the overall standard deviation and can be computed as follows (Schmitz et al., 2006):

$$\sigma_{tot} = \sqrt{D_{hr}{}^2\left(\sigma_{sun}{}^2 + \sigma_{bp}{}^2 + \sigma_{ast}{}^2 + \sigma_t{}^2\right)} \tag{6.90}$$

$$\sigma_{ast} = \frac{\sqrt{0.5\left(H_t^2 + W_s^2\right)}}{4D_{hr}} \tag{6.91}$$

$$H_t = d\left|\frac{D_{hr}}{f} - \cos\,\omega\right| \tag{6.92}$$

$$W_s = d\left|\frac{D_{hr}}{f}\cos\,\omega - 1\right| \tag{6.93}$$

where $f = D_{hr}$, and d is the square root of the total heliostat area and can be computed according to Fig. 6.16:

$$d = \sqrt{LW \times LH} \tag{6.94}$$

Tower height can be estimated using following equation (Schmitz et al., 2006):

$$H_t = 29.1 + 0.5113P_{th,SF} - 0.00887P_{th,SF}^{1.5} + 32801.719P_{th,SF}^{-2} \tag{6.95}$$

The thermal power input to the receiver can be determined by the following equation (Schmitz et al., 2006):

$$P_{th,SF} = \dot{m}_f \Delta H \tag{6.96}$$

6.6.2 Central receiver modeling

The energy balance equation for receiver is as follows (Li, Yang, et al., 2018):

$$Q_{rec} = Q_{hel} - Q_{rec,loss} \tag{6.97}$$

$$Q_{rec,loss} = Q_{ref} + Q_{rad} + Q_{conv} \tag{6.98}$$

where Q_{rec} is the heat energy received by the receiver; $Q_{rec,\ loss}$ is the receiver thermal loss; Q_{ref} is the thermal loss from the receiver tube surface due to reflectivity; Q_{rad} is the radiation loss; and Q_{conv} is the convection loss.

$$Q_{ref} = (1 - \alpha)Q_{hel} \tag{6.99}$$

$$Q_{rad} = \sum \varepsilon\sigma_0 A_r\left(T_{wall,i}^4 - T_{amb}^4\right) \tag{6.100}$$

$$Q_{conv} = \sum h_{conv}A_r\left(T_{wall,i} - T_{amb}\right) \tag{6.101}$$

where A_r is the lateral surface of the receiver tube; $T_{wall,i}$ is the inner wall temperature; T_{amb} is the ambient temperature; and h_{conv} is the convective heat transfer coefficient.

$$h_{conv} = 0.557 \times 10^{-6} \left(\frac{T_{wall,i} - T_{amb}}{H_t} \right)^{0.25} \tag{6.102}$$

Heat absorbed by HTF in the receiver is as follows (Li, Yang, et al., 2018):

$$Q_{HTF} = \dot{m}_{HTF} (h_o - h_i) \tag{6.103}$$

where $\dot{m}_{HTF}$ is the mass flow rate of HTF; and h_o and h_i is the specific enthalpy of HTF at the outlet and inlet of the receiver, respectively.

6.7 Heat transfer fluid

The thermal efficiency and performance of STP plant depend upon the heat transfer fluid (HTF), hence the characteristics of HTF become an absolute importance (Zaharil & Hasanuzzaman, 2020). HTFs must, in general, be efficient with storage materials and storage media, work within the appropriate temperature range, absorb and transfer heat effectively, and circulate well in an enclosed space. As a result, in order to ensure high efficiency and the previously mentioned properties, the key characteristics for solar HTFs have been established, and they can be classified as follows (Ho & Iverson, 2014):

- High thermal stability and a broad operating temperature range: This entails a low melting point to minimize heat tracing criteria (heat tracing is often used to prevent HTF solidification in pipes) and a high upper temperature limit to enable effective thermodynamic cycles to be used.
- Since the difference in temperature between the receiving surface and the HTF will be low, excellent heat transfer characteristics will be critical, enabling the fluid to attain high temperatures without putting too much thermal stress on the material. To improve the heat transfer coefficient, the HTF should have a high thermal conductivity. It is advantageous to have a low viscosity to minimize pressure drop and pumping capacity. Direct thermal storage is achievable with a high specific heat capacity, but alternative methods with a secondary medium are also available.
- Because of the minimal working pressure, thin tube walls may be used. This lowers the temperature difference on the wall and, as a result, the temperature-induced mechanical stress.
- Safety and corrosion actions are significant operational considerations. Low hazard products and good material compatibility are also highly recommended, as are cost-effective products and low maintenance costs.

6.7.1 Available heat transfer fluid

In the last few decades, different types of HTF have been studied for thermal performance enhancement and to design a cost-effective CSP plant. HTF is categorized into two main parts: traditional and nontraditional HTF. The traditional HTF includes water/steam, and synthetic oil. The nontraditional HTF includes nanofluids, molten salts, ionic liquid, liquid metals, and gaseous fluid. Fig. 6.15 shows different types of HTF available for solar thermal application. Table 6.5 shows thermal and physical properties of thermal oil used in CSP.

Fig. 6.16 shows some physical and thermal properties of molten salt used as HTF in CSP. The thermophysical properties of the solar salt can be estimated from the correlations reported in Table 6.6. The properties of Hitec can be determined using Table 6.9 correlations.

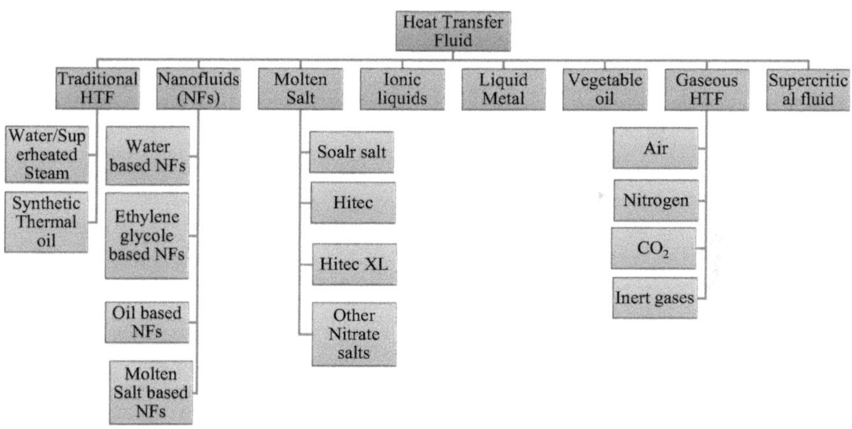

FIGURE 6.15 Classification of available heat transfer fluid (Vignarooban et al., 2015).

TABLE 6.5 Thermal and physical properties of thermal oils.

Thermal oils	Thermal conductivity (W/mK)	Melting point (°C)	Stability (°C)	Cost ($/kg)
Mineral oil	~0.1	−20	300	0.3
Synthetic oil	~0.1	−20	350	3
Silicon oil	~0.1	−20	400	5
Xceltherm 600 (paraffinic mineral oil)	~0.1	NA	315	N/A

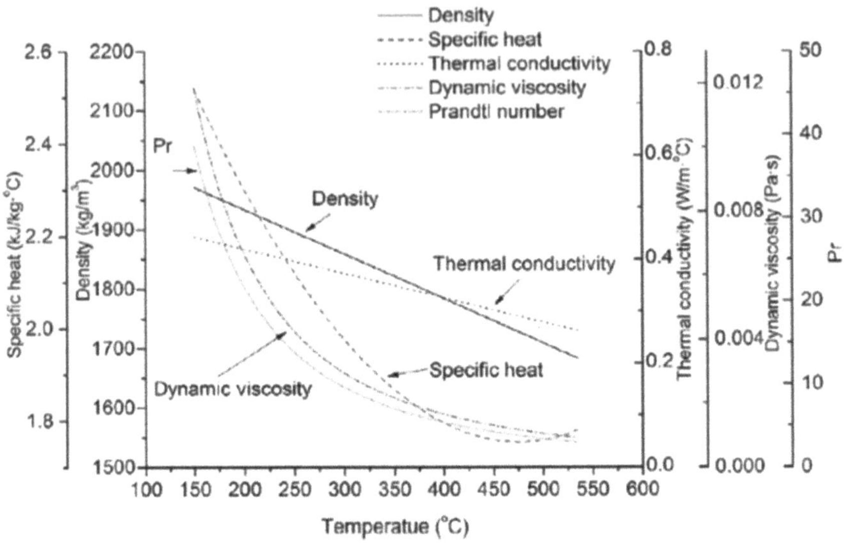

FIGURE 6.16 Properties of Hitec (Xiao et al., 2015).

Table 6.7 lists some important characteristics of liquid metals. We can deduct from Table 6.7 that liquid metals are more expensive than air (0$/kg), water (0$/Kg), oil (0.3–5 $/kg), and molten salt HTFs (0.93–1.3$/Kg).

A comparison between liquid and gaseous HTF properties is listed in Table 6.8.

6.7.2 Correlations for thermophysical property of HTFs

Table 6.9 shows the equations required to determine thermophysical properties of HTFs as a function of temperature.

6.8 Site selection and mapping

The selection of the site in the entire life cycle of the CSTPPs is highly important from the insight of sustainability. First, since it would have a necessary effect on the cost, the optimal location will decide the funded project feasibility. Second, an inappropriately chosen location may have a hostile impact on public life. Finally, the chosen location that does not meet environmental standards may not be environmentally friendly, but rather harmful. It is well recognized that the selection of the site decides the future electricity generation and socioeconomic values of the power plant and it needs to pay more emphasis.

TABLE 6.6 Thermophysical properties of commonly used molten salts in CSP (Vignarooban et al., 2015).

Molten salts	Composition (wt%)	Specific heat capacity (kJ/kgK)	Thermal conductivity (W/mK)	Melting point (°C)	Viscosity	Stability (°C)	Cost ($/kg)
Solar salts	$NaNO_3$ (60)–KNO_3 (40)	1.1	0.55	220	0.00326	600	0.5
Hitec	$NaNO_3$ (7)–KNO_3 (53)–$NaNO_2$ (40)	1.56 @ 300°C	0.2 @ 300°C	142	0.00316 @ 300°C	535	0.93
Hitec XL	$NaNO_3$ (7) KNO_3 (45)–$Ca(NO_3)_2$ (48)	1.45 @ 300°C	0.52 @ 300°C	120	0.00637 @ 300°C	550	1.1
Na–K–Li nitrates	$NaNO_3$ (28)–KNO_3 (52)–$LiNO_3$ (20)	1.091	NA	130	0.03 @ 300°C	600	1.1
Li–Na–K carbonates	Li_2CO_3 (32.1)–Na_2CO_3 (33.4)–K_2CO_3 (34.5)	1.4–1.5	NA	400	0.0043 @ 800°C	800–850	1.2–1.3
Sandia mix	$NaNO_3$ (9–18)–KNO_3 (40–52)–$LiNO_3$ (13–21)–$Ca(NO_3)_2$ (20–27)	1.16–1.44 @ 247°C	0.654 @ 250°C	<95	0.005 –0.007 @ 300°C	500	0.62–0.81
Halotechnics SS-500	$NaNO_3$ (6)–KNO_3 (23)–$LiNO_3$ (8)–$CsNO_3$ (44)–$Ca(NO_3)_2$ (19	1.22 @ 150°C	N/A	65	N/A	500	N/A

TABLE 6.7 Properties of liquid metals (Vignarooban et al., 2015).

Liquid metals	Composition (wt%)	Specific heat capacity @ 600°C (kJ/kgK)	Thermal conductivity @ 600°C (W/mK)	Melting point (°C)	Viscosity @ 600°C (Pa s)	Stability (°C)	Cost ($/kg)
Na	-	1.25	46	98	0.00021	883	2
Pb–Bi	Pb (44.5%): Bi (55.5%)	0.15	12.8	125	0.00108	1533	13
Na–K	Na (22.2%): K (77.8%)	0.87	26.2	−12	0.00018	785	2

TABLE 6.8 Physical property comparison between liquid and gaseous HTF (Vutukuru, Pegallapati, & Maddali, 2019).

HTF	Toxicity	Flammability	Corrosivity	Temperature stability
Therminol VP1	Yes	Yes	Cu corrosive	Until 400°C
Dowtherm Q	Yes	Yes	Noncorrosive	Until 400°C
Hitec XL	No	No	Low corrosivity	Until 550°C
CO_2	No	No	Noncorrosive	>550°C
He	No	No	Noncorrosive	>550°C

6.8.1 Method related to sites selection of solar power plants

Multicriteria decision-making (MCDM) and geographical information system (GIS) approaches have been widely used in the field of energy to select an appropriate site, project evaluation, and equipment evaluation. Although GIS can be used as a primary resource for the loading, use and evaluation of spatial data from solar energy, MCDM approaches could enable the determination of locations of prospective solar power plants and the evaluation of alternative locations, considering important indicators in the process of site selection (Shorabeh et al., 2019). MCDM is a combination of theories, strategies, and techniques implemented in a systematic and organized way to assist decision-makers to take complex decisions. MCDM ranking techniques involves Analytic Hierarchy Process (AHP), Elimination et Choix Traduisant la Realité (ELECTRE), PROMETHEE, Technique for Order Preference by Similarity to Ideal Solution (TOPSIS), Analytic Network Process (ANP), and Vlsekriterijumska Optimizacija IKompromisno Resenje (VIKOR) to identify different site domains. Table 6.10 shows the advantages and disadvantages of different MCDM ranking methods.

However, the reliability of STPP site assessment will be decreased by some issues with the MCDM methods. First, in the assessment of the STPP site, qualitative data cannot be accurately interpreted. Under linguistic circumstances, human beings are actively making choices. Evaluations of the output of alternatives on qualitative parameters by means of linguistic terms but instead numerical values are more fitting. Linguistic words such as “good,” “fair,” and “poor” are generally used when measuring the taste of food, for instance. Besides that, in some countries in southeast Asia, due to cultural variations, most managers have poorer probabilistic reasoning than their English peers. Managers prefer “I don’t know” to convey their views of

TABLE 6.9 HTFs thermophysical property correlations (Benoit, Spreafico, Gauthier, & Flamant, 2016).

Fluid	Equations
Therminol VP-1	Specific heat $\left[\frac{J}{KgK}\right]$ $C_p = 2.82T + 716$
	Thermal Conductivity[W/mK] $k = 1.73 \times 10^{-7}T^2 + 7.62 \times 10^{-6}T + 0.14$
	Density kg/m^3 $\rho = -7.61 \times 10^{-4}T^2 - 2.24 \times 10^{-1}T + 1191$
	Dynamic viscosity [Pa s] $\mu = \left(-23 \times 10^{-5}T^3 + 5.61 \times 10^{-3}T^2 - 19.89T + 1822\right)^{-1}$
Solar salt	$C_p = 1443 + 0.172(T - 273.15)$
	$k = 0.443 + 1.9 \times 10^{-4}(T - 273.15)$
	$\rho = 2090 - 0.636(T - 273.15)$
	$\mu = 2.2714 \times 10^{-2} - 1.2 \times 10^{-4}(T - 273.15) + 2.281 \times 10^{-7}(T - 273.15)^2 - 1.474 \times 10^{-10}(T - 273.15)^3$
Hitec salt	$C_p = 1.56 \times 10^3$
	$k = 0.411 + 4.36 \times 10^{-4}(T - 273.15) - 1.54 \times 10^{-6}(T - 273.15)^2$
	$\rho = 2084 - 0.74(T - 273.15)$
	$\mu = 10^{2.7374}(T - 273.15)^{-2.104}$
Hitec XL	$C_p = 1.56 \times 10^3$
	$k = 0.519$
	$\rho = 2240 - 0.827(T - 273.15)$
	$\mu = 10^{6.1374}(T - 273.15)^{-3.36406}$

Liquid metals Sodium (Na)	$C_p = 1658.2 - 0.84790T + 4.4541 \times 10^{-4}T^2 - 2.99 \times 10^6 T^{-2}$
	$k = 124.67 - 0.11381T + 5.5226 \times 10^{-5}T^2 - 1.1842 \times 10^{-8}T^3$
	$\rho = 219 + 275.32\left(1 - \frac{T}{2503.7}\right) + 511.58\left(1 - \frac{T}{2503.7}\right)^{0.5}$
	$\ln(\mu) = -6.4406 - 0.3958\ln(T) + \frac{556.835}{T}$
Lead-bismuth eutectic	$C_p = 159 - 2.72 \times 10^{-2}T + 2.12 \times 10^{-6}T^2$
	$k = -1.741 \times 10^{-6}T^2 + 1.517 \times 10^{-2}T + 3.61$
	$\rho = 11096 - 1.3236T$
	$\mu = 0.000494\exp\left(\frac{6270}{RT}\right)$
Air	$C_p = 0.1805T + 950.1$
	$k = -9.72 \times 10^{-9}T^2 + 7.14 \times 10^{-5}T + 6.22 \times 10^{-3}$
	$\rho = \frac{P}{r_{air}T}$
	$\mu = \mu_r\left(\frac{T}{T_r}\right)^{1.54}\frac{T_r+S_{air}}{T+S_{air}}$ Where, $T_r = 273.15$ K is the reference temperature, $\mu_r = 1.716$ Pa s is the reference viscosity and $S_{air} = 110.4$ K is the Sutherland's constant for air
CO_2	$C_p = -3.32 \times 10^{-4}T^2 + 0.918T + 651$
	$k = -1.57 \times 10^{-8}T^2 + 9.74 \times 10^{-5}T - 1.10 \times 10^{-2}$
	$\rho = \frac{P}{r_{CO_2}T}$
	$\mu = -1.23 \times 10^{-11}T^2 + 5.3 \times 10^{-8}T + 5.94 \times 10^{-7}$

Continued

TABLE 6.9 HTFs thermophysical property correlations (Benoit, Spreafico, Gauthier, & Flamant, 2016).—cont'd

Fluid	Equations
He	$C_p = -2.58 \times 10^{-6} T^2 + 8.97 \times 10^{-3} T + 5183$
	$k = -3.91 \times 10^{-8} T^2 + 3.33 \times 10^{-4} T + 7.08 \times 10^{-2}$
	$\rho = \frac{P}{r_{He} T}$
	$\mu = -4.7 \times 10^{-12} T^2 + 4.23 \times 10^{-8} T + 8.64 \times 10^{-6}$
H_2	$C_p = 1.72 \times 10^{-3} T^2 - 1.72 T + 14994$
	$k = 3.6 \times 10^{-8} T^2 + 4.32 \times 10^{-4} T + 5.94 \times 10^{-2}$
	$\rho = \frac{P}{r_{H_2} T}$
	$\mu = -3.08 \times 10^{-12} T^2 + 1.97 \times 10^{-8} T + 3.69 \times 10^{-6}$

TABLE 6.10 **The MCDM ranking approaches advantages and disadvantages.**

Approaches	Advantages	Disadvantages
AHP	**a)** Hierarchical ordered structures **b)** It is an effective method to determine the evaluation process consistency	**a)** Data are collected from experience **b)** More requirements and alternatives are needed for the large sets of comparisons
ELECTRE	**a)** Decision is made by thresholds of indifference and preference, and outranking method **b)** It is best suitable when incomparable alternatives are present	**a)** As it requires complex computational methods **b)** Possible alternatives may not be identified
PROMETHEE	**a)** Required group level decision **b)** Decision can be made from different properties and attributes	The objective doesn't get proper structure
TOPSIS	**a)** Very easy to understand and utilize **b)** It works with fundamental ranking process	**a)** Failure to consider the distinction between positive and negative values **b)** The relative value of these distances fails to be taken into account
ANP	**a)** Act with suggestions and interdependency. **b)** Comfortably apply relevant softwares	**a)** More matrices of pairwise comparison are needed **b)** With one variable in isolation, it cannot be measured
VIKOR	**a)** Present the subjective preferences of DMs' **b)** Include a benefit level to a comparative approach	**a)** Difficult under contradictory conditions

uncertainty on behalf of the Southeast Asian culture pattern, whereas most English managers take the probability appraisal to explain their perception of uncertainty. Unfortunately, this sort of information cannot be managed by most MCDM methods used in the collection of STPP locations, such as AHP and TOPSIS. So the conventional approach that uses the numerical value to replace the linguistic words would lead to the loss or deviation of decision data.

Second, in fact, the separate statement of the MCDM methods is unlikely to satisfy. The obvious implication in most MCDM methods is that the criteria are distinct from one another. For example, one of the common aggregation techniques of the MCDM methods is the weighted average method, whose foundation is the additivity phenomenon that is distinguished by an independence axiom. It is well understood, however, that things are usually related. Despite trying to eliminate the dependent requirements, we can still not make sure that the parameters are completely independent of each other.

The fuzzy set theory (FST) is integrated with the MCDM method in order to represent the subjective preferences of specialists more precisely. The incorporation of the FST allowed fuzzy numbers to be frequently used to demonstrate uncertainty in the context of site selection. To some level, however, certain degrees of uncertainty for the evaluation are shown by decision-making (DM) as they rely on intuition and knowledge to determine the complex indicators of active site during the DM process. Intuitionistic fuzzy set (IFS) is more robust and practical when solving uncertainty issues than conventional fuzzy set, since IFS concurrently contains membership, nonmembership, and hesitation degrees. As a modified form of IFN, triangular intuitionistic fuzzy numbers have gained considerable interests as they articulate the views of DM more specifically through the use of a triangular fuzzy number to replace a linguistic variable.

The MCDM method involves three major phases:

Phase-I: Identification of alternative sites and evaluation criteria
Phase-II: Determination of weight vector of evaluation criterion for FST or probabilistic linguistic theory
Phase-III: Selection of alternative using MCDM ranking approaches.

6.8.2 Evaluation criteria

Evaluation criteria play a vital role in the site selection and project implementation of CSTPP. To identify a content location for CSTPP, it is definitive for us to develop a technical and comprehensive evaluation criteria framework. Considering sustainability aspect, evaluation criteria consist six main criterions:

- Resource criterion (C1),
- Economy criterion (C2),
- Land criterion (C3),
- Infrastructure and construction criterion (C4),
- Environmental criterion (C5), and
- Social criterion (C6).

The main criteria are interconnected with some subcriteria that are presented in Fig. 6.17:

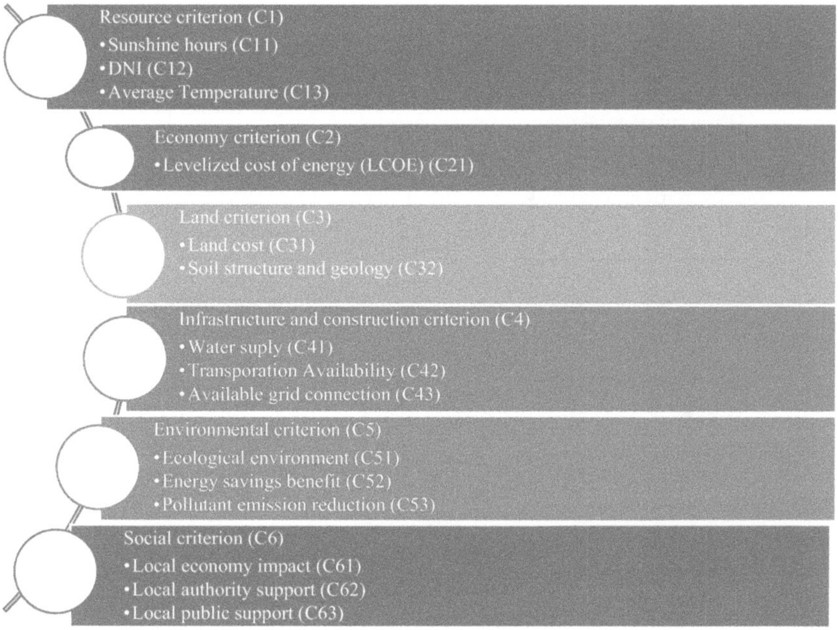

FIGURE 6.17 Evaluation criteria and associated subcriteria (Aragonés-Beltrán, Chaparro-González, Pastor-Ferrando, & Pla-Rubio, 2014).

6.8.3 AHP and ANP ranking approach

The reasoning behind the application of AHP/ANP approach for site selection are: (a) They facilitate decision-makers to use a structured approach to evaluate complicated DM issues, breaking down the key question into simplified and realistic subproblems, (b) ANP can be used if there are interrelationships between sets of elements (criteria and alternatives), (c) the thorough review of goals and interrelationships between the components of the clusters forces the DM to focus critically on its project priority strategy and on the issue of DM itself, resulting in a clearer understanding of the problem and a more accurate final decision. To select a site for the STP plant project using ANP/AHP approach, three broader phases should follow: The following steps can be taken to solve an MCDN problem using AHP approach. Fig. 6.18 shows the method scheme used for AHP approach.

i) The problem of DM is organized as a hierarchy and is subdivided into many layers. The highest level of the hierarchy is the primary objective of the problem of choice. The tangible and/or intangible criteria and subcriteria contributing to the target are the lower levels. The lowest level consists of the options to quantify in terms of the criteria.

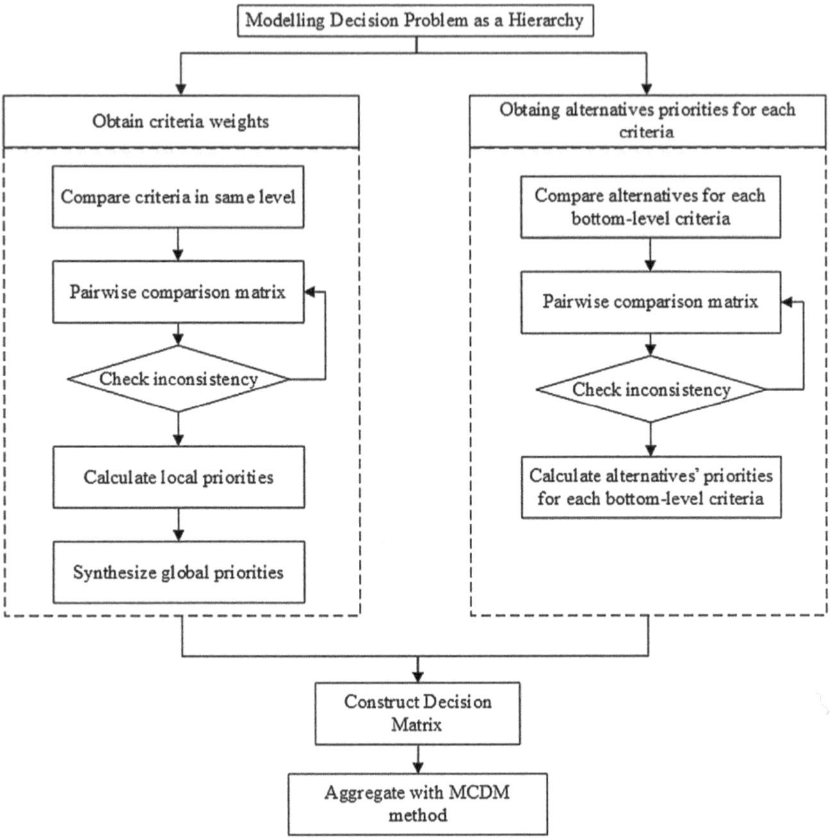

FIGURE 6.18 AHP method scheme (Aragonés-Beltrán et al., 2014).

ii) The weight for criteria is obtained using (Saaty, 1990) discrete 1–9 scale, the n criteria is compared at same level. Based on the DM's judgments a_{ij}, a pairwise comparison matrix A is created at each level.

$$A = \begin{bmatrix} 1 & a_{21} & \cdots & a_{1n} \\ a_{21} & 1 & \ldots & a_{2n} \\ \vdots & \vdots & \ddots & \vdots \\ a_{n1} & a_{n2} & \ldots & 1 \end{bmatrix} \tag{6.104}$$

A certain degree of inconsistency is anticipated and permitted. A consistency ratio (CR) of the matrix A is used to check the inconsistencies of judgment. Saaty (1990) suggested measuring the consistency index (CI) to see if the degree of inconsistency is "rational" using $CI = \frac{\lambda_{max} - n}{n-1}$,

TABLE 6.11 RI values (Saaty, 2000).

N	1	2	3	4	5	6	7	8	9	10	11	12
RI	0	0	0.52	0.89	1.11	1.25	1.35	1.40	1.45	1.49	1.51	1.54

where λ_{max} is the maximum eigenvalue. The consistency ratio can be estimated as:

$$CR = \frac{CI}{RI} \tag{6.105}$$

Here *RI* (Table 6.11) is the random index value that depends on the size of the matrix *n* and evaluated from experimental data.

Saaty (2000) proposed a threshold value for *CR* for different size of the matrix A. For $n = 3$, 4 and $n \geq 5$, the threshold values are 0.05, 0,08, and 0.1 respectively. If CR is under a certain threshold, the matrix is assumed to have reasonable consistency, and the comparison matrix's defined priorities are appropriate. If CR is over the threshold value, the matrix A judgments must be reconsidered. The matrix A yields the local priorities vector P= (p_1, p_2, p_i, … … p_n). Since these are the priorities of entities at the same level of the hierarchy, they are called local priorities (Aragonés-Beltrán et al., 2014). The local priorities of all criteria are combined to decide the global priority of all criteria, g_i, i = 1,2, …, N^H, where N^H is the number of criteria and subcriteria in the hierarchy, by multiplying the element's local priority to the element's global priority.

iii) For each criterion, the evaluation of alternatives is collected.

iv) The bottom-level parameters and alternatives are used to construct the decision matrix.

v) An MCDM approach is used to aggregate the alternative priorities and criteria priorities.

While AHP is conceptually straightforward to use, it's rigorous hierarchical structure can't handle the intricacies of many real-world challenges. Saaty (1990) suggested the ANP model, a general type of AHP, as a solution. ANP depicts a DM problem as a clustered network of parameters and alternatives (all referred to as elements). Within and between clusters, a network should provide feedback and interdependent links, and any of the parts in the network can be connected in any way. This allows for more precise simulation of complex environments. A super matrix can be used to describe the effect of network elements on other elements in the same network. The stages involved the ANP approach are depicted in Fig. 6.19.

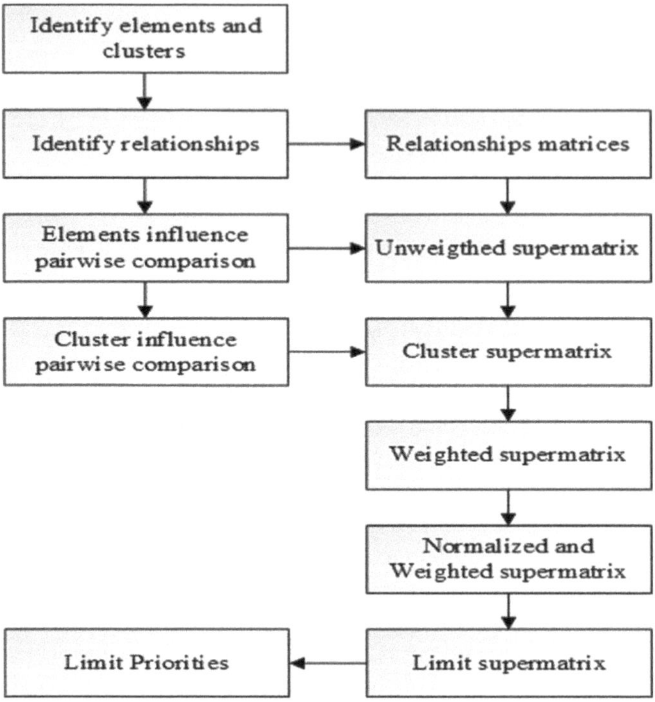

FIGURE 6.19 Stages in ANP approach (Aragonés-Beltrán et al., 2014).

6.8.4 Site selection principle

The national and regional solar energy use strategies, land use policy, thermodynamic and power system strategies, local construction and infrastructure strategies, combined thermodynamic and energy loads, and solar energy policy would all be used to choose solar power plant sites. Quicksand, swamps, wetlands, Salt Lake basins, and other undesirable areas with a slope greater than 1% can be eliminated as potential sites employing a GIS system when taking into account the proximity between the power plant and the power grid. The placement of high-temperature, high-pressure thermal storage tanks must take into account (1) the necessary heat transfer distance that can be achieved and (2) the impact on plant area protection. Lengthy thermal transmission suffers from significant thermal losses because thermal storage is a high-temperature operation. As a result, thermal storage tanks should be placed as close as possible to the turbo-generator unit. CSP technology only uses DNI for solar-electrical transformation. This condition allows the option of a restricted geographical region. The site with largest DNI offer the most fascinating possibilities for the installation of new power plant. Table 6.12 shows the spectrum accumulated annual solar DNI requirement for site selection of STP plant.

TABLE 6.12 Annual solar direct normal irradiance for site selection.

DNI (kWh/m²/year))	Comments
DNI<1600 kWh/(m².a)	Not suitable
DNI = 1600–2000	Suitable
DNI>2000	Recommended

Since the degree of irradiance is approximately proportional to a plant's electricity production, calculating the DNI with the highest possible accuracy is of greatest importance. These data are already conclusive during the implementation of the CSP project with the goal of identifying a certain installation site and anticipating the grounds for commercial appropriateness. Until now, the available solar irradiance (DNI) in a site is only accessible when a reasonably highly accurate local meteorological station is installed, equipped with required instruments. The calculated data must be coupled with longer satellite historical data sets and numerical weather prediction models to obtain all the necessary information with adequate accuracy for the project implementation. The meteorological data are used during the plant operation to assess the plant's performance, i.e., to detect any malfunctioning or deterioration or performance degradation due to the unpredictable operating schemes. Several instruments make up the meteorological station including sun-tracking pyrheliometers or rotating shadow band irradiometer. The pyrheliometer includes solar position tracker, a solar radiation-sensing unit, a collimated light tube and sighting system with solar radiation, as well as the connector socket and the incoming light safety shield. The sensing element consists of thermopile and a sensing surface. As solar radiation is sensed by the sensing surface, the thermopile produces thermoelectromotive force proportional to the radiation energy absorbed. The meteorological stations also use shading ball mounting pyranometers to estimate two other related components of radiation known as global horizontal irradiance and diffuse horizontal irradiance.

6.8.5 Land selection and site elevation

The following prerequisites must be met while selecting a land (Wang, 2019):

- Emphasis should be provided to the use of desert mountains, deserts, wastelands, and other difficult-to-use land that is undesirable for agricultural, ecological, and industrial development; attempt should be taken to avoid occupying farmland, or if that is not feasible, to occupy less farmland while constructing CSP generation projects.

- Land for the field of solar collector, the electricity generation area, the residential area, and permanent roads are all required to be incorporated in the CSP generation projects.
- The site altitude must be higher than the 100-year flood mark. If the elevation falls below this mark, the plant area must be built with adequate flood-control structures and constructed in its fullest during the early stages of the project. Land surface protection must be considered when introducing flood-control strategies in the concentration area.

6.9 Design of thermal collector for power plant

Solar thermal collector is the device that absorbs thermal energy of the incident solar radiation into the surface of the collector. The collector is broadly classified as nontracking also known as fixed or stationary and tracking collectors. Fig. 6.20 shows classifications of solar thermal collectors.

6.9.1 Flat plate solar collectors

Flat plate solar collectors (FPSCs) are the widely used solar energy collectors that consist of glazing covers, absorber plates coated with high absorptive coatings (black coatings), tubes or channels to facilitate HTFs, insulation materials, and frames. To determine the thermal performance of FPSCs, steady-state condition is considered and energy balance equation is applied to the thermal resistance network of typical single FPSCs (Fig. 6.21). Fig. 6.22 shows sketch of a flat-plate solar collector. The HTFs inside the tubes or channels absorbed radiant solar energy when rays of solar hit the collector. The

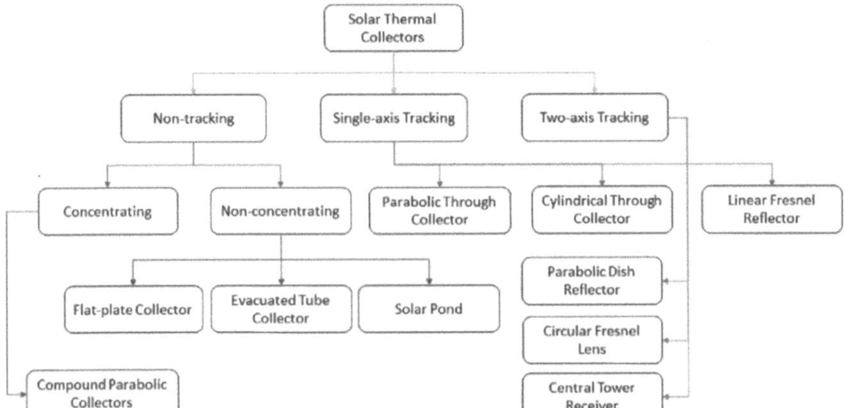

FIGURE 6.20 Classifications of solar thermal collectors (Gorjian et al., 2020).

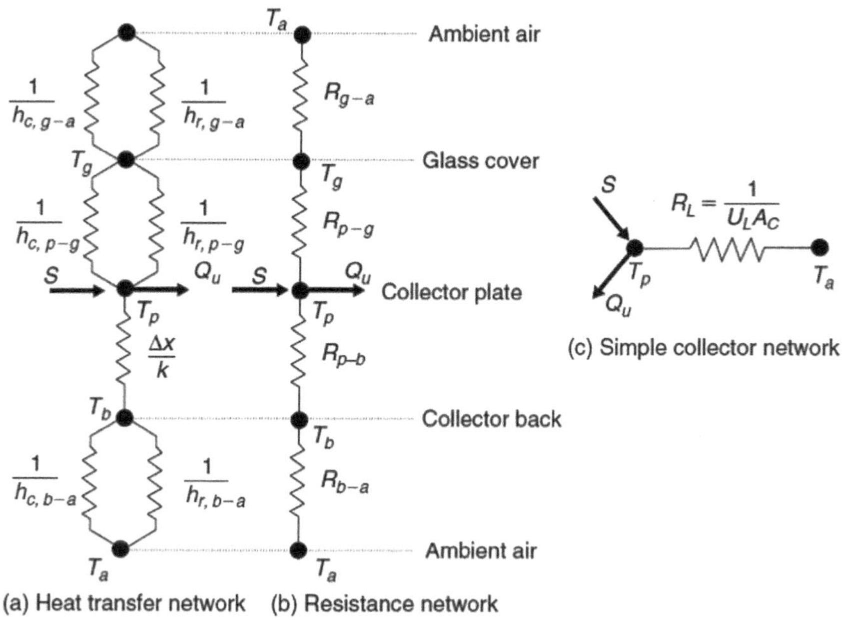

FIGURE 6.21 Thermal resistance circuit for a single FPSC, (A) conduction, convection, and radiation units, (B) resistance of plates' and (C) a simple collector (Sakhaei & Valipour, 2019).

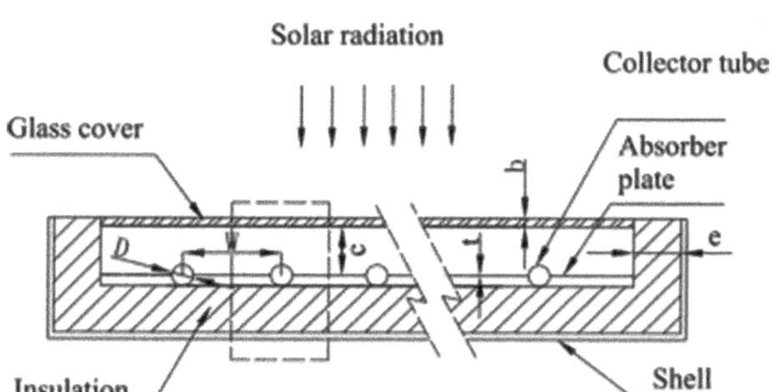

FIGURE 6.22 Sketch of a flat-plate solar collector (Jiandong et al., 2015).

useful solar energy (Q_u) gained by the collector is the difference between the heat absorbed by the HTFs and the amount of heat lost from the collector to the air and can be determined according to the following equation (Gorjian et al., 2020; Klein, 1975; Sakhaei & Valipour, 2019):

$$Q_u = A_{ab} F_R [DNI - U_L (T_{ab} - T_{amb})] \tag{6.106}$$

where A_{ab} is the collector absorber surface area, T_{ab} is the collector absorber temperature, and U_L is the overall heat loss coefficient.

F_R is estimated as (Klein, 1975):

$$F_R = \frac{\dot{m}_{HTF} C_p}{A_{ab} U_L} \left(1 - \exp\left[\frac{U_L F' A_c}{\dot{m}_{HTF} C_p} \right] \right) \tag{6.107}$$

where C_p is the specific heat of HTF at constant pressure, F' is the collector efficiency factor which decreases with an increase in the overall heat loss coefficient U_L, and A_c is the collector area. The collector efficiency factor F' is a function of the thickness of the absorber plate (δ) and calculated by Hottel-Whiller-Bliss model which is given by Sakhaei and Valipour (2019):

$$F' = \left[\frac{\frac{1}{U_L}}{W\left[\left(\frac{1}{U_L[D_o + (W - D_o)F]}\right) + \frac{1}{C_b} + \frac{1}{\pi D_i h_{fi}}\right]} \right]$$

where F is the fin efficiency factor and $F = \frac{\tanh\left[\frac{m(W - D_o)}{2}\right]}{\left[\frac{m(W - D_o)}{2}\right]}$ and $m = \sqrt{U_L / k_{ab}\delta_{ab}}$ and C_b is the bond conduction.

The overall heat loss coefficient U_L can be determined by Sakhaei and Valipour (2019):

$$U_L = U_{top} + U_{bottom} + U_{edge} \tag{6.108}$$

Convection loss

Heat loss from the absorber plate to the glass by convection can be calculated by Hollands et al. (1976):

$$Q_{conv,ab-g} = A_{ab} h_{c,ab-g} (T_{ab} - T_{amb}) \tag{6.109}$$

The convective heat transfer coefficient, $h_{c,p\text{-}g}$ for a tilt angle 60 degrees referred to Fig. 6.23 can be calculated by (Hollands, Unny, Raithby, & Konicek, 1976):

$$Nu = \frac{h_{c,ab-g} L}{k} = 1 + 1.446 \left[1 - \frac{1708}{Ra \times \cos(\theta)} \right]^{+} \left\{ 1 - \frac{1708[\sin(1.8\theta)]^{1.6}}{Ra \times \cos(\theta)} \right\} + \left\{ \left[\frac{Ra \times \cos(\theta)}{5830} \right]^{0.333} - 1 \right\}^{+} \tag{6.110}$$

Here + sign indicates that only positive values are accepted. The Rayleigh number is given by (Hollands et al., 1976):

$$Ra = \frac{g\beta Pr}{\vartheta^2} (T_{ab} - T_g) L^3 \tag{6.111}$$

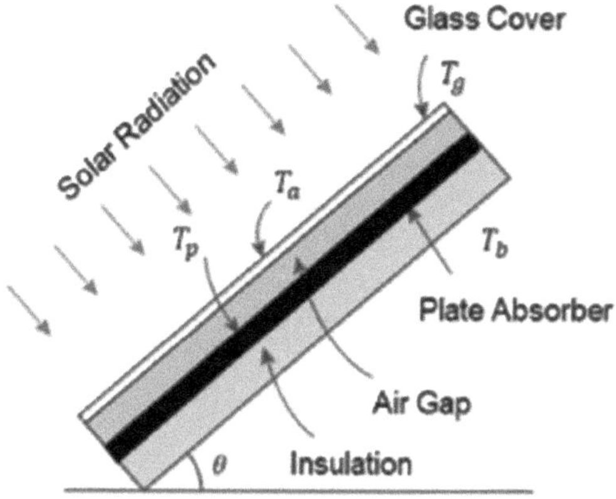

FIGURE 6.23 Heat loss from inclined FPSCs (Sakhaei & Valipour, 2019).

The properties of the HTF are calculated at the mean temperature $(\frac{T_{ab}+T_g}{2})$.

For a vertical collector, convective heat transfer is given by Shewen, Hollands and Raithby (1996):

$$Nu = \frac{h_{c,ab-g}L}{k} = \left[1 + \left(\frac{0.0665Ra^{0.333}}{1+\left(\frac{9600}{Ra}\right)^{0.25}}\right)^2\right]^{0.5} \tag{6.112}$$

Conduction loss

The heat loss by conduction in FPSCs appears through the insulation in the collector bottom, and the edges of the collector.

$$U_{bottom,\ c} = \frac{k_{bottom}}{t_{bottom}} \tag{6.113}$$

$$U_{edge,c} = \frac{k_{edge}}{t_{edge}} \tag{6.114}$$

Radiation loss

In the FPSCs, the radiation heat transfer happens between the absorber plate and the glass cover, and between the glass cover wall and the sky. The radiation heat transfer coefficient between absorber plate and glass cover is given by Sakhaei and Valipour (2019):

$$h_{r,ab-g} = \frac{\sigma\left(T_{ab}+T_g\right)\left(T_{ab}^2+T_g^2\right)}{\left(\frac{1}{\varepsilon_{ab}}\right)+\left(\frac{1}{\varepsilon_g}\right)-1} \tag{6.115}$$

The heat transfer by radiation between the absorber plate and the glass cover is given by Sakhaei and Valipour (2019):

$$Q_{r,ab-g} = \frac{A_{ab}\sigma\left(T_{ab}^4 - T_g^4\right)}{\left(\frac{1}{\varepsilon_{ab}} + \frac{1}{\varepsilon_g}\right) - 1} \tag{6.116}$$

The radiation heat transfer coefficient between the glass cover and is the sky given by Sakhaei and Valipour (2019):

$$h_{r,g-amb} = \varepsilon_g\sigma\left(T_g + T_{amb}\right)\left(T_g^2 + T_{amb}^2\right) \tag{6.117}$$

The heat dissipated from the bottom of the collector composed of convection and radiation heat transfer to the ambient and can be given by Sakhaei and Valipour (2019):

$$U_{bottom} = \frac{1}{\frac{t_{bottom}}{k_{bottom}} + \frac{1}{h_{c,bottom-amb}}} \tag{6.118}$$

Heat removed from the edge of the collector is given by:

$$U_{edge} = \frac{1}{\frac{t_{edge}}{k_{edge}} + \frac{1}{h_{c,edge-amb}}} \tag{6.119}$$

Heat transfer coefficient from the top is given by:

$$R_{top} = R_{ab-g} + R_{g-amb} = \frac{1}{U_{top}A_{ab}} \tag{6.120}$$

$$R_{ab-g} = \frac{1}{A_{ab}\left(h_{c,ab-g} + h_{r,ab-g}\right)} \tag{6.121}$$

$$R_{g-amb} = \frac{1}{A_{ab}\left(h_{c,g-amb} + h_{r,g-amb}\right)} \tag{6.122}$$

For multiple cover system (Klein, 1975):

$$Q_{top} = \frac{(T_{ab} - T_{amb})A_{ab}}{\frac{N_g}{\left(C/T_{ab}\right)\left[\frac{T_{ab}-T_{amb}}{N_g+f}\right]^{0.33}} + \frac{1}{h_w}} + \frac{\sigma\left(T_{ab}^4 - T_{amb}^4\right)A_{ab}}{\frac{1}{\varepsilon_{ab} + 0.05N_g(1-\varepsilon_{ab})} + \frac{2N_g + f - 1}{\varepsilon_g} - N_g} \tag{6.123}$$

$$f = \left(1 - 0.04h_w + 0.0005h_w^2\right)\left(1 + 0.091N_g\right) \tag{6.124}$$

$$C = 365.9\left(1 - 0.00883s + 0.0001298s^2\right) \tag{6.125}$$

where N_g is the number of covers, h_w is the convection coefficient between the top glass cover and the sky, s is the collector tilt (for horizontal collector, $s = 0$). Here, h_w is a function of wind velocity (V) and the length (L) of the cover which can be given by Sakhaei and Valipour (2019):

$$h_w = \frac{8.6V^{0.6}}{L^{0.4}} \tag{6.126}$$

The solar energy to thermal energy conversion efficiency (η_{FPC}) is given by:

$$\eta_{FPC} = \frac{Q_u}{DNI \times A_c}$$

where Q_u can also be calculated as:

$$Q_u = \dot{m}_{HTF} C_p (T_o - T_i) \tag{6.127}$$

6.9.2 Evacuated tube collectors

Evacuated tube collectors (ETCs) are composed of multiple long and parallel tubes of glass or plastic which contain a copper heat pipe and coated absorber plate (Fig. 6.24). There are various models of ETCs and a general classification is shown in Fig. 6.25. In the ETCs, the HTF flows upward by natural convection to heat exchanger that is located at the manifold area and heat is transferred to another fluid to operate any thermal process or to store in a

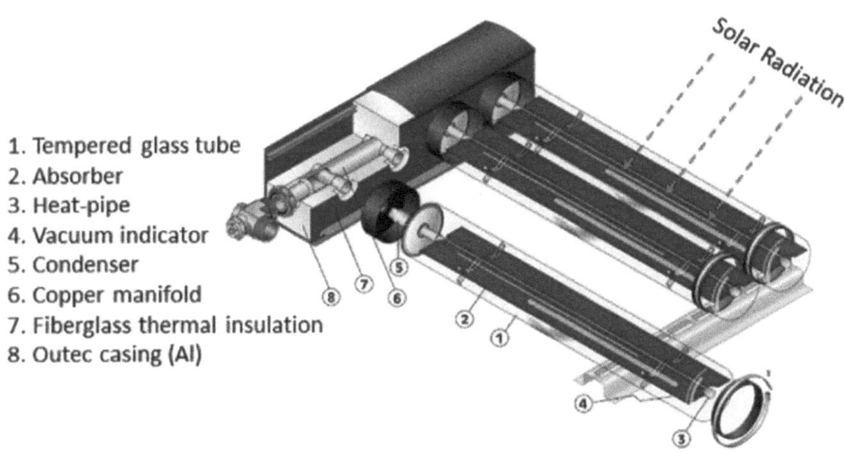

FIGURE 6.24 View of an ETC and its main components (Gorjian et al., 2020).

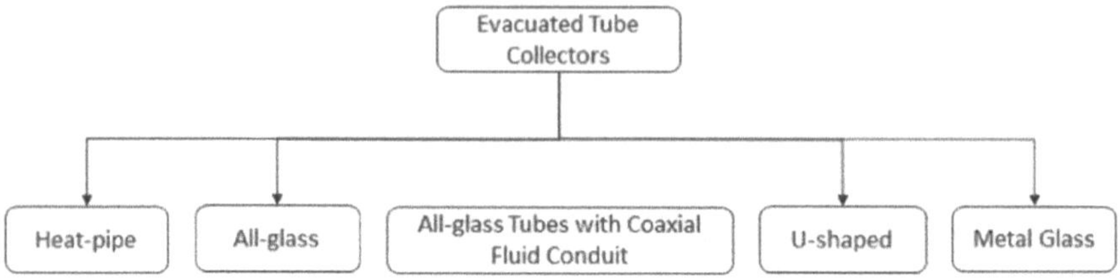

FIGURE 6.25 Classifications of ETCs (Ataee & Ameri, 2015).

storage system for later application. Steady state and quasi dynamic methods are used to calculate the energetic efficiency of the ETCs. At steady-state conditions, the heat transfer balance of the ETCs is (Daghigh & Shafieian, 2016):$Q_u = Q_{ab} + Q_l$, where Q_{abs} is the solar energy (solar heat) strikes the inner space of the heat pipe across the glass cover and Q_l is the lost heat.

Q_{abs} is calculated as (Daghigh & Shafieian, 2016):

$$Q_{abs} = \tau_{g,o}\tau_{g,i}\alpha_c A_{ab} N_{he} DNI \quad (6.128)$$

The amount of lost energy from the absorber to the sky due to the temperature gradient can be given by (Daghigh & Shafieian, 2016):

$$Q_l = \frac{T_{ab} - T_{amb}}{R_{ab-gi} + R_{gi} + R_{gi-go} + R_{go} + R_{go-amb}} \quad (6.129)$$

The resistance element in the above-mentioned equation can be calculated from the thermal resistance circuit between the absorber surface and the sky is shown in Fig. 6.26.

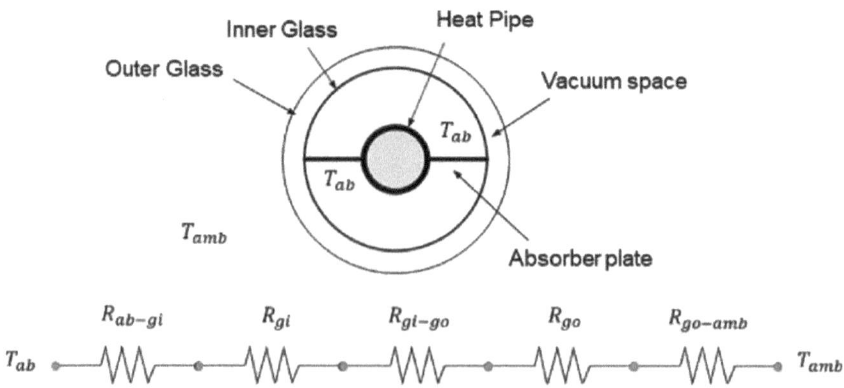

FIGURE 6.26 Cross section of a single tube and thermal resistance circuit (Daghigh & Shafieian, 2016).

Resistance for convection and radiation between absorber and inner glass is given by Daghigh and Shafieian (2016):

$$R_{ab-gi} = \frac{1}{h_{air}\left(\frac{A_{ab}N_{hp} + A_{gi}}{2}\right)} + \frac{\frac{1-\varepsilon_{ab}}{\varepsilon_{ab}A_{ab}N_{hp}} + \frac{1}{A_{ab}N_{hp}F_{ab-g}} + \frac{1-\varepsilon_{gi}}{\varepsilon_{gi}A_{gi}}}{\sigma(T_{ab} + T_{gi})\left(T_{ab}^2 + T_{gi}^2\right)} \quad (6.130)$$

Inner glass resistance for conduction:

$R_{gi} = \frac{t_{gi}}{k_{gi}A_{gi}}$, where t_{gi} is the thickness of the inner glass.

Resistance for convection and radiation between ambient and outer glass is given by:

$$R_{go-amb} = \frac{1}{A_g}\left(\frac{1}{h_{amb}} + \frac{1}{\sigma\varepsilon_{go}(T_{go} + T_{amb})\left(T_{go}^2 + T_{amb}^2\right)}\right) \quad (6.131)$$

Conduction resistance of outer glass:

$$R_{go} = \frac{t_{go}}{k_{go}A_{go}} \quad (6.132)$$

At the evacuated space, radiation resistance that exists can be given by:

$$R_{gi-go} = \frac{\frac{1-\varepsilon_{gi}}{\varepsilon_{gi}A_{gi}N_{hp}} + \frac{1}{A_{gi}N_{hp}F_{gi-go}} + \frac{1-\varepsilon_{go}}{\varepsilon_{go}A_{go}}}{\sigma(T_{gi} + T_{go})\left(T_{gi}^2 + T_{go}^2\right)} \quad (6.133)$$

When all of the radiation energy leaves the absorber and reaches the inner and outer glasses, $F_{ab\text{-}g}$ and $F_{gi\text{-}go}$ are equal to one (Daghigh & Shafieian, 2016).

The amount of heat absorbed by heat pipe can be modeled from a multiple thermal resistance that consists evaporator and condenser wall ($R_{e,p}$), internal resistance ($R_{e,i}$), and wick conduction ($R_{e,wick}$) that can be given by Daghigh and Shafieian (2016). Fig. 6.27 shows the thermal resistance network of heat pipe.

$$\sum R = R_{e,p} + R_{e,wick} + R_{e,i} + R_v + R_{c,i} + R_{c,p} \quad (6.134)$$

From 6.31 the following can be evaluated (Daghigh & Shafieian, 2016):

$$k_{wick} = \frac{k_l[k_l + k_s - (1-\varepsilon_{wick})(k_l - k_s)]}{k_l + k_s + (1-\varepsilon_{wick})(k_l - k_s)} \quad (6.135)$$

$$h_e = \frac{k_l}{t_{wick}} \quad (6.136)$$

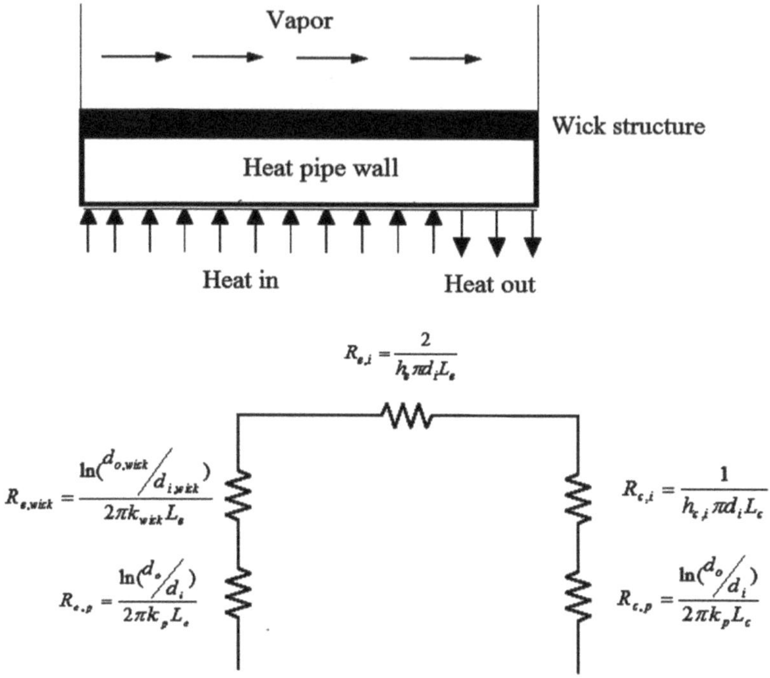

FIGURE 6.27 Thermal resistance network of heat pipe (Daghigh & Shafieian, 2016).

$$h_{c,i} = 0.728 \left[\frac{g \sin(\theta) \rho_l (\rho_l - \rho_v) k^3 h_{fg}}{d_i \mu_i \Delta T_i} \right]^{0.25} \tag{6.137}$$

The collector efficiency is: $\eta_{ETC} = \frac{Q_u}{A_c DNI}$, where Q_u can also be calculated as:

$$Q_u = \dot{m}_{HTF} C_p (T_o - T)_i \tag{6.138}$$

6.10 Solar aided power generation

The solar aided power generation (SAPG) concept is a hybrid solar power system that uses low-grade solar thermal energy to substitute the high-grade heat of extraction steam in a regenerative Rankine cycle (RRC) power plant for feedwater preheating. The main characteristic of SAPG is that solar heat conveyed by HTF or steam is only used to replace extraction steam by preheating feedwater approaching the boiler, rather than directly entering the turbine for generating power. As a result, the recovered extraction steam could be used to produce power in the turbine. The power produced by the solar thermal input is applied to the energy produced by the saved/displaced steam. Once the thermal sink temperature is set, the maximum temperature of the solar thermal source limits or caps the thermal efficiency of any STP plant. The highest temperature of solar thermal input limits the full performance of a traditional STP system. The solar to power efficiency of the SAPG method, on the other hand, is no longer constrained or restricted by the temperature of solar thermal input, but by the maximum temperature of the entire plant, i.e., the combustion temperature, which is typically much higher than the temperature of solar heat input (Qin, Hu, & Li, 2020). The advantages of an SAPG system are listed below (Hu et al., 2010):

- Over a traditional coal-fired power plant and a solar-only power plant, the SAPG technology offers better thermodynamic first and second law efficiencies, i.e., energy and exergy efficiencies.
- SAPG technology will use the current infrastructure of fossil fuel fired power plants, resulting in a low installation cost and high social, environmental, and economic advantages.
- The SAPG can be used to alter existing power plants as well as new power plants, with little or no risk to the function of the existing power plants.
- It is not mandatory to use the thermal storage system, which is still theoretically in its infancy. Another advantage of SAPG technology is its simplicity.
- The SAPG is adaptable in its use. SAPG can be introduced to a power station in phases, based on the capital available.
- Solar collectors with a low temperature range, such as evacuated tubes and flat plate collectors, can be implemented in the SAPG system. Non-concentrating collectors have technological advantages over concentrating collectors in SAPG plants, according to studies.

6.11 Hybridization of power cycle with solar thermal energy resources

6.11.1 Hybridizing with Rankine cycle

In particular, there are two methods (low temperature solar heat and high temperature solar heat) for integrating solar thermal energy into RRC power plants, as shown in Fig. 6.28. Both of these solutions could be combined in a single power plant. Table 6.13 enlisted qualitative advantages and disadvantages of hybridizing approach in an RC power station.

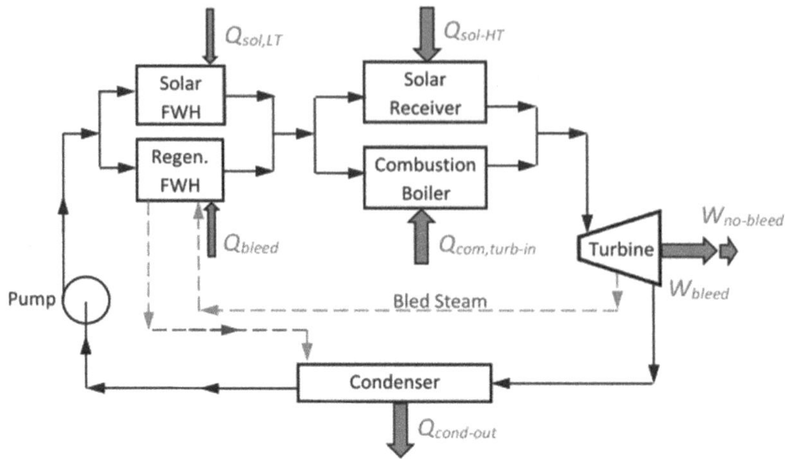

FIGURE 6.28 Basic options to hybridize an RRC with solar thermal energy (Nathan et al., 2018).

TABLE 6.13 Qualitative advantages and disadvantages of hybridizing approach.

Option	Temperature range	Where to integrate	Advantages	Disadvantages
Low temperature solar heat, $Q_{sol,\ LT}$	<300°C	To heat the feedwater to the boiler	Low cost Thermodynamic leverage Low temperature solar heating Simple and robust	Low solar share Low efficiency
High temperature solar heat, $Q_{sol,\ HT}$	>300°C			

6.12 Performance analysis

6.12.1 Overall efficiency

For any hybrid solar power conversion system, the ratio of electricity output to overall energy consumption defines the overall efficiency. The overall energy consumption includes heat from the combustion of fossil fuels as well as total solar irradiation obtained by the solar collectors.

$$\eta_{overall} = \frac{W_{net}}{Q_{fossil} + Q_{solar}} \tag{6.139}$$

where W_{net} is the total power output from the cycle, Q_{fossil} is the heat input by fossil fuels, and Q_{solar} is the overall solar energy collected by the solar collector field.

6.12.2 Fuel-based efficiency

The ratio of electricity provided by a power plant to energy input from only fossil energy sources is known as fuel-based efficiency. Since solar power is unlimited, any hybrid STP plant's fuel-based efficiency is provided by:

$$\eta_{fuel-based} = \frac{W_{net}}{Q_{fossil}} \tag{6.140}$$

6.12.3 Solar-to-electric efficiency

The solar-to-electric efficiency is used to evaluate the output of converting solar energy from the solar field. It can be defined as:

$$\eta_{solar} = \frac{W_{net} - W_{ref}}{Q_{solar}} \tag{6.141}$$

where W_{ref} is the reference thermodynamic cyclic net power output.

6.12.4 Solar heat fraction

The sum of thermal energy produced by the solar field divided by the total energy input to the power plant is the fraction of solar heat input.

$$\chi_{solar} = \frac{Q_{solar}}{Q_{fossil} + Q_{solar}} \tag{6.142}$$

6.12.5 Fuel saving fraction

The quantity of fuel saved in comparison to the reference power conversion system for the same capacity when the plant is operating in fuel saving mode (constant output) is given by:

$$\chi_{saving} = \frac{Q_{fossil,ref} - Q_{fossil}}{Q_{fossil,ref}} \tag{6.143}$$

where $Q_{fossil,ref}$ is the heat input to the reference power plant system at the same capacity.

References

Aragonés-Beltrán, P., Chaparro-González, F., Pastor-Ferrando, J. P., & Pla-Rubio, A. (2014). An AHP (Analytic Hierarchy Process)/ANP (Analytic Network Process)-based multi-criteria decision approach for the selection of solar-thermal power plant investment projects. *Energy, 66*, 222–238. https://doi.org/10.1016/j.energy.2013.12.016

Assael, M. J., Gialou, K., Kakosimos, K., & Metaxa, I. (2004). Thermal conductivity of reference solid materials. *International Journal of Thermophysics, 25*(2), 397–408. https://doi.org/10.1023/B:IJOT.0000028477.74595.d5

Ataee, S., & Ameri, M. (2015). Energy and exergy analysis of all-glass evacuated solar collector tubes with coaxial fluid conduit. *Solar Energy, 118*, 575–591. https://doi.org/10.1016/j.solener.2015.06.019

Behar, O. (2018). Solar thermal power plants - a review of configurations and performance comparison. *Renewable and Sustainable Energy Reviews, 92*, 608–627. https://doi.org/10.1016/j.rser.2018.04.102

Behar, O., Khellaf, A., & Mohammedi, K. (2015). A novel parabolic trough solar collector model - validation with experimental data and comparison to engineering equation solver (EES). *Energy Conversion and Management, 106*, 268–281. https://doi.org/10.1016/j.enconman.2015.09.045

Behar, O., Sbarbaro, D., & Morán, L. (2020). A practical methodology for the design and cost estimation of solar tower power plants. *Sustainability, 12*(20), 1–16. https://doi.org/10.3390/su12208708

Bellos, E., & Tzivanidis, C. (2018). Thermal analysis of parabolic trough collector operating with mono and hybrid nanofluids. *Sustainable Energy Technologies and Assessments, 26*, 105–115. https://doi.org/10.1016/j.seta.2017.10.005

Benoit, H., Spreafico, L., Gauthier, D., & Flamant, G. (2016). Review of heat transfer fluids in tube-receivers used in concentrating solar thermal systems: Properties and heat transfer coefficients. *Renewable and Sustainable Energy Reviews, 55*, 298–315. https://doi.org/10.1016/j.rser.2015.10.059

Bishoyi, D., & Sudhakar, K. (2017). Modeling and performance simulation of 100 MW PTC based solar thermal power plant in Udaipur India. *Case Studies in Thermal Engineering, 10*, 216–226. https://doi.org/10.1016/j.csite.2017.05.005

Collado, F. J., & Guallar, J. (2012). Campo: Generation of regular heliostat fields. *Renewable Energy, 46*, 49–59. https://doi.org/10.1016/j.renene.2012.03.011

Collado, F. J., & Guallar, J. (2013). A review of optimized design layouts for solar power tower plants with campo code. *Renewable and Sustainable Energy Reviews, 20*, 142–154. https://doi.org/10.1016/j.rser.2012.11.076

Daghigh, R., & Shafieian, A. (2016). Energy-exergy analysis of a multipurpose evacuated tube heat pipe solar water heating-drying system. *Experimental Thermal and Fluid Science, 78*, 266–277. https://doi.org/10.1016/j.expthermflusci.2016.06.010

Dudley, E., Kolb, J., Mahoney, A., Mancini, T., M, S., & Kearney, D. (1994). *Test results: SEGS LS-2 solar collector.* Sandia National Laboratory. https://doi.org/10.2172/70756. Report: SAND94- 1884. 140.

Forristall, R. (2003). *Heat transfer analysis and modeling of a Parabolic Trough solar receiver implemented in engineering equation solver* (Issue October). NREL/TP-550-34169.

Gaul, H., & Rabl, A. (1980). Incidence-angle modifier and average optical efficiency of parabolic through collectors. *Journal of Solar Energy Engineering, 102*, 16–21. https://doi.org/10.1115/1.3266115

Gnielinski, V. (1976). New equations for heat and mass transfer in turbulent pipe and channel flow. *International Chemical Engineering, 16*(2), 359–368.

Gorjian, S., Ebadi, H., Calise, F., Shukla, A., & Ingrao, C. (2020). A review on recent advancements in performance enhancement techniques for low-temperature solar collectors. *Energy Conversion and Management, 222*, 113246. https://doi.org/10.1016/j.enconman.2020.113246

Gorjian, S., & Ghobadian, B. (2015). Solar thermal power plants: Progress and prospects in Iran. *Energy Procedia, 75*, 533–539. https://doi.org/10.1016/j.egypro.2015.07.447

Ho, C. K., & Iverson, B. D. (2014). Review of high-temperature central receiver designs for concentrating solar power. *Renewable and Sustainable Energy Reviews, 29*, 835–846. https://doi.org/10.1016/j.rser.2013.08.099

Hollands, K. G. T., Unny, T. E., Raithby, G. D., & Konicek, L. (1976). Free convective heat transfer across inclined air layers. *Journal of Heat Transfer, 98*(2), 189–193. https://doi.org/10.1115/1.3450517

Hu, E., Yang, Y. P., Nishimura, A., Yilmaz, F., & Kouzani, A. (2010). Solar thermal aided power generation. *Applied Energy, 87*(9), 2881–2885. https://doi.org/10.1016/j.apenergy.2009.10.025

Jiandong, Z., Hanzhong, T., & Susu, C. (2015). Numerical simulation for structural parameters of flat-plate solar collector. *Solar Energy, 117*, 192–202. https://doi.org/10.1016/j.solener.2015.04.027

Key world energy statistics 2020. (2020). OECD. https://doi.org/10.1787/295f00f5-en

Islam, M. K., Hasanuzzaman, M., & Rahim, N. A. (2015). Modelling and analysis on the effect of different parameters on a parabolic-trough concentrating solar system. *RSC Advances, 5*(46), 36540. https://doi.org/10.1039/C4RA12919A, 36456.

Klein, S. A. (1975). Calculation of flat-plate collector loss coefficients. *Solar Energy, 17*(1), 79–80. https://doi.org/10.1016/0038-092X(75)90020-1

Krishna, Y., Faizal, M., Saidur, R., Ng, K. C., & Aslfattahi, N. (2020). State-of-the-art heat transfer fluids for parabolic trough collector. *International Journal of Heat and Mass Transfer, 152*. https://doi.org/10.1016/j.ijheatmasstransfer.2020.119541

Kuehn, T. H., & Goldstein, R. J. (1976). Correlating equations for natural convection heat transfer between horizontal circular cylinders. *International Journal of Heat and Mass Transfer, 19*(10), 1127–1134. https://doi.org/10.1016/0017-9310(76)90145-9

Leon, M. A., & Kumar, S. (2007). Mathematical modeling and thermal performance analysis of unglazed transpired solar collectors. *Solar Energy, 81*(1), 62–75. https://doi.org/10.1016/j.solener.2006.06.017

Lippke, F. (1995). *Simulation of the part-load behaviour of a 30 MW SEGS plant, SAND-95-1293*. https://doi.org/10.2172/95571

Li, C., Yang, Z., Zhai, R., Yang, Y., Patchigolla, K., & Oakey, J. E. (2018). Off-design thermodynamic performances of a solar tower aided coal-fired power plant for different solar multiples with thermal energy storage. *Energy, 163*, 956–968. https://doi.org/10.1016/j.energy.2018.08.186

Li, C., Zhai, R., Liu, H., Yang, Y., & Wu, H. (2018). Optimization of a heliostat field layout using hybrid PSO-GA algorithm. *Applied Thermal Engineering, 128*, 33–41. https://doi.org/10.1016/j.applthermaleng.2017.08.164

Li, C., Zhai, R., & Yang, Y. (2017). Optimization of a heliostat field layout on annual basis using a hybrid algorithm combining particle swarm optimization algorithm and genetic algorithm. *Energies, 10*(11). https://doi.org/10.3390/en10111924

Mohammad, S. T., Al-Kayiem, H. H., Assadi, M. K., Sabir, O., & Khlief, A. K. (2018). An integrated program of a stand-alone parabolic trough solar thermal power plant: Code description and test. *Case Studies in Thermal Engineering, 12*, 26–37. https://doi.org/10.1016/j.csite.2018.02.006

Nathan, G. J., Jafarian, M., Dally, B. B., Saw, W. L., Ashman, P. J., Hu, E., & Steinfeld, A. (2018). Solar thermal hybrids for combustion power plant: A growing opportunity. *Progress in Energy and Combustion Science, 64*, 4–28. https://doi.org/10.1016/j.pecs.2017.08.002

Qin, J., Hu, E., & Li, X. (2020). Solar aided power generation: A review. *Energy and Built Environment, 1*(1), 11–26. https://doi.org/10.1016/j.enbenv.2019.09.003

Quezada-García, S., Sánchez-Mora, H., Polo-Labarrios, M. A., & Cázares-Ramírez, R. I. (2019). Modeling and simulation to determine the thermal efficiency of a parabolic solar trough collector system. *Case Studies in Thermal Engineering, 16*, 100523. https://doi.org/10.1016/j.csite.2019.100523

Saaty, T. L. (1990). *Multicriteria decision making: The analytic hierarchy process: Planning, priority setting resource allocation* (2nd ed.). RWS Publications.

Saaty, T. L. (2000). *Fundamentals of decision making and priority theory with the analytic hierarchy process* (Vol. 6). RWS publications.

Sakhaei, S. A., & Valipour, M. S. (2019). Performance enhancement analysis of the flat plate collectors: A comprehensive review. *Renewable and Sustainable Energy Reviews, 102*, 186–204. https://doi.org/10.1016/j.rser.2018.11.014

Schmitz, M., Schwarzbözl, P., Buck, R., & Pitz-Paal, R. (2006). Assessment of the potential improvement due to multiple apertures in central receiver systems with secondary concentrators. *Solar Energy, 80*(1), 111–120. https://doi.org/10.1016/j.solener.2005.02.012

Shawky, A. A., & Khalil, A. K. H. (2020). Design and performance analysis of small-scale parabolic trough solar collectors using sustainable materials. In *Encyclopedia of renewable and sustainable materials*. Elsevier Ltd. https://doi.org/10.1016/b978-0-12-803581-8.11136-1

Shewen, E., Hollands, K. G. T., & Raithby, G. D. (1996). Heat transfer by natural convection across a vertical air cavity of large aspect ratio. *Journal of Heat, 118*, 993–995. https://doi.org/10.1115/1.2822603

Shorabeh, S. N., Firozjaei, M. K., Nematollahi, O., Firozjaei, H. K., & Jelokhani-Niaraki, M. (2019). A risk-based multi-criteria spatial decision analysis for solar power plant site selection in different climates: A case study in Iran. *Renewable Energy, 143*, 958–973. https://doi.org/10.1016/j.renene.2019.05.063

Sadik, K., Shah, K., & Aung, W. (1987). *Handbook of single-phase convective heat*. John Wiley & Sons.

Sparrow, E. M., Abraham, J. P., & Tong, J. C. K. (2004). Archival correlations for average heat transfer coefficients for non-circular and circular cylinders and for spheres in cross-flow. *International Journal of Heat and Mass Transfer, 47*(24), 5285–5296. https://doi.org/10.1016/j.ijheatmasstransfer.2004.06.024

Türk Togrul, I., & Onat, E. (1999). Study for estimating solar radiation in Elazig using geographical and meteorological data. *Energy Conversion and Management, 40*(14), 1577–1584. https://doi.org/10.1016/S0196-8904(99)00035-7

Vignarooban, K., Xu, X., Arvay, A., Hsu, K., & Kannan, A. M. (2015). Heat transfer fluids for concentrating solar power systems - a review. *Applied Energy, 146*, 383–396. https://doi.org/10.1016/j.apenergy.2015.01.125

Vutukuru, R., Pegallapati, A. S., & Maddali, R. (2019). Suitability of various heat transfer fluids for high temperature solar thermal systems. *Applied Thermal Engineering, 159*, 113973. https://doi.org/10.1016/j.applthermaleng.2019.113973

Wang, Z. (2019). *Design of solar thermal power plants* (1st ed.). Academic Press. https://doi.org/10.1016/C2017-0-03007-0

Xiao, P., Guo, L., & Zhang, X. (2015). Investigations on heat transfer characteristic of molten salt flow in helical annular duct. *Applied Thermal Engineering, 88*, 22–32. https://doi.org/10.1016/j.applthermaleng.2014.09.021

Yılmaz, İ. H., & Mwesigye, A. (2018). Modeling, simulation and performance analysis of parabolic trough solar collectors: A comprehensive review. *Applied Energy, 225*, 135–174. https://doi.org/10.1016/j.apenergy.2018.05.014

Yilmaz, I. H., & Söylemez, M. S. (2014). Thermo-mathematical modeling of parabolic trough collector. *Energy Conversion and Management, 88*, 768–784. https://doi.org/10.1016/j.enconman.2014.09.031

Zaharil, H. A., & Hasanuzzaman, M. (2020). Modelling and performance analysis of parabolic trough solar concentrator for different heat transfer fluids under Malaysian condition. *Renewable Energy, 149*, 22–41. https://doi.org/10.1016/j.renene.2019.12.032

Zarza-Moya, E. (2018). Concentrating solar thermal power. In *A comprehensive guide to solar energy systems*. Elsevier Inc. https://doi.org/10.1016/b978-0-12-811479-7.00007-5

Zuo, L., Liu, Z., Ding, L., Qu, N., Dai, P., Xu, B., & Yuan, Y. (2020). Performance analysis of a wind supercharging solar chimney power plant combined with thermal plant for power and freshwater generation. *Energy Conversion and Management, 204*, 112282. https://doi.org/10.1016/j.enconman.2019.112282

Chapter 7

Solar thermal energy storage

Zabir Mahmud[1] and Md Hasanuzzaman[2]
[1]*Environmental Systems Graduate Program, School of Engineering, University of California Merced, Merced, CA, United States;* [2]*Higher Institution Centre of Excellence (HICoE), UM Power Energy Dedicated Advanced Centre (UMPEDAC), University of Malaya, Kuala Lumpur, Malaysia*

Chapter outline

7.1 Introduction

Worldwide, fossil fuels are used as a primary means of producing electricity and the burning of fossil fuels is creating a disaster for us. If this trend continues, there will be fossil fuel shortages in the immediate future. We must continually find some tools that can replace the current method of generating electricity in order to resolve the threatening situation. It is possible to accept renewable resources such as solar and wind as some of the viable alternatives for this fossil fuel. For example, if we look around the world right now, we will find some excellent uses of various renewable technologies such as power generation, pool water heating, underfloor heating or radiators, cooling technologies, dryers for agricultural goods, and water desalination, and so on. These tools are normal but not adequately consistent. Consequently, there is a lack of synchronization between the demand for these services and their

Technologies for Solar Thermal Energy. https://doi.org/10.1016/B978-0-12-823959-9.00002-7

availability. This has become the world's biggest challenge, precluding the transition to renewable energy. In addition, for people all over the world, these technologies are very new, so it will take time to develop them as the ideal substitute for fossil fuels. Most of these use rely on weather conditions that cause the supplied energy to intermittently and fluctuate. Many researchers around the world are trying to fill these gap situations and one of the solutions could be the solar thermal energy storage system.

For designing such a storage, it requires to combine the system with solar collectors in order to store and retrieve thermal energy whenever important. This device functions according to the user's need for a number of charging and discharging cycles. During charging, a receiver from the sun absorbs the solar energy from the system and this energy is stored. This storage falls into practice and fulfills the demand by discharging the accumulated energy if the renewable electricity fails to satisfy the peak demand due to unfavorable weather conditions or other unforeseen and inevitable circumstances. Although it seems very simple and straightforward, in fact, there are a lot of nuances in this whole system.

The entire world is looking forward to zero carbon emissions in power generation for the time being and has already initiated numerous initiatives to do so. The EU has already targeted zero carbon emissions by 2050, and the US state of California has set a target of zero carbon electricity by 2045. They have already begun working according to their plans, and these plans include the development of solar thermal energy storage as one of the main steps (Sanchez Perez, Bailey, Kurtz, & Mahmud, 2020). In addition, it would be a great opportunity for any developing nation to make an advanced step toward great evolution by efficiently exploiting its renewable resources. Therefore, this whole system is a very critical aspect to consider in order to address the erratic and unpredictable existence of solar irradiation.

7.2 Types of solar thermal energy storage system

Solar thermal energy storage can be of two types if we consider the thermal process (Sarbu & Sebarchievici, 2018).

(a) Sensible thermal energy storage (STES)
(b) Latent thermal energy storage (LTES)

7.2.1 Sensible solar thermal energy storage

This storage is carried out without altering its phase by increasing or decreasing a material's temperature. For supporting the load like space heating or domestic hot water, heat is extracted from storage. Liquid or solid may be the material. With its lower cost and higher specific heat, the most common material used in this process is water. Molten salts, oils, and liquid metals have

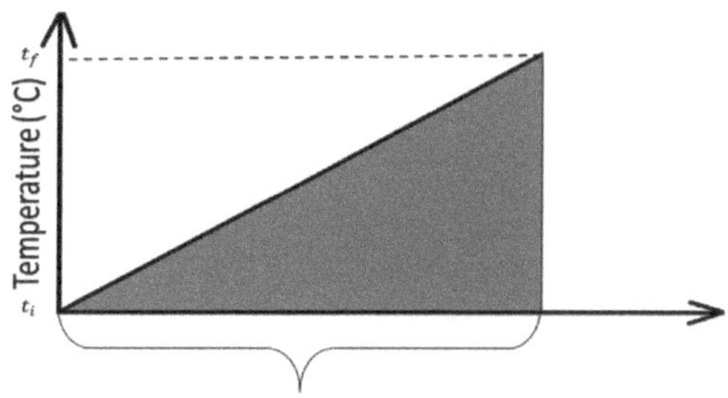

FIGURE 7.1 Sensible solar thermal energy storage (STES) system.

high boiling point, so they perform better above 100°C processes. There are three drivers for the stored energy in STES systems: mass, heat capacity, input and output temperature difference of storage medium (Dincer, 2002) (Fig. 7.1).

For STES systems:

$$Q_s = \int_{t_i}^{t_f} mc_p dt = mc_p(t_f - t_i) \quad (7.1)$$

where Q_s = heat stored (J).
m = mass of storage medium (kg).
c_p = specific heat (J/kg K).
t_f = final temperature and t_i = initial temperature (°C).

Temperature gap between storage and the environment works as a driver of heat loss from an STES system. The rate of heat release and extraction works as a function of thermal diffusivity. Besides, one of the material characteristics, thermal conductivity, impacts energy-in and energy-out of the storage.

$$\lambda = \rho c_p \alpha \quad (7.2)$$

where λ = thermal conductivity (W/m.K).
ρ = density (kg/m^3).
α = thermal diffusivity (m^2/s).

7.2.2 Latent solar thermal energy storage

The LTES system has become more powerful than the STES system because of its lower overnight cost, high-level storage density, and reduction in environmental losses. During charging and discharging, they generally work on short temperature ranges. The downside, however, involves corrosive effects on metallic walls and the possibility of supercooling due to phase segregation

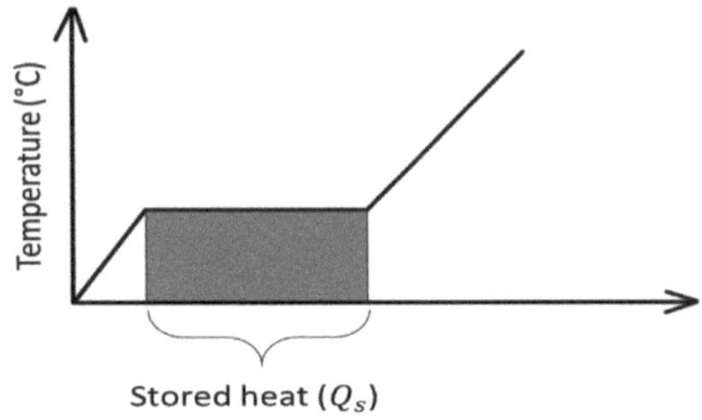

FIGURE 7.2 Latent solar thermal energy storage (LTES) system.

of materials (Pielichowska & Pielichowski, 2014; Saha, Ruslan, Morshed, & Hasanuzzaman, 2020). It must include three fundamental components:

(a) phase change materials (PCMs) and its range in temperature
(b) storage control system
(c) mechanism of transferring energy (Fig. 7.2)

The energy from LTES can be calculated by integrating the heat flow over a period, *t*. The heat flow is calculated by the following Eq. (7.3) (Kuravi, Trahan, Goswami, Rahman, & Stefanakos, 2013):

$$Q_s = ml_f \tag{7.3}$$

$$\dot{Q}_{HTF} = \dot{m}_{HTF} C_{p,HTF} \left(T_{HTF,i} - T_{HTF,f} \right) \tag{7.4}$$

where $\dot{m}_{HTF}$ = mass flow rate of HTF (Kg/s).
l_f = latent heat of fusion (J/Kg).

Now the heat flow is integrated for a certain time to determine the total energy stored by Eq. (7.4).

$$Q_{HTF,total} = \int_0^t \dot{Q}_{HTF} dt \tag{7.5}$$

After setting the maximum energy deliverable, the required PCM volume can be determined through Eq. (7.5) (Kuravi et al., 2013). While operating under similar conditions, in order to make a comparison of the behavior of an LTES systems with a standard two-tank STES unit, Eq. (7.6) is useful to figure out the two-tank system size (Kuravi et al., 2013).

$$V_{LHS,total} = \frac{Q_{HTF,total}}{\rho_s c_{ps} \left(T_{m1} - T_{HTF,cold} \right) + \rho_l h_{ls} + \rho_l c_{pl} \left(T_{HTF,hot} - T_{m2} \right)} \tag{7.6}$$

$$V_{SHS} = \frac{Q_{HTF,total}}{\rho_l c_{pl}\left(T_{HTF,hot} - T_{HTF,cold}\right)} \quad (7.7)$$

7.3 Corresponding solar thermal energy storage technologies

7.3.1 Two-tank direct system

There are two tanks in this system: one at high temperature and low temperature shown in Fig. 7.3 (Cocco & Serra, 2015). To pass heat and store energy, the same material is used. The liquid flows to the collector from the low temperature tank where it warms up by directly absorbing the solar energy from the sun and increment in temperature occurs before entering in the high temperature tank. The flow continues through a heat exchanger and produces steam. Temperature of the fluid drops down and it reaches the tank with lower thermal state. This system was used in California Solar Two Power Tower and Solar Electric Generating Station I. For transferring heat and storing energy, mineral oil is used in the former one and the later one uses molten salt (Cocco & Serra, 2015).

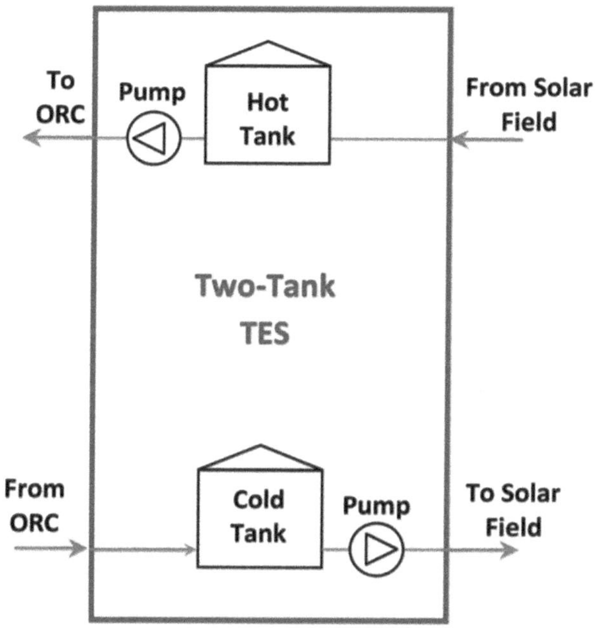

FIGURE 7.3 Two-tank direct storage.

Volume of the storage is determined by:

$$V = \frac{E_{ST}}{\left[\rho_s c_{ps}(1-\varepsilon) + \rho_o c_{po}\varepsilon\right](T_H - T_C)} \tag{7.8}$$

where E_{ST} = energy storage capacity
ρ_s = density (solid).
ρ_o = density (thermal oil).
ε = bed void fraction (For two-tank direct system, $\varepsilon = 1$).

7.3.2 Two-tank indirect system

The operating theory is different from the previous one. Two different fluids are used: one for transferring heat and another one as storage medium. The heat transfer fluid (HTF) flows to the collector with a lower temperature and the temperature increases as the solar energy consumption becomes hot. This hot fluid flows to a heat exchanger, where its thermal energy is transferred to the storage fluid that comes as a cold from another tank, and the temperature increases by absorbing the heat from the HTF. Finally, the HTF comes back to the tank and the loop continues to be maintained. The storage fluid goes back to the storage tank at high temperature and releases steam as it was in the direct one. The overall cost of the system increases as it needs a supplementary heat exchanger. Several parabolic towers of Spain and the United States are planning to include this system. Organic oil would be used as HTF and molten salt as the fluid of storage. A two-tank indirect solar thermal storage system is shown in Fig. 7.4 (Li, Xu, Song, Wang, & Guofeng, 2017).

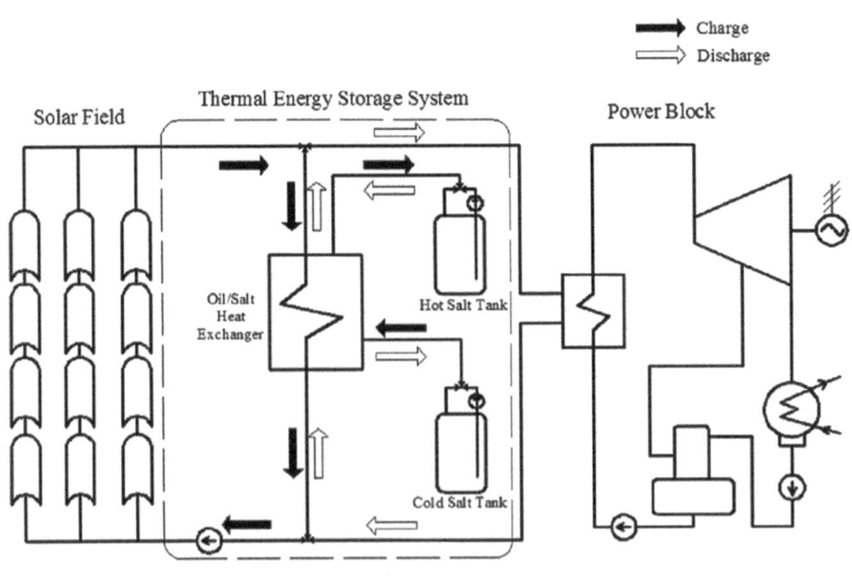

FIGURE 7.4 Two-tank indirect storage.

7.3.3 Single tank thermocline system

This system stores energy in a solid medium (such as silica sand) shown in Fig. 7.5 (Cocco & Serra, 2015). Inside a single tank, there is a temperature difference; one part has higher temperature, and the other part has lower. A temperature gradient (known as the thermocline) separates these two regions. The high-temperature HTF flows thermocline top and, temperature gets down while going out of the tank. The thermocline is shifted downward and upward while charging and discharging. The fluid inside the tank gets stratified so that it becomes easy to stabilize and retain the thermocline against the effect of buoyancy. This scheme was implemented at Solar One power tower. It uses steam and mineral oil as HTF and storage fluid, respectively (Cocco & Serra, 2015).

In particular, a work-based 2-phase 1D model initially allows prediction of the coefficient of convective heat transfer along the two mediums, i.e., the solid material and the thermal oil (Schumann, 1929). The convective coefficient of heat transfer h_v between solid particles and oil and heat transfer coefficient α is determined by following equations:

$$h_v = \frac{6(1-\varepsilon)}{d}\alpha \tag{7.9}$$

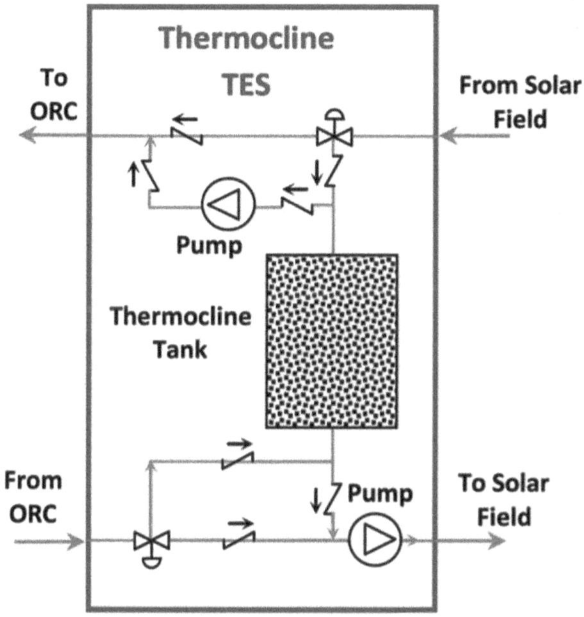

FIGURE 7.5 Single tank thermocline.

$$\alpha = \frac{Nu.K}{d} \tag{7.10}$$

where Nusselt number $= f$ (Reynolds number, Prandtl number)

$$Nu = 2 + 1.1\left(Re_d^{0.6}Pr^{0.33}\right) \tag{7.11}$$

Finally, the heat losses Q_w along the walls of the tank are calculated as follows:

$$Q_w = UA_w\Delta T_w \tag{7.12}$$

where U = coefficient of heat transfer (global).

A_w = area of walls

ΔT_w = difference of temperature between internal oil and air

For thermocline tank, U is a function of conductive thermal resistance of wall insulation and the coefficients of internal and external convective heat transfer whereas for traditional oil tank, it only considers the external one. The conductive thermal resistance of metal walls is disregarded for both calculations.

The internal heat transfer coefficient, particularly for thermocline tank, is assumed to be 0.8 times heat transfer coefficient α (Ismail & Stuginsky Jr, 1999). The external heat transfer coefficient is determined using the following equation:

$$Nu_{H_B} = 0.1Ra_{HB}^{\frac{1}{3}} \tag{7.13}$$

7.4 Effectiveness analysis of thermal storage

According to Belusko and Bruno (Belusko & Bruno, 2008), an STES system can function like a heat exchanger. Transfer of heat takes place between HTF and PCM at phase change temperature. The effectiveness (ε) of and STES system can be calculated similarly as it is for the heat exchanger. The effectiveness-Number of Transfer Units (ε-NTU) technique can be a feasible way to determine the efficiency of storage system. A cylindrical tank loaded with PCM is considered to establish a mathematical model using ε-NTU method. In this model, the HTF flows through tubes inside the tank (Bruno, Belusko, & Tay, 2010). The results could be used as a validation as a research tool for designing an STES unit with PCMs.

For shifting phase in a heat exchanger as well as a storage system (El-Dessouky & Al-Juwayhel, 1997), the effectiveness is expressed by Eqs. (7.7) and (7.8):

$$\in\ = 1 - e^{-NTU} \tag{7.14}$$

$$NTU = \frac{UA}{\dot{m}c_p} = \frac{1}{R_T\,\dot{m}\,c_p} \tag{7.15}$$

Convection is ignored during the transmission of thermal energy, so heat can only be transferred into the PCM by conduction. When a crossflow style tube is surrounded by a volume of PCM in a storage device, the thermal resistance R_T reduces the rate of heat transfer. A variety of factors influence tolerance (i.e., thermal conductivity, diameter and thickness of the pipe, etc.) and is given by Eq. (7.9).

$$R_T = R_{HTF} + R_{WALL} + R_{PCM} = \frac{1}{2\pi\, R_i\, L\, h_f} + \frac{\ln\left(\frac{R_o}{R_i}\right)}{2\pi\, K_w\, L} + \frac{1}{S\, K_{PCM}} \quad (7.16)$$

where R_{HTF} = HTF resistance
R_{WALL} = Tube wall resistance
R_{PCM} = PCM resistance
R_i = Tube inner radius and R_o = Tube outer radius while
L = Length of tube
h_f = coefficient of heat transfer (HTF).
K_w = Thermal conductivity (wall).
K_{PCM} = Thermal conductivity (PCM).

The phase shift mechanism of PCMs entails displacement of the boundary between stages, which is known as the Stefan or shifting boundary state. Two thermal processes include solidification and melting in LTES systems. A solid PCM absorbs heat energy during melting, which reflects fundamental energy deposit. The PCM's latent heat is restored by solidification of the solvent (Sparrow, Larson, & Ramsey, 1981). The heat transfer processes may be regulated by conduction, convection (mostly natural convection), or both conduction and convection at same time. The Stephan condition is the most challenging part of addressing liquid–solid phase shift problems (Stefan, 1891). Numerical solutions for melting problems can be classified into deforming and fixed grid systems based on how the moving boundary state is treated numerically (Voller, 1996).

The transition can be explained by the following (Sarbu & Dorca, 2019):

$$\rho L \frac{ds(\tau)}{d\tau} = \lambda_s \frac{\delta t_s}{\delta\tau} - \lambda_l \frac{\delta t_l}{\delta\tau} \quad (7.17)$$

The phase shift fraction is the fraction of substance that has yet to shift phase and is closely related to the line of solid to liquid inside PCM to prevent complexities. Instead of time, this can be used as the distinguishing variable (Belusko & Bruno, 2008) (Fig. 7.6). A square volume of PCM surrounds a tube and the phase change fraction (δ) is defined as follows:

$$\delta = \frac{Z^2 - \pi r_o^2}{W^2 - \pi r_o^2} \quad (7.18)$$

$$Z = \sqrt{\delta\left(W^2 - \pi R_o^2\right) + \pi R_o^2} \quad (7.19)$$

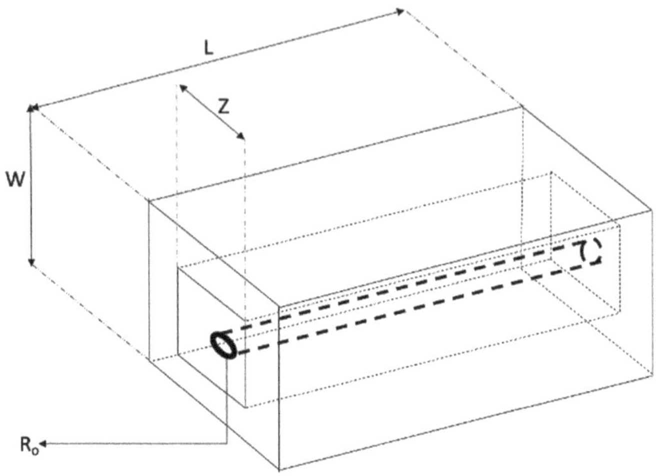

FIGURE 7.6 Phase change fraction.

Eq. (7.10) shows the formula for the shape factor, S found in the above equation. Figure expresses the continuous phase change scenario of a PCM where "*Z*" dimension is the changing width of the PCM during phase change while the "*W*" dimension is the maximum width of the PCM after phase change.

$$S = \frac{2\pi L}{\ln\left(\frac{0.5Z}{R_o}\right)} \tag{7.20}$$

7.5 Design of a solar thermal energy storage system

In agricultural product processing, drying consumes the most energy. There are a number of dryers used in this purpose, one is Hohenheim type. Here we will represent a method to design such a solar thermal energy storage system (Escobar-Ochoa, Cuervo-Andrade, & Rincon-Prat, 2018). In this dryer, the heat transfer is calculated for internal and external surface individually by Eqs. (7.13) and (7.14), respectively. The internal and external heat transfer coefficient is as below:

$$h_{int} = \frac{Nu_{int} K_{PCM}}{d_{int}} \tag{7.21}$$

$$h_{ext} = \frac{Nu_{ext} K_a}{D_{ext}} \tag{7.22}$$

Finally, *U* can be determined.

$$\frac{1}{U} = \frac{1}{h_{ext}} + \frac{1}{h_{int}} \tag{7.23}$$

The sizing of the storage cylinder depends on actual heat transfer rate, $\dot{Q}_{real}$ and desired heat transfer rate, $\dot{Q}_d$.

$$\dot{Q}_{max} = C_{min}\Delta T_{max} \tag{7.24}$$

$$\dot{Q}_{real} = \varepsilon \dot{Q}_{max} \tag{7.25}$$

$$\dot{Q}_d = \dot{m}c_{p,air@Tav}(T_o - T_F) \tag{7.26}$$

where T_o = Inlet temperature

T_F = Outlet temperature (desired)

T_{av} = Average temperature

The actual heat transfer rate is expected to be more than the desired heat flow. If it doesn't fulfill the condition, geometrical reshaping is conducted.

7.6 Energy storage material thermal properties & selection

The compatibility between the PCM and the material of the container is extremely dominant to evaluate the stability of the LTES system. A list of compatibility between material of the tank and PCM is presented in Table 7.1 (Khan, Khan, & Ghafoor, 2016). The thermochemical properties of inorganic substances are shown in Table 7.2.

7.7 Example and solution

Example 7.1. For solar thermal plants using concentric tubes, an LTES of high temperature was planned. Solar salt and lead were used as the respective HTF and PCM. The warm HTF drops down while charging (melting), and flow reverses during discharging (solidification), as shown in Fig. 7.11.

The HTF engages in the system at 565°C and the HTF is at 287°C inlet temperature during discharging. 10 h value was chosen for charging and discharging. Physical properties are given here:

Properties	Unit	HTF (solar salt, 40% KNO_3 and 60% NO_3)	PCM (lead)
Dynamic viscosity	kg/m.s	0.00326	-
Melting point	°C	-	327
Density (solid)	kg/m^3	-	11,340
Density (liquid)	kg/m^3	1820	10,640
Specific heat capacity (solid)	J/kg.K	-	133
Specific heat capacity (liquid)	J/kg.K	1533	155
Thermal conductivity (solid)	W/m.K	-	29
Thermal conductivity (liquid)	W/m.K	0.52	16
Latent heat of fusion	J/kg	-	26,284
Melting point (lower)	°C	-	324
Melting point (upper)	°C	-	330

TABLE 7.1 PCM compatibility with material.

Phase change material	References	Brass	Copper	Aluminum	Stainless steel	Carbon steel
$Zn(NO_3)_2.6H_2O$	Das, Bordoloi, Muigai and Kalita (2020)	N	N	N	Y	N
	Cabeza et al. (2001)	N	N	N	Y	N
$Na_2HPO_4.12H_2O$	Das et al. (2020)	Y	Y	N	Y	N
	Cabeza et al. (2001)	Y	C	N	Y	C
$CaCl_2.6H_2O$	Das et al. (2020)	Y	Y	N	C	N
	Cabeza et al. (2001)	Y	Y	C	Y	C
$Na_2S_2O_3.5H_2O$	Moreno et al. (2014)	–	N	C	Y	C
S10 ($Na_2SO4 + NH_4Cl$ + sepiolite)	Moreno et al. (2014)	–	N	–	Y	N
C10 ($Na_2SO4 + H_2O$ + additives)	Moreno et al. (2014)	–	N	Y	Y	N
$ZnCl_2.3H_2O$	Moreno et al. (2014)	–	Y	N	Y	N
$NaOH.15H_2O$	Moreno et al. (2014)	–	N	N	Y	C
$K_2HPO_4.6H_2O$	Moreno et al. (2014)	–	C	N	Y	N
$MgSO_4.7H_2O$	Moreno et al. (2014)	–	N	Y	Y	N
$Zn(NO_3).4H_2O$	Moreno et al. (2014)	–	N	N	Y	N
$K_3PO_4.7H_2O$	Moreno et al. (2014)	–	N	N	Y	Y

For the above table, *C*, Caution; *N*, No; *Y*, yes; -, Not available.

TABLE 7.2 Thermochemical properties of inorganic substances (Zalba, Marın, Cabeza, & Mehling, 2003).

Compound	Melting temperature	Heat of fusion	Thermal conductivity	Density
H_2O	0	333	0.612	998
$LiClO_3.3H_2O$	8.1	253	–	1720
$ZnCl_2.3H_2O$	10	–	–	–
$KF.4H_2O$	18.5	231	–	1447
$CaCl_2.6H_2O$	29	190.8	0.54	1562
$LiNO_3.3H_2O$	30	296	–	–
$Na_2SO_4.10H_2O$	32.4	254	0.544	1485
$Na_2CO_3.10H_2O$	32–36	246.5	–	1442
$CaBr_2.6H_2O$	34	115.5	–	1956
$Zn(NO_3)_2.6H_2O$	36	146.9	0.464	1828
$Na_2S_2O_3.5H_2O$	48	201	–	1600
$Na(CH_3COO).3H_2O$	58	264	–	1450
NaOH	64.3	227.6	–	1690
$Mg(NO_3)_2.6H_2O$	89	162.8	0.49	1550
$MgCl_2.6H_2O$	117	168.6	0.57	1450
$NaNO_3$	307	172	0.5	2260

Continued

TABLE 7.2 Thermochemical properties of inorganic substances (Zalba, Marın, Cabeza, & Mehling, 2003).—cont'd

Compound	Melting temperature	Heat of fusion	Thermal conductivity	Density
KNO_3	333	266	0.5	2.11
KOH	380	149.7	0.5	2044
$MgCl_2$	714	452	–	2140
NaCl	800	492	5	2160
Na_2CO_3	854	466.7	–	2533
KF	857	275.7	2	2370

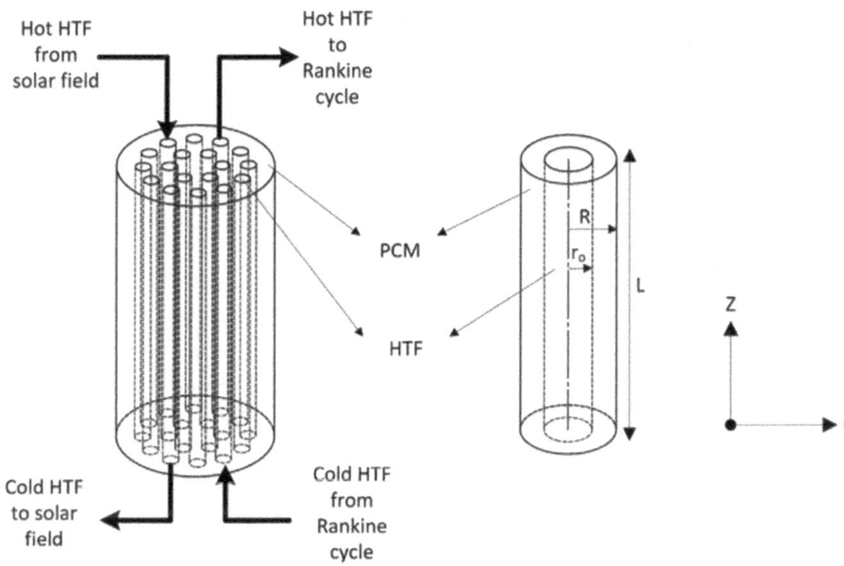

FIGURE 7.7 Generalized system for cylindrical LTES (left) and unit (right).

Determine:

i. The volume needed for the latent heat solar thermal storage
ii. Compare the volume with a conventional sensible heat solar thermal storage.

Solution. For the following calculation, the assumptions are as follows:
nondimensional length (L/d) = 10
nondimensional radius (R/r_o) = 1.3
length (L) = 1m

To calculate mass flow rate of molten salt, a simulation code of Rankine cycle was considered to use (Tehrani, Taylor, Saberi, & Diarce, 2016).

The amount of thermal energy required to operate the power cycle over the 10 h of storage had to be computed in order to determine the volume of PCM.

Eq. (7.4) was used to calculate the hourly thermal energy needed for Rankine cycle operation, while Eq. (7.5) was used to calculate the total stored energy.

$$\dot{Q}_{HTF} = 7.14 \times 1553 \times (565 - 287) = 3082580.76 \text{ W} = 3.083 \text{ MW}$$

$$Q_{HTF,total} = 3082580.76 \times 10 \times 3600 \text{ W} = 1.109729 \times 10^{11} \text{ W} = 110.97 \text{ GJ}$$

From the amount of stored energy, we can calculate the volumes for both latent and sensible heat storage by Eqs. (7.6) and (7.7).

Volume occupied by latent heat storage:

$$V_{LHS,total} = \frac{1.109729 \times 10^{11}}{11340 \times 133 \times (324 - 287) + 10640 \times 4770 + 10640 \times 155 \times (565 - 330)} = 224.6 \text{ m}^3$$

Volume needed for conventional sensible heat storage:

$$V_{SHS} = \frac{1.109729 \times 10^{11}}{10640 \times 155 \times (565 - 287)} = 242.05 \text{ m}^3$$

Therefore, it is clear that the volume for the LTES is less than the STES for accumulating same amount of energy.

The nondimensional radius and the nondimensional length also affect the stored energy for both charging and discharging process, which can be determined by further simulation. For your better understanding, we are providing a figure to show the impact of those variables on the stored energy (Tehrani et al., 2016).

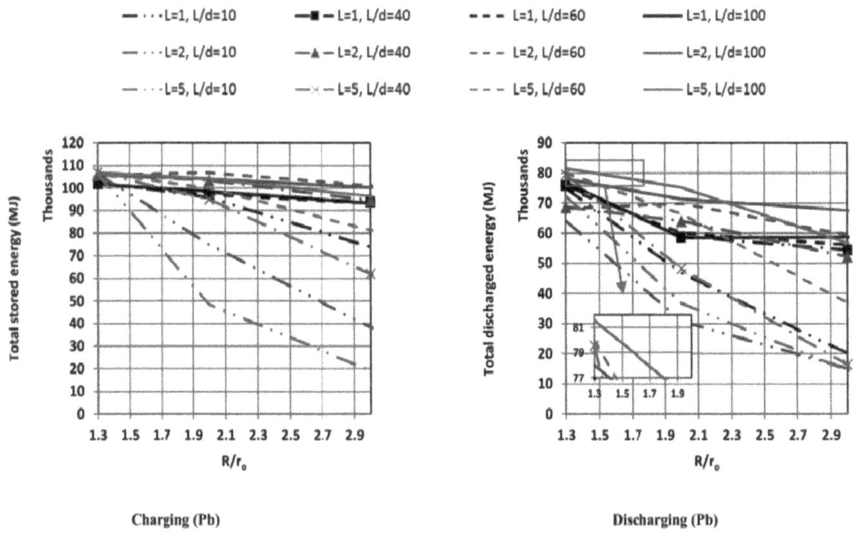

Example 7.2. The business "X" is trying to work out the efficiency of the conservation of solar thermal energy. A cylindrical tank with PCM inside it and HTF flowing through tubes inside the tank was used for storing energy. By changing its phase from solid to liquid, the PCM can store heat. In the $\epsilon - NTU$ system, considering all resistances during the heat transfer process through the storage tank, find the efficiency of the storage.

Characteristics	Unit	Value
Inner radius (tube)	M	0.02
Outer radius (tube)	M	0.022
Length (tube)	M	1.5
Specific heat (paraffin)	J/kg.K	2490
Thermal conductivity (paraffin)	W/m.K	0.22
Thermal conductivity (copper)	W/m.K	385
Heat transfer coefficient (water)	W/m.K	100

Solution. Thermal resistance will be exerted by HTF, PCM, and tube wall. They could be determined as the following:

$$R_{HTF} = \frac{1}{2\pi R_i L h_f} = \frac{1}{2\pi \times 0.02 \times 1.5 \times 100} = 0.053\ \text{K/W}$$

$$R_{WALL} = \frac{\ln\left(\frac{R_o}{R_i}\right)}{2\pi K_w L} = \frac{\ln\left(\frac{0.022}{0.02}\right)}{2\pi \times 385 \times 1.5} = 2.63 \times 10^{-5}\text{K/W}$$

$$R_{PCM} = \frac{1}{SK_{PCM}} = \frac{1}{\frac{2\pi L}{\ln\left(\frac{0.5Z}{R_o}\right)} K_{PCM}} = \frac{1}{\frac{2\pi \times 1.5}{\ln\left(\frac{0.5 \times 0.6}{0.022}\right)} \times 0.22} = 0.15\ \text{K/W}$$

After finding the above three resistance, we calculate the thermal resistance of the tube using Eq. (7.16).

$$R_T = R_{HTF} + R_{WALL} + R_{PCM} = 0.053 + 2.63 \times 10^{-5} + 0.15 = 0.203\ \text{K/W}$$

$$NTU = \frac{1}{R_T \dot{m} c_p} = \frac{1}{0.203 \times 0.01 \times 2490} = 0.1978$$

$$\epsilon = 1 - e^{-NTU} = 1 - e^{-0.1978} = 17.95\%$$

Example 7.3. A solar thermal energy storage is designed for a Hohenheim type solar tunnel dryer and a cross-flow heat exchanger was involved to heat the air driving from the collector. The air velocity is set at $u = 0.8$ m/s. Paraffin is used as the PCM and tubes are made of Copper. The thermochemical properties of the PCM and the dimensionless numbers required are given in a table.

Properties	Unit	Value
Phase change temperature (PCM)	K	60
Specific heat (PCM liquid state)	J/kg.K	2490
Specific heat (PCM solid state)	J/kg.K	59,000
Density (PCM liquid state)	kg/m^3	782
Density (PCM solid state)	kg/m^3	893
Density ratio (liquid/solid)	—	0.88
Thermal conductivity (PCM)	W/m.K	0.22
Latent heat of fusion (PCM)	J/kg	189,000
Thermal diffusivity (PCM liquid state)	m^2/s	8.5×10^{-8}
Thermal expansion coefficient	K^{-1}	7.78×10^{-4}
Kinematic viscosity (PCM liquid state)	m^2/s	5.23×10^{-6}
Dynamic viscosity (PCM liquid state)	kg/m.s	4.84×10^{-3}
Prandtl number	—	61.53
Prandtl number of the tube surface	—	338,018
Number of archimedes	—	1.3×10^5
Stefan number	—	6.6×10^{-3}
Heat transfer area	m^2	0.0413
Reynolds number	—	8620

Find:

i. The internal heat transfer coefficient of the tube
ii. The external heat transfer coefficient for the tube
iii. The overall heat transfer coefficient of the tube
iv. The amount of PCM in the storage cylinder
v. Whether there is a need of resizing or not.

Solution. For circular cylinders in cross flow, $C = 0.26$ is assumed for Reynolds number between 1000 and 2×10^5.

Correlations proposed by Bareiss et al. (Hossain, Woods, & Bala, 2005) are used for analyzing heat transfer in tube core. Expressions include:

Average dimensionless melting time:

$$\tau_a = 2.49\left(\frac{Ste}{Pr.Ar}\right)^{0.25} \acute{\rho}^{-0.75}\left(\frac{1}{1+C}\right) = 2.49\left(\frac{6.6 \times 10^{-3}}{61.53 \times 1.3 \times 10^5}\right)^{0.25}$$

$$\times\, 0.88^{-0.75}\left(\frac{1}{1+0.26}\right) = 0.01166$$

Average Nusselt number:

$$Nu_{int} = 0.2\left(\frac{Pr.Ar}{Ste}\right)^{0.25} \acute{\rho}^{-0.75}\,(1+C) = 0.2\left(\frac{61.53 \times 1.3 \times 10^5}{6.6 \times 10^{-3}}\right)^{0.25}$$

$$\times\, 0.88^{-0.75}(1+0.26) = 51.75$$

For staggered configuration, the external Nusselt number, Nu_{ext} can be estimated. Reynold number, Re_D is calculated considering maximum air velocity in the setup.

$$Nu_{ext} = 0.35\left(\frac{S_T}{S_L}\right)^{0.2} Re_d^{0.6} Pr^{0.36}\left(\frac{Pr}{Pr_s}\right)^{0.25} = 0.35(1)^{0.2} \times 8620^{0.6}$$

$$\times 61.53^{0.36}\left(\frac{61.53}{338018}\right)^{0.25} = 41.16$$

The Fourier number and the phase change duration is then determined.

$$F_o = \frac{\tau_a}{Ste} = \frac{0.01166}{6.6 \times 10^{-3}} = 1.77$$

$$t = \frac{F_o\, r_{int}^2}{\alpha} = \frac{1.77 \times 0.019^2}{8.5 \times 10^{-8}} = 7517.3 \text{ s} = 125.29 \text{ min}$$

The internal and external heat transfer coefficient are determined by Eqs. (7.21) and (7.22). Then U is derived using Eq. (7.23).

$$h_{int} = \frac{Nu_{int}\, K_{PCM}}{d_{int}} = \frac{51.75 \times 0.22}{0.019 \times 2} = 299.6 \text{ W/m.K}$$

The internal heat transfer coefficient of the tube = 299.6 W/m.K.

$$h_{ext} = \frac{Nu_{ext}\, K_a}{D_{ext}} = \frac{41.16 \times 0.026}{0.019 \times 2} = 28.22 \text{ W/m.K}$$

The external heat transfer coefficient of the tube = 28.22 W/m.K.

$$\frac{1}{U} = \frac{1}{h_{ext}} + \frac{1}{h_{\text{int}}} = \frac{1}{28.22} + \frac{1}{299.6}$$

$$U = 25.79 \text{ W/m.K}$$

The overall heat transfer coefficient of the tube = 25.79 W/m.K. $\dot{Q}_d$ was calculated by using Eq. (7.26) and it was found $\dot{Q}_d = 1608.24$ W. So, the amount of PCM is calculated.

$$m_{PCM} = \frac{\dot{Q}_d t}{\lambda} = \frac{1608.24 \times 7517.3}{189000} = 64 \text{ kg}$$

To find out the necessity of any resizing, we determine the actual heat transfer rate and compare it with the desired heat flow.

$$\dot{Q}_{max} = C_{min}\Delta T_{max} = 321.65 \times 5.5 = 1769.075 \text{ W}$$

The $\in$ and NTU was calculated using Eqs. (7.14) and (7.15),

$$\in = 0.73$$

$$\mathrm{NTU} = 1.32$$

$$\dot{Q}_{real} = \in \dot{Q}_{max} = 0.73 \times 1769.075 = 1291.4 \text{ W}$$

Here, $\dot{Q}_{real} < \dot{Q}_{max}$ and it means that resizing occurs.

To reach the desired value, we can increase the NTU and also the effectiveness by following steps (Escobar-Ochoa et al., 2018):

- Increase in U.
- Increase in A_T.
- Simultaneous increase in U and A_T if the above two doesn't work

Example 7.4. In a two-storage tank, the storage capacity of 200 kWh is targeted. The HTF used in the system is thermal oil ($\rho_o = 2750 \text{ kg/m}^3$ and $c_{po} = 900 \text{ J/kg K}$). The oil inlet and outlet temperatures are 204°C and 305°C. What should be the size of the tank to get the desired storage capacity?

Solution. Assume, for the two-tank direct system, $\varepsilon = 1$.

$$\begin{aligned} V &= \frac{E_{ST}}{\left[\rho_s c_{ps}(1-\varepsilon) + \rho_o c_{po}\varepsilon\right](T_H - T_C)} \\ &= \frac{200 \times 1000 \times 3600}{[0 + 2750 \times 900 \times 1](305 - 204)} \\ &= 2.88 \text{ m}^3 \end{aligned}$$

The size of the tank should be 2.88 m^3.

References

Belusko, M., & Bruno, F. (2008). *Design methodology of PCM thermal storage systems with parallel plates.*

Bruno, F., Belusko, M., & Tay, N. S. (2010). *Design of PCM thermal storage systems using the effectiveness-NTU method.* International Solar Energy Society.

Cabeza, L., Illa, J., Roca, J., Badia, F., Mehling, H., Hiebler, S., & Ziegler, F. (2001). Middle term immersion corrosion tests on metal-salt hydrate pairs used for latent heat storage in the 32 to 36° C temperature range. *Materials and Corrosion, 52*(10), 748–754.

Cocco, D., & Serra, F. (2015). Performance comparison of two-tank direct and thermocline thermal energy storage systems for 1 MWe class concentrating solar power plants. *Energy, 81*, 526–536. https://doi.org/10.1016/j.energy.2014.12.067

Das, D., Bordoloi, U., Muigai, H. H., & Kalita, P. (2020). A novel form stable PCM based bio composite material for solar thermal energy storage applications. *Journal of Energy Storage, 30*, 101403.

Dincer, I. (2002). On thermal energy storage systems and applications in buildings. *Energy and Buildings, 34*(4), 377–388.

El-Dessouky, H., & Al-Juwayhel, F. (1997). Effectiveness of a thermal energy storage system using phase-change materials. *Energy Conversion and Management, 38*(6), 601–617.

Escobar-Ochoa, M., Cuervo-Andrade, S., & Rincon-Prat, S. (2018). Methodology for the design of a thermal energy storage module for a solar tunnel dryer using phase change materials (PCM). *Revista UIS Ingenierías, 17*(1), 9–20.

Hossain, M., Woods, J., & Bala, B. (2005). Optimisation of solar tunnel drier for drying of chilli without color loss. *Renewable Energy, 30*(5), 729–742.

Ismail, K. A. R., & Stuginsky, R., Jr. (1999). A parametric study on possible fixed bed models for pcm and sensible heat storage. *Applied Thermal Engineering, 19*(7), 757–788. https://doi.org/10.1016/S1359-4311(98)00081-7

Khan, Z., Khan, Z., & Ghafoor, A. (2016). A review of performance enhancement of PCM based latent heat storage system within the context of materials, thermal stability and compatibility. *Energy Conversion and Management, 115*, 132–158.

Kuravi, S., Trahan, J., Goswami, D. Y., Rahman, M. M., & Stefanakos, E. K. (2013). Thermal energy storage technologies and systems for concentrating solar power plants. *Progress in Energy and Combustion Science, 39*(4), 285–319.

Li, X., Xu, E., Song, S., Wang, X., & Guofeng, Y. (2017). Dynamic simulation of two-tank indirect thermal energy storage system with molten salt. *Renewable Energy, 113*, 1311–1319. https://doi.org/10.1016/j.renene.2017.06.024

Moreno, P., Miró, L., Solé, A., Barreneche, C., Solé, C., Martorell, I., & Cabeza, L. F. (2014). Corrosion of metal and metal alloy containers in contact with phase change materials (PCM) for potential heating and cooling applications. *Applied Energy, 125*, 238–245.

Pielichowska, K., & Pielichowski, K. (2014). Phase change materials for thermal energy storage. *Progress in Materials Science, 65*, 67–123.

Saha, S., Ruslan, A. R. M., Morshed, A. K. M. M., & Hasanuzzaman, M. (2020). Global prospects and challenges of latent heat thermal energy storage: A review. *Clean Technologies and Environmental Policy.* https://doi.org/10.1007/s10098-020-01997-7

Sanchez Perez, P. A., Bailey, R., Kurtz, S., & Mahmud, Z. (2020). *Method for quantifying value of storage toward reaching 100% renewable electricity.*

Sarbu, I., & Dorca, A. (2019). Review on heat transfer analysis in thermal energy storage using latent heat storage systems and phase change materials. *International Journal of Energy Research, 43*(1), 29–64.

Sarbu, I., & Sebarchievici, C. (2018). A comprehensive review of thermal energy storage. *Sustainability, 10*(1), 191. Retrieved from http://www.mdpi.com/2071-1050/10/1/191.

Schumann, T. E. W. (1929). Heat transfer: A liquid flowing through a porous prism. *Journal of the Franklin Institute, 208*(3), 405–416. https://doi.org/10.1016/S0016-0032(29)91186-8

Sparrow, E. M., Larson, E., & Ramsey, J. (1981). Freezing on a finned tube for either conduction-controlled or natural-convection-controlled heat transfer. *International Journal of Heat and Mass Transfer, 24*(2), 273–284.

Stefan, J. (1891). Über die Theorie der Eisbildung, insbesondere über die Eisbildung im Polarmeere. *Annalen der Physik und Chemie, 42*, 269–286.

Tehrani, S. S. M., Taylor, R. A., Saberi, P., & Diarce, G. (2016). Design and feasibility of high temperature shell and tube latent heat thermal energy storage system for solar thermal power plants. *Renewable Energy, 96*, 120–136.

Voller, W. (1996). Anoverview of numerical methods for solving phase change problems. *Advances in Numerical Heat Transfer, 1*, 341.

Zalba, B., Marın, J. M., Cabeza, L. F., & Mehling, H. (2003). Review on thermal energy storage with phase change: Materials, heat transfer analysis and applications. *Applied Thermal Engineering, 23*(3), 251–283.

Chapter 8

Solar drying system

Md Zillur Rahman[1] and Md Hasanuzzaman[2]

[1]*Department of Mechanical Engineering, Ahsanullah University of Science and Technology, Dhaka, Bangladesh;* [2]*Higher Institution Centre of Excellence (HICoE), UM Power Energy Dedicated Advanced Centre (UMPEDAC), University of Malaya, Kuala Lumpur, Malaysia*

Chapter outline

Technologies for Solar Thermal Energy. https://doi.org/10.1016/B978-0-12-823959-9.00010-6

8.1 Fundamental conception of the solar drying system

Drying is the process of removing moisture from a product to preserve it or to reach the levels of moisture referred to as the equilibrium condition suitable for storage without spoilage. Drying is one of the most energy-intensive processes because of the high latent heat of vaporisation and the inherently inefficient use of warm air as the most commonly used drying medium (Kamarulzaman, Hasanuzzaman, & Rahim, 2021; Mujumdar, 2014). For instance, the drying product necessitates the removal of more than 80% of its mass from its moisture content. It is necessary to invest the energy of 9.04 MJ (= 2.69 kWh) for approximately 1 kg of dry product from 5 kg of the fresh product, and the energy demand is generally 3.59 kWh/kg of the dry product for a dryer with an efficiency of 75% (Nair, Espinosa-Santana, Guerrero-Martínez, López-Ortiz, & Nair, 2020). The drying process occurs in two steps. In the first step, the energy from the heating source is transferred to the product, and the water content is then carried from inside the product to the surface and dried in the air in the form of water vapor at a constant drying rate. A slow drying process occurs in the second step, and this process involves drying the properties of the product (Chauhan, Kumar, & Tekasakul, 2015; Ekechukwu, 1999; El-Sebaii & Shalaby, 2012). A decrease in the moisture content through drying can extend a product's lifetime by delaying or preventing the action of the spore germinations, chemical reactions (enzymatic/nonenzymatic), and the formation of microscopic organisms (which cause rot or fermentation), namely enzymes, bacteria, fungi, yeasts, and molds. It reduces the weight of the product to facilitate transportation and storage capacity (Belessiotis & Delyannis, 2011; Djebli, Hanini, Badaoui, Haddad, & Benhamou, 2020). The drying process can also preserve the nutritive benefits and flavor of the product (Djebli et al., 2020).

Drying is a complex phenomenon related to the transfer of thermal energy and moisture content, accompanied by some rate processes (i.e., physico-chemical transformations) that can induce variations in the product quality and the mechanisms associated with the heat and mass transfer. Physical changes such as puffing, shrinkage, decomposition of the product, crystallization, and glass transition may happen. Also, on a few occasions, chemical or biochemical reactions can occur, which may lead to variations in color, odor, texture, or other properties (e.g., acidity, toxicity, flammability, and so on) of the product (Mujumdar, 2014).

Open sun drying (OSD) is the most commonly used process for drying horticultural products such as vegetables, fruits, and grains, especially in tropical and subtropical areas around the world. In this process, the products are typically spread outdoor, often over the prepared ground or mat, and exposed directly to sun and wind. As a result, this process suffers from some drawbacks: the damage of the product due to the infestation of birds, animals, insects, and rodents, the degradation of the product caused by direct exposure to rain, solar radiation, wind, and dew, contamination due to wind-blown dirt, dust, debris, and environmental pollution, probable cracking of the substance, growth of microorganisms, and loss of germination ability (Imre, 2014; Sandali, Boubekri, & Mennouche, 2019). Open sun drying may also result in an overdried or often undried product with relatively high water content due to its slow and intermittent drying (Norton, 2017), thus exacerbating the quality of the final product.

Postharvest losses (i.e., degradation in quantity and quality of a crop from harvest to consumption) can be more than 30%, and a significant reduction of losses can be made by adequate drying of products (Imre, 2014). Solar drying, however, can be used as an alternative to the conventional drying processes and a remedy to deal with the problems of OSD. In the solar drying system, the conversion of solar energy into heat energy happens in the solar collector, and this heat energy is subsequently used to remove water content from the product directly (convection), indirectly (conduction), by radiation, or a combination of any of them. In contrast, the transportation of heat energy to the product occurs continuously or intermittently by convection, conduction, radiation, or electromagnetic fields (Mujumdar, 2014). Applying the solar dryer can not only decrease postharvest losses but also significantly enhance the dried product quality when compared to the OSD. The solar energy-based dryer is not as expensive as electricity or fossil fuel-powered dryers, and the latter is also responsible for environmental pollution due to emissions of CO_2, SO_2, and NO_X. The underlying factors for the higher use of the solar dryer are: (1) CO_2 emission and pollution due to the use of fossil fuels, (2) depleting of conventional energy sources, (3) the certainty of life cycle system costs, and (4) enabling more rapid drying than an open sun dryer (Norton, 2017).

Typically, agricultural products are very susceptible to drying conditions. It is essential to perform drying in a way so that drying does not have any adverse effects on their flavor, texture, color, or nutrients. Consequently, the selection of drying parameters, especially temperature, is crucial (Kalogirou, 2014). The drying temperature usually lies in the range of 35−45°C, depending on the product. Temperatures range from 37 to 71°C, destroy bacteria and cause inactivation of enzymes, even though these temperatures tend to induce surface hardening of the crops (Norton, 2017). An excessive acceleration of product surface dehydration may also cause case hardening (wherein the outer layers of the product are overdried along with product shrinkage), and the subsequent mold formation in the internal core of the product.

The physical structure of the product and transfer mechanisms of water content (i.e., mass) differ during the drying (Mujumdar & Beke, 2003). Over 99% of the dryers involve in extracting the water content from the product. As mentioned earlier, the heat transfer to the substance may occur by convection (direct dryers), conduction (indirect dryers), radiation, or electromagnetic field. Most heat transfer modes provide heat energy at the drying product's boundaries, but the thermal energy diffuses into the produce mainly through conduction. The moisture of the interior of the product has to transport to the product's boundary before it is moved away by a drying medium such as hot air, direct combustion gas for convective type dryer, or vacuum for non-convective type dryer. Heat and mass transfer processes are illustrated in Fig. 8.1. Transfer of water content inside the material may happen through any or a combination of the following mass transfer mechanisms (Mujumdar, 2014).

- Liquid diffusion (if the wet solid temperature is below the liquid's boiling point)
- Vapor diffusion (if the liquid is vaporised inside the product)
- Knudsen diffusion (if drying occurs at very low pressures and temperatures (e.g., in freeze-drying))
- Surface diffusion may also occur
- Hydrostatic pressure differences (if internal vaporisation rates surpass the vapor transport rate via solid to the surrounding environment)
- Combinations of mechanisms as mentioned above

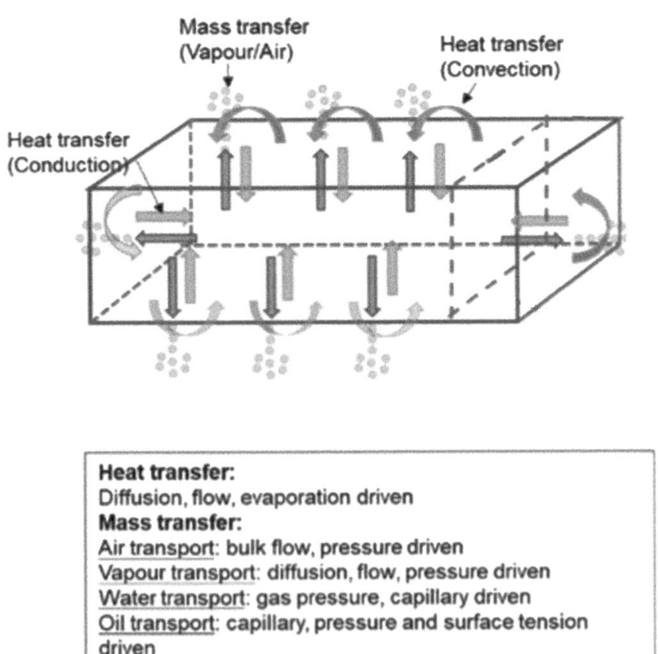

FIGURE 8.1 Heat and mass transfer in drying (Xu et al., 2020).

8.2 Classification of solar dryers

The solar drying system can generally be classified according to the type of drying or the mode of air circulation. Four kinds of solar dryers, such as direct, indirect, mixed-mode, and hybrid, have successfully been employed to dry crops. Such a dryer's airflow mechanism is mainly on the basis of the mode of natural convection or forced air circulation. These dryers are explained in detail in the following section.

8.2.1 Direct solar dryer

In a direct solar dryer, the product is heated directly by the sun's radiation, and moisture of the succulent foods is removed by natural convection. A solar dryer may have a black, absorbent surface that absorbs and transforms the sun's radiation spectrum into heat. The products that need to be dried are spread directly over the black surface. These dryers can also consist of glass covers, enclosures, and vents to improve performance (Pardhi & Bhagoria, 2013). A direct solar dryer is shown in Fig. 8.2. Examples of direct solar dryers are the tunnel dryer, cabinet dryer, rack dryer, tray and tent dryer, multirack dryer, and greenhouse dryer.

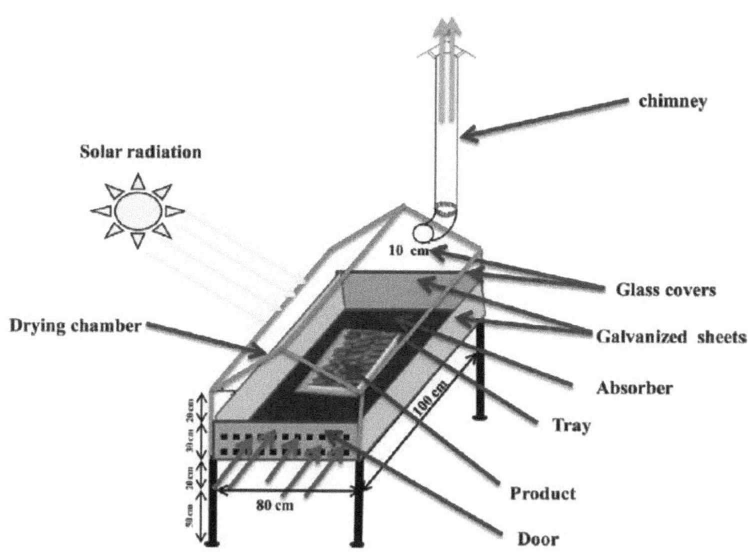

FIGURE 8.2 Schematic diagram of the direct solar dryer (Mennouche et al., 2014; Sandali, Boubekri, Mennouche, & Gherraf, 2019).

8.2.2 Indirect solar dryer

The indirect solar dryer comprises of mainly two parts: a solar thermal energy-absorbing unit known as a solar collector and a crop dehydration unit known as a drying chamber, as illustrated in Fig. 8.3. In this dryer, the solar collector warms the ambient air instead of directly exposing the product to the intensity of solar radiation. The heated air is then circulated over the fresh product in the drying chamber, taking moisture from the product, and exits upwards often through a chimney (Chauhan et al., 2015). Convective heat transfer between the product and nearby heated drying air dehydrates the fresh product (Ssemwanga, Makule, & Kayondo, 2020). Notably, in forced convection solar dryer, the hot drying air is flowed either by using fans or blowers (i.e., active force convection dryer) or by natural convection or buoyancy force because of wind pressure (i.e., passive force convection dryer) (Ssemwanga et al., 2020). The heat exchange rate between the drying air and the products increases due to forced air circulation. Some parameters can have a significant influence on the performance of the drying system, namely the ambient temperature, drying temperature, the amount of useful thermal energy gained from the solar thermal collector, the relative humidity of heated air, and the air velocity (El-Sebaii & Shalaby, 2012). This system can protect the crops or other materials from contamination caused by wind-blown debris, dust, and dirt. This system can also prevent infestation by

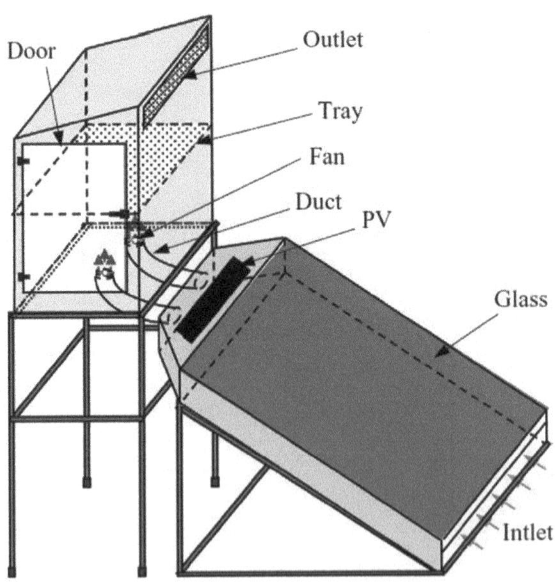

FIGURE 8.3 Indirect solar dryer with forced convection (Djebli et al., 2020).

birds, insects, or animals. Direct sunlight may chemically change some foods and make them less appetizing (Weiss & Buchinger, 2012). This system is useful in the preparation of those crops whose vitamin content cannot be degraded by sunlight (Pardhi & Bhagoria, 2013).

8.2.3 Mixed-mode solar dryer

The mixed-mode solar dryer is composed of direct and indirect solar dryers. This dryer collects solar thermal energy using the solar collector and drying chamber roof, as shown in Fig. 8.4. The coupled effects of the incident solar radiation over the drying product and the heated drying air from the solar collector offer the heat needed for the drying process (Pardhi & Bhagoria, 2013).

8.2.4 Hybrid solar dryer

The hybrid uses the features of both the direct and indirect solar dryer systems simultaneously to ensure appropriate drying conditions (see Fig. 8.5). The drying operation is not solely dependent on solar radiation intensity. The solar photovoltaic powered blower may be applied in this system for air circulation. The coupled effect of the incident direct solar heat flux and preheated air in the solar collector generates the resultant heat flux and consequently enhances the

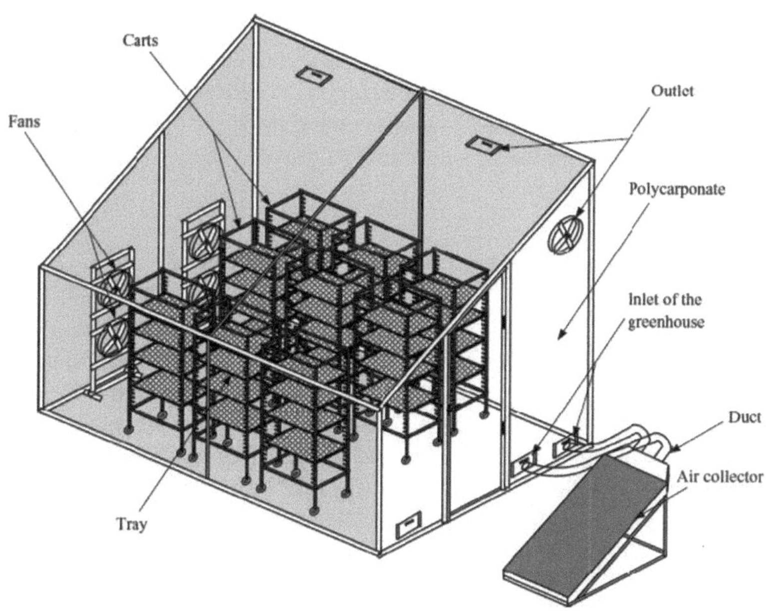

FIGURE 8.4 Schematic diagram of the mixed-mode solar dryer (Djebli et al., 2020).

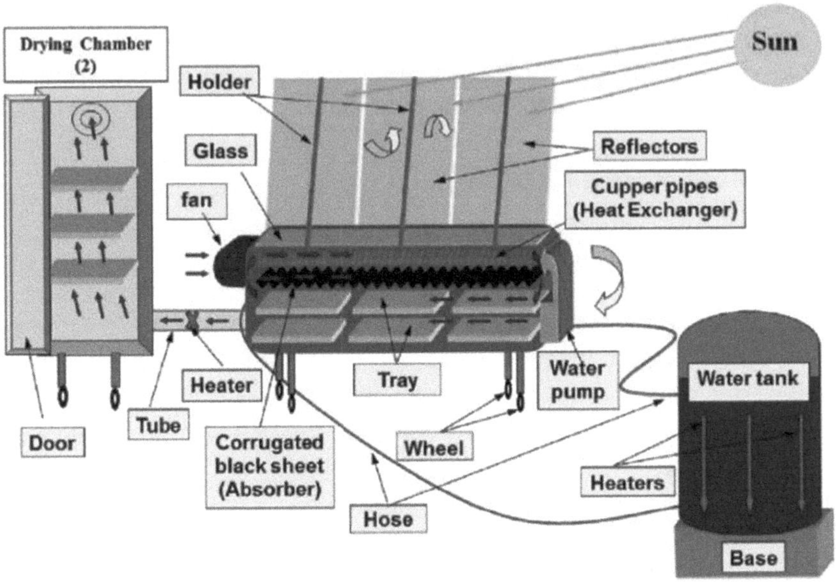

FIGURE 8.5 Schematic diagram of the integrated hybrid solar dryer system (Amer, Gottschalk, & Hossain, 2018).

drying operation (Ssemwanga et al., 2020). This system can control the removal of moisture from any crops efficiently and ensure the proper quality of the produce (Chauhan et al., 2015).

Solar energy is not always available, especially at night and during the rainy periods. In such cases, hybrid solar dryers where solar thermal energy is coupled with alternative energy sources, including solid biomass fuels, fossil fuels, and electricity, are used as alternatives to solar energy (Pirasteh, Saidur, Rahman, & Rahim, 2014).

8.3 Components and selection of a solar dryer

8.3.1 Components of a solar dryer

A solar dryer may be considered as it comprises of three main components — a drying chamber, a solar collector, and some type of airflow system, as illustrated in Fig. 8.6. In the drying chamber, drying takes place, and the material is spread on the chamber to get dehydrated, whereas the solar collector converts the solar radiation spectrum into heat. A solar dryer may also have the following components or accessories.

- Equipment of heat transfer for transporting thermal energy to the air or product
- Device for keeping the circulation of drying air

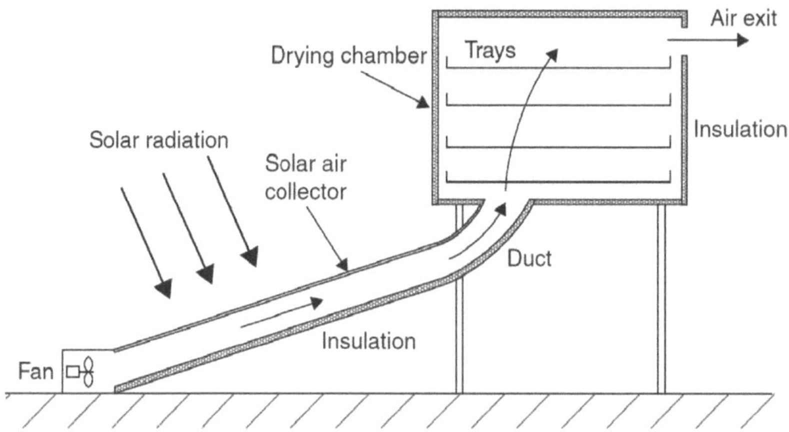

FIGURE 8.6 Typical components of a solar dryer (Kalogirou, 2014).

- Thermal energy storage system (non-compulsory)
- An alternative energy source (non-compulsory)
- Measurement and control equipment (non-compulsory)
- Transparent cover, absorber, trays, ducts, pipes, circuits, solar chimney (optional), and other appliances

8.3.2 Selection of a solar dryer

More than 400 types of dryers have previously been mentioned in the literature, while more than 100 distinct types are usually available (Mujumdar, 2014). It is not easy to choose a drier from a large variety of solar dryers, and the choice is generally subjective that depends heavily on local conditions and drying products. The best configuration of a solar dryer for each local area is different. Thus a successful food drying typically involves a period of testing and modifications according to the location requirements (Green & Schwarz, 2001). Since a variety of solar dryers are available for drying a wide range of produces, the choice of one over the others also depends on some attributes such as ease of handling of materials, their drying behavior, the final quality of the product, and the resources available (Mujumdar, 2014).

8.4 Solar drying design considerations for a product

The procedure for designing a solar drying system is associated with collecting some parameters such as location, solar radiation, humidity, wind speed, and so on. Some other parameters such as the amount of moisture to be removed from the crops, the required useful energy for drying crops, drying rate, absorbed energy by the solar collector, and unit cost the drying need to be estimated. The dryer size depends on the total area required to dry each unit of the product. It is possible to obtain up to 2 kg of dried product per m^2 of dryer/collection area at a dryer having an efficiency of 75% (Nair et al., 2020). The drying temperature is determined based on the maximum temperature that should not have any adverse effects on the taste, flavor, texture, and color of the resulting produce. Drying of a product takes place at temperature ranges from 35 to 45°C based on its type. The technical requirements for the development of solar dryers are well-designed compact collectors, high-performance and versatile storage systems, and long-life drying systems (El-Sebaii & Shalaby, 2012). For conducting the drying of a product, the predictions of the characteristics of various units of solar drying systems can be made using mathematical considerations/models. The distributions of temperature and water content in the substance can be obtained through simulations. It is also feasible to predict the estimation of thermodynamic parameters (e.g., drying time, energy consumption, and thermal efficiency), along with some other parameters such as the area of collector, surface exchange rate, wind speed, and so on.

8.4.1 Mathematical considerations of the triple-pass solar collector

The solar collector acts as a preliminary source of energy to a solar dryer, along with performing energy conversion and transfer. It converts the direct and diffused solar radiation intensity into heat, and the transformation of the energy occurs in the collector's absorber. The absorber material or coating has a high solar radiation absorption coefficient. The absorbed radiation enhances the internal energy and temperature of the absorber. The solar collectors are available for solar drying in the temperature range of 80 to 240°C (Schweiger et al., 2000). Table 8.1 demonstrates three types of solar collectors that can be applied for drying purposes, such as flat plate, stationary compounded, and evacuated tube. The collector efficiency relies on the differences in temperature between ambient and inlet to the collector. The collector efficiency and performance increase with increasing irradiation levels (Motevali, Minaei, & Khoshtagaza, 2011). A high-performance collector is required to obtain high temperature and efficiency. The parabolic solar collector can deliver a maximum temperature of 250°C in an application of process heating (Pirasteh et al., 2014).

Energy balance analysis can be carried out on different parts of the conventional solar collector (i.e., single-pass), double-pass solar collector, and triple-pass solar collector. However, the triple-pass type is more efficient because of the three times heating of the drying air in the collector. The triple-pass solar collector, along with associated heat transfer coefficients at its different surfaces, is shown in Fig. 8.7 (Khouya, 2020).

TABLE 8.1 Characteristics of various stationary solar collectors (Pirasteh et al., 2014).

Type of stationary collector	Indicative temperature range (°C)	Efficiency irradiation levels (W/m^2)		Concentration ratio
		500	1000	
Flat plate	30–80	0.71–0.75	0.72–0.75	1
Compound parabolic	60–240	0.45–0.73	0.58–0.72	1–5
Evacuated tube	50–200	0.44–0.82	0.62–0.82	1

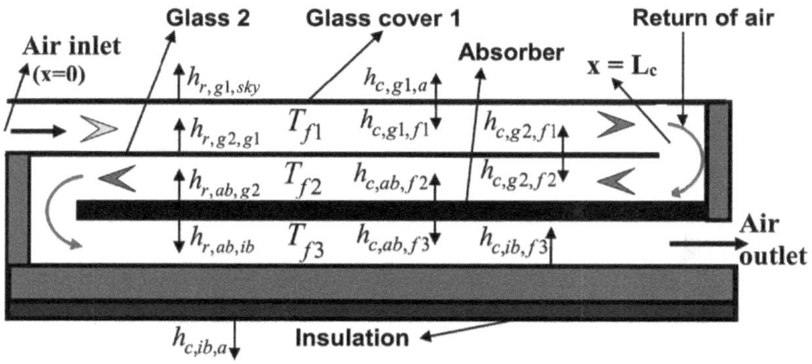

FIGURE 8.7 Triple-pass solar collector (Khouya, 2020).

The thermal energy balance equations in the various parts of the multipass collector are made considering certain assumptions (Aboul-Enein, El-Sebaii, Ramadan, & El-Gohary, 2000; Khouya, 2020; Jain, 2005) such as (1) The transfer of heat is unidirectional, (2) The thermal capacity of the transparent cover, absorbing surface, and the insulator is negligible, (3) There is no thermal gradient across the thickness of the collector's glass material, (4) There is no vertical thermal stratification of the airflow in the ducts, (5) The volumetric shrinkage of the substance during drying is negligible, and (6) The solar collector is correctly sealed using an insulating material, and no collector air leakage occurs.

Energy balance equations

The energy balance equations for the triple-pass solar collector are described below, as outlined in (Aboul-Enein et al., 2000; Jain, 2005; Khouya, 2020).

For glass 1

A portion of the solar energy incident is absorbed by glass cover 1, while a significant portion is transmitted to glass 2. The remaining energy is lost to the surroundings by convection and radiation. The equation of the energy balance for glass cover 1 can be expressed as follows.

$$A_{co}I\alpha_{g1} + h_{r,g1,g2}A_{co}\left(T_{g2} - T_{g1}\right) = \\ h_{c,g1,f1}A_{co}\left(T_{g1} - T_{f1}\right) + h_{c,g1,a}A_{co}\left(T_{g1} - T_{amb}\right) + h_{r,g1,a}A_{co}\left(T_{g1} - T_{sky}\right) \tag{8.1}$$

where A_{co} is the collector area (m^2), I is the solar intensity (W/m^2), α is the absorptivity coefficient, h is the heat transfer coefficient (W/m K), T is the temperature (K), and subscripts $g1$, $g2$, $f1$, a, c, r, and amb are the glass cover 1, glass cover 2, airstream 1, drying air, convection, radiation, and ambient, respectively.

For the airstream between glass 1 and 2

The energy balance for the airstream (exchanges heat by convection with both glasses) between glass 1 and 2 is as follows.

$$\dot{m}_a C_{p,a}\frac{dT_{f1}}{dx} = lh_{c,g1,f1}\left(T_{g1} - T_{f1}\right) + lh_{c,g2,f1}\left(T_{g2} - T_{f1}\right) \tag{8.2}$$

where $\dot{m}_a$ is the mass flow rate (kg/s), $C_{p,a}$ is the specific heat (kJ/kg K) of drying air, l is the collector width (m), and x is the abscissa (m).

For glass 2

A part of solar energy is absorbed by glass 2, and the transmission of heat energy occurs by convection to the airstream between two glasses, and between glass 2 and absorber. A heat transfer also occurs by radiation between glass 1 and 2, and then with the absorber. The energy balance can be formulated as

$$A_{co}I\tau_{g1}\alpha_{g2} + h_{r,ab,g2}A_{co}\left(T_{ab} - T_{g2}\right) + h_{c,g2,f2}A_{co}\left(T_{f2} - T_{g2}\right) = \\ h_{r,g1,g2}A_{co}\left(T_{g2} - T_{g1}\right) + h_{c,g2,f1}A_{co}\left(T_{g2} - T_{f1}\right) \tag{8.3}$$

where τ is the transmission factor and subscripts ab and $f2$ are the absorber and airstream 2, respectively.

For the airstream between glass 2 and absorber

The drying air exchanges heat with glass 2 and absorber through convection. The heat balance equation is expressed as

$$\dot{m}_a C_{p,a}\frac{dT_{f2}}{dx} = lh_{c,g2,f2}\left(T_{g2} - T_{f2}\right) + lh_{c,ab,f2}\left(T_{ab} - T_{f2}\right) \tag{8.4}$$

For the absorber plate

The absorber plate absorbs all the heat transported through the glass 2. It exchanges thermal energy by convection with airstream between absorber and glass 2, and between absorber and insulator. The absorber also exchanges thermal energy by radiation with glass 2 and the insulator. The thermal balance of the absorber plate is

$$\begin{aligned} A_{ab} I \tau_{g1} \tau_{g2} + h_{c,ab,f2} A_{ab} \left(T_{f2} - T_{ab}\right) + h_{c,ab,f3} A_{ab} \left(T_{f3} - T_{ab}\right) = \\ h_{r,ab,g2} A_{ab} \left(T_{ab} - T_{g2}\right) + h_{r,ab,ib} A_{ab} (T_{ab} - T_{ib}) \end{aligned} \tag{8.5}$$

where subscripts $f3$ and ib are the airstream 3 and insulating background, respectively.

For the airstream between absorber and insulator

At this stage, the airstream exchanges heat with the absorber and insulator by convection. The energy balance is expressed as

$$\dot{m}_a C_{p,a} \frac{dT_{f3}}{dx} = l h_{c,ib,f3} \left(T_{ib} - T_{f3}\right) + l h_{c,ab,f3} \left(T_{ab} - T_{f3}\right) \tag{8.6}$$

For the insulator

The heat transfer happens by radiation with absorber and convection with airstream 3 and atmospheric air. The thermal balance is written as

$$T_{ib} = \frac{h_{c,ib,a} T_{amb} + h_{c,ib,f3} T_{f3} + h_{r,ab,ib} T_{ab}}{h_{c,ib,a} + h_{r,ab,ib} + h_{c,ib,f3}} \tag{8.7}$$

Heat transfer coefficient

Heat transfer coefficient inside the solar collector

For using in Eqs. (8.1)–(8.7), the coefficient of convective heat transfer inside the solar thermal collector may be estimated as

$$h = \frac{Nu \lambda_a}{D_H} \tag{8.8}$$

The Nusselt number (Nu) can be calculated as follows.

$$Nu = \begin{cases} 0.666 \mathrm{Re}^{0.5} \mathrm{Pr}^{1.3} & \text{for laminar flow } \left(\mathrm{Re} < 3 \times 10^5\right) \\ 0.035 \mathrm{Re}^{0.5} \mathrm{Pr}^{1.3} & \text{for turbulent flow } \left(\mathrm{Re} > 3 \times 10^5\right) \end{cases} \tag{8.9}$$

The Reynolds number (Re) of Eq. (8.9) may be determined by the following Eq. (8.10).

$$Re = \frac{U_f D_H}{\nu_f} \tag{8.10}$$

where U_f is the velocity of the heat transfer fluid (i.e., gas, which is a mixture of air and water vapor) flow (m/s), ν_f is the viscosity of the heat transfer fluid flow (m^2/s), and D_H is the hydraulic diameter of the collector (m).

Heat transfer coefficient from glass 1 to outside air

The convective heat transfer coefficient from glass 1 to wind can be estimated using Eq. (8.11), and this equation is valid for wind speed in the range of $0 \leq U_{amb} \leq 7$ m/s.

$$h_{c,g1,a} = h_{c,ib,a} = 2.8 + 3U_{amb} \tag{8.11}$$

where U_{amb} is the ambient wind speed (m/s).

Radiative heat transfer coefficient

The radiative heat transfer coefficient between glass cover and sky, between glass cover and absorber, and between the absorber and interior face of the insulator can be calculated using the following expression.

$$h_{r,x,y} = \frac{\sigma\left(T_x^2 + T_y^2\right)\left(T_x + T_y\right)}{\frac{1}{\varepsilon_x} + \frac{1}{\varepsilon_y} - 1} \tag{8.12}$$

where T is the surface temperature (K), subscripts x and y are the sky, glass, absorber, and insulator (i.e., (glass 1, sky)/(glass 1, glass 2)/(absorber, insulator)), ε is the emissivity coefficient of the surface, and σ is the Stephan Boltzmann constant (5.67×10^{-8} W/m^2 K^4).

The equivalent sky temperature is determined as

$$T_{sky} = 0.0552(T_{amb})^{1.5} \tag{8.13}$$

Thermal energy analysis

The air is heated by a solar thermal collector, and the warm air is then used as a traditional fluid in the solar drying system. The quantity of the extracted solar thermal energy is vital to decide whether a sufficient amount of heat energy is available or not to heat the air. The useful energy which is gained by working fluid to be transferred to the drying chamber. The total useful energy required for moisture evaporation can be estimated considering the fluid mass flow rate

$\dot{m}_a$, its specific heat $C_{p,a}$, and the inlet-outlet temperature difference of the solar collector.

$$Q_u = \int_0^{t_{dr}} \dot{m}_a C_{p,a} \left(T_{f3}(x = L_c) - T_{amb} \right) dt \tag{8.14}$$

where L_c is the collector length (m).

The solar energy which the collector captures during the drying operation may be computed using the following mathematical expression.

$$Q_{solar} = \int_0^{t_{dr}} I A_{co} dt \tag{8.15}$$

where I is the total incident radiation on the horizontal surface during the drying period (kJ/m^2).

It is necessary to estimate the performance (i.e., efficiency) of the solar collector, which can be obtained considering the ratio of useful energy transferred to the heat transfer fluid and the solar thermal energy collected by the solar collector.

$$\eta_c = \frac{Q_u}{Q_{solar}} \tag{8.16}$$

where η_c is the collector efficiency (typically in the range of 30%–50%) (Tonui, Mutai, Mutuli, Mbuge, & Too, 2014).

8.4.2 Mathematical considerations of the drying chamber

In general, convective drying is used for preserving and storing the substance for more prolonged periods by extracting its moisture content. This drying is a complex operation, during which simultaneous heat and mass transfer occur between drying medium and product. The heat and mass transfer equations are mentioned in this section for the forced convection solar drying system.

Energy balance equations

Mass balance for drying air

Conservation of drying air mass represents that the product loses the moisture, while the drying air absorbs it. It is stated as (Lamrani, Khouya, & Draoui, 2019)

$$m_a \frac{dW}{dt} + \dot{m}_a (W - W_0) = -M_0 \frac{dX}{dt} \tag{8.17}$$

where m_a is the mass of drying air (kg), $\dot{m}_a$ is the mass flow rate of drying air (kg/s), W is the absolute humidity of the air inside the drying chamber (kg water/kg dry air), W_0 is the absolute humidity of the air at the inlet of the drying chamber (kg water/kg dry air), M_0 is the product mass (dry basis) (kg), X is the moisture content of the product (kg water/kg dry product), and t is the time (s).

Energy balance for drying air

The energy balance of drying air implies that the loss of heat energy from the drying air, while the product gains it. It is expressed as (Khouya, 2020)

$$m_a C_{pr,a}\left(\frac{dT_a}{dt} + U_a\frac{dT_a}{dx}\right) = h_{th}A_p\left(T_p - T_a\right) \tag{8.18}$$

where T_a is the drying air temperature (K), T_p is the product temperature (K), U_a is the drying air velocity (m/s), h_{th} is the heat transfer coefficient (W/m^2 K), and A_p is the exchange surface area of the product (m^2).

The specific heat of the drying air at constant pressure $C_{pr,a}$ (J/kg K) can be calculated using the following equation.

$$C_{pr,a} = 1835 - 0.734(T_a - 273.15) \tag{8.19}$$

Amount of air required for drying

The quantity of air needed for drying can either be based on the volumetric flow rate (Eq. 8.20) or mass flow rate of air (Eq. 8.21) (Tonui et al., 2014).

$$V_a = \frac{W_a}{t_d} \tag{8.20}$$

where V_a is the volumetric flow rate of air (m^3/s), W_a is the amount of air required (m^3), and t_d is the total drying time (s).

$$\dot{m}_a = \rho_a V_a \tag{8.21}$$

where $\dot{m}_a$ is the mass flow rate of air (kg/s) and ρ_a is the density of air (kg/m^3).

Mass conservation of a product

A product's mass conservation refers to the quantity of moisture it loses in the drying air during the drying process. The moisture content of a product can be estimated either by wet basis (kg water/kg wet matter) or dry basis (kg water/kg dry matter). The amount of moisture removal from the product (by wet basis) per unit of the dried product is expressed as (Kumar & Kandpal, 2005)

$$X_w = \frac{M_i - M_f}{1 - M_i} \tag{8.22}$$

where M_i and M_f are the initial and final moisture contents (in fraction) of the product, respectively.

The moisture content by wet basis is mostly used in industry, whereas academicians and researchers use the moisture content by dry basis. The dry-based moisture content (kg water/kg dry matter) is formulated as

$$X_d = \frac{M_i - M_f}{M_f} \tag{8.23}$$

Energy conservation of a product

The heat transfer rate for a product from the drying air during the drying process can be found as similar to that of drying air energy conservation, and the heat transfer rate is written as (Khouya, 2020)

$$m_p C_{pr,p} \frac{dT_p}{dt} = h_{th} A_p (T_a - T_p) - M_0 \Delta H \frac{dX}{dt} \tag{8.24}$$

where m_p is the mass of the product (kg), $C_{pr,p}$ is the heat transfer coefficient of the product at constant pressure (kJ/kg K), and ΔH is the specific latent heat (kJ/kg).

The specific latent heat can be determined as (Khouya, 2020)

$$\Delta H = 4.186(597 - 0.56T_p) \tag{8.25}$$

Amount of energy required for drying

The required useful energy to dry a unit quantity of a product on a dry mass basis can be calculated using the following relation (Kumar & Kandpal, 2005).

$$E_{dry} = \left(\frac{1 - M_f}{1 - M_i}\right) s(T_d - T_{amb}) + \left(\frac{M_i - M_f}{1 - M_i}\right) h_{fg}$$

$$E_{dry} = E_{sh} + Eem \left(\text{Where } E_{sh} = \left(\frac{1 - M_f}{1 - M_i}\right) s(T_d - T_{amb}), \right. \tag{8.26}$$

$$\left. E_{em} = \left(\frac{M_i - M_f}{1 - M_i}\right) h_{fg} \right)$$

where E_{sh} is the useful energy (W) needed for the sensible heating of the wet product from ambient temperature (T_{amb}) to drying temperature (T_d), E_{em} is the useful energy (W) needed for moisture evaporation in the product, h_{fg} is the water evaporation enthalpy at drying temperature (kJ/kg), s is the specific heat of the product by wet basis (kJ/kg K), which can be estimated using the following Eq. (8.27) (ASHRAE, 1974).

$$s = (0.80M_i + 0.20)4.1868 \times 10^{-3} \tag{8.27}$$

The useful energy needed to dry a unit quantity of a crop on a wet basis is expressed as (ASHRAE, 1974).

$$E_{dry} = \left(\frac{M_i - M_f}{1 - M_f}\right) h_{fg} + s(T_d - T_{amb}) \tag{8.28}$$

Relative humidity and drying rate

The equilibrium moisture content of the product depends on the temperature and relative humidity (RH) of drying air during the drying process. The relative humidity within the product decreases as the drying rate increases. However, the drying rate and drying process are affected by ambient weather conditions that include temperature, sunlight hours and solar radiation intensity, cloud cover, and wind speed (Ssemwanga et al., 2020). The drying rate can be estimated as (Fuller, 2000).

$$\frac{dX}{dt} = \frac{h_{ht} A_p (T_a - T_{wb})}{L_e} \tag{8.29}$$

where $\frac{dX}{dt}$ is the drying rate (kg/s), T_a is the dry bulb temperature of the drying air (K), T_{wb} is the wet-bulb temperature of the drying air at the surface of the product (K), and L_e is the latent heat of evaporation of water (kJ/kg), which can be written as follows.

$$L_e = 4186.8(597 - 0.56T_a) \tag{8.30}$$

The relative humidity can be calculated within and around the dryer by the following equation (Khouya, 2020; Lamrani et al., 2019; Simo-Tagne, Zoulalian, Rémond, & Rogaume, 2018).

$$RH = \frac{WP_{atm}}{(W + 0.622)P_{sat}(T_a)} \tag{8.31}$$

where W is the absolute humidity of the air inside the drying chamber (kg water/kg dry air), P_{atm} is the atmospheric pressure (Pa), and P_{sat} is the saturated pressure (Pa) at a drying air temperature (T_a) calculated as

$$P_{sat}(T) = 10^{(7.625T_a/(T_a - 29.212)) - 3} \text{ for } 273\text{ K} \leq T_a \leq 413\text{ K} \tag{8.32}$$

Heat transfer coefficient

For a solar dryer system, knowing the convective heat transfer coefficient between the drying air and product inside the dryer is essential. The drying rate rises with increasing the coefficient of convective heat transfer. The coefficient can be determined using the correlation as follows (Khouya, 2020).

$$h_{ht} = \begin{cases} \frac{\lambda_a}{D_H}\left(0.066\mathrm{Re}^{0.5}\mathrm{Pr}^{0.33}\right) \text{for laminar flow } (\mathrm{Re} \leq 2100) \\ \frac{\lambda_a}{D_H}\left(0.023\mathrm{Re}^{0.8}\mathrm{Pr}^{0.3}\right) \text{ for turbulent flow } (\mathrm{Re} > 2100) \end{cases} \tag{8.33}$$

where Re is the Reynolds number, Pr is the Prandtl number, D_H is the hydraulic diameter (m) given by $D_H = 2e$ (where e is the stick thickness), λ_a is the air thermal conductivity (W/m K), which can be calculated according to the drying air temperature (T_a) as

$$\lambda_a = 4 \times 10^{-5} T_a + 0.0246 \tag{8.34}$$

Thermal efficiency

It is important to know the overall thermal efficiency of the drying operation. The efficiency can be calculated by taking into account the ratio of the energy necessary for the drying (i.e., to evaporate the moisture from the initial moisture content of the product until it reaches the equilibrium moisture content) to the thermal energy provided during the drying operation (Khouya & Draoui, 2019).

$$\eta_d = \frac{m_w \Delta H + m_w C_{p,w}\left(T_{w,f} - T_{w,i}\right)}{Q_u} \tag{8.35}$$

where η_d is the thermal efficiency of the solar dryer, m_w is the mass of water (kg), $C_{p,w}$ is the specific heat of water (kJ/kg K), $T_{w,f}$ is the final water temperature (K), and $T_{w,i}$ is the initial water temperature (K).

Unit cost of the solar drying

The unit expense of the solar product drying $\left(UC_{dry}\right)$ is the total expense over the year of a solar dryer divided by the quantity of dried crops by the dryer ($/kg), as expressed in Eq. (8.36) (Kandpal & Garg, 2003; Kumar & Kandpal, 2005).

$$UC_{dry} = \frac{f_{es} UE_{dry} p_c \left(CRF_{d,T} + m\right)}{365 CUF I \eta_d} \tag{8.36}$$

where p_c is the solar dryer cost per unit aperture area ($/m^2), m is the annual operation and maintenance costs of the solar dryer as a fraction of its capital cost, f_{es} is the fraction of useful energy requirement for drying supplemented by solar energy, CUF is the capacity utilization factor of the solar dryer, I is the design value of the daily solar radiation, UE_{dry} is the useful energy needed for drying a unit quantity of crop and is computed based on the dry mass

(Eq. 8.26) or wet mass (Eq. 8.28) basis, and $CRF_{d,T}$ is the capital recovery factor, which is formulated as

$$CRF_{d,T} = \frac{d(1+d)^T}{(1+d)^T - 1} \tag{8.37}$$

where d is the discount rate in fraction and T is the useful lifespan of the solar dryer (years).

Notably, the aperture area of a solar dryer to dry Q_{dry} quantity of wet product can be computed following Eq. (8.38) (Kumar & Kandpal, 2005).

$$A_a = \left(\frac{f_{es} Q_{dry} UE_{dry}}{365 ICUF \eta_d}\right) \tag{8.38}$$

where A_a is the aperture area of the solar dryer (m^2) and Q_{dry} is the potential quantity of a product for solar drying (kg).

Problem 8.1. For drying grapes raisin in a solar dryer, the required fraction of useful energy of 1 and the required useful energy of 1.91 MJ/kg for drying on a dry basis, the capacity utilization factor of 0.25, the cost of \$15/m^2 of aperture area, annual operation and maintenance cost of 1% of the capital investment of \$980, the average daily solar radiation of 19.8 MJ/m^2, the dryer thermal efficiency of 79%, and expected to be operational for 10 years. Estimate the unit expense of the solar product drying. Take a discount rate of 5%.

Solution. The capital recovery factor is given by

$$CRF_{d,T} = \frac{d(1+d)^T}{(1+d)^T - 1}$$

$$CRF_{d,T} = \frac{0.05(1+0.05)^{10}}{(1+0.05)^{10} - 1}$$

$$CRF_{d,T} = 0.13$$

The unit expense of the solar product drying is

$$UC_{dry} = \frac{f_{es} UE_{dry} p_c \left(CRF_{d,T} + m\right)}{365 CUFI \eta_d}$$

$$UC_{dry} = \frac{1 \times 1.91 \text{ MJ/kg} \times 15 \text{ \$/m}^2 (0.13 + 9.8\$)}{365 \times 0.25 \times 19.8 \text{ MJ/m}^2 \times 0.79}$$

$$UC_{dry} = 0.20 \text{ \$/kg}$$

8.4.3 Software tools in solar drying calculations

Simulation is an essential tool for designing and controlling the process of a drying system. Determining optimal parameters for optimum design and operating conditions can be performed via simulation. Various simulation tools such as CFD (Fluent/Ansys), Matlab, and Comsol Multiphysics are very useful in the development of mathematical models intending to predict air temperature, product temperature, water content, water evaporation, drying kinetics, and crop color for different solar drying systems. At the same time, Origin Pro, Sigma Plot, and SPSS can be beneficial for fitting the parameters of drying kinetics. One additional software named TRNSYS is a simulation program in the areas of solar energy, and the use of this program is to create and explain the drying characteristics of products in various types of dryers. The design parameters of a solar dryer system can be optimized using simulation tools well before the development begins, and they also save time and capital investment in solar drying systems (Chauhan et al., 2015).

8.5 Characteristics of the solar drying

8.5.1 Drying curves

The changes in moisture content can be obtained over time and are expressed as a curve in Fig. 8.8A. This curve illustrates how a substance reduces its moisture content, and the drying rate can be estimated from the curve. The drying rate curve varies from one material to another, owing to variations in material types, structures, and constituents. These drying rate curves play an essential role in understanding the drying characteristics of a product. Fig. 8.8B and C display the drying rate curve of a product over the drying time and water content, respectively. The characterizations of the drying process

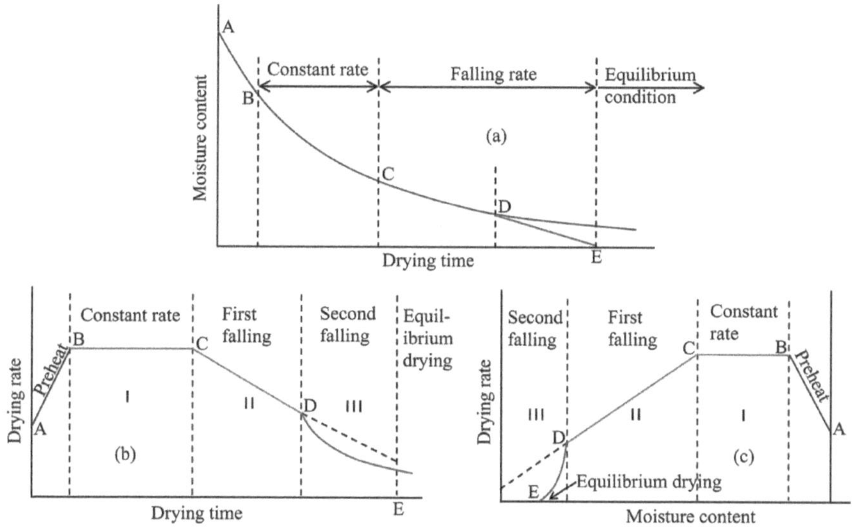

FIGURE 8.8 Drying curves.

can be performed, taking into account the different periods in which the rate of drying is different. Generally, three phases (or stages) can be identified where the rate of drying shows dissimilarities significantly due to the mechanisms associated with the internal process. In Fig. 8.8B and C, the initial phase (segment BC, phase I) is the constant rate period (drying occurs at a constant rate), the subsequent phase (segment CE, phase II) is the first falling rate period (drying rate reduces almost linearly), and the last phase (segment ED, phase III) is the second decreasing rate period. Note that section AB is the initial time needed to heat the substance to drying temperature. The drying periods in the drying of a product are explained as follows (Babalis et al., 2017; Belessiotis & Delyannis, 2011).

- Constant rate period (phase I): During this period, the surface of the product stays saturated with moisture, and the flow of moisture occurs smoothly through diffusion to the surface at a constant rate. At the end of this phase, the surface starts to be no longer saturated with humidity as the moisture comes to the critical point (i.e., point C in Fig. 8.8B and C, critical moisture content). The first falling rate period (phase II) then begins.
- First falling rate period (phase II): At this stage, the product moisture is in critical condition of moisture content, and no adequate amount of moisture is transferred to the surface with the help of capillary action to evaporate quickly from the inside of the product. Hence, the product surface is slowly dried, and the product's drying rate curve continuously falls.
- Second falling rate period (phase III): During this stage, the surface of the substance reaches nearly dry. The moisture circulation occurs at a decreased rate from the product's inside to the unsaturated surface by diffusion subjected to the concentration gradient. The gradient is produced by the depletion of water at the surface or as a result of capillary action. However, the water content continuously reduces until an equilibrium condition is reached and drying ends.

8.5.2 Drying kinetics

Drying kinetics can be interpreted as the dependency of attributes influencing the drying and drying rate. It explains the variation of a drying product's average water content as well as temperature over time, while drying dynamics represent the moisture content and temperature profiles in the drying substance. It is feasible to estimate the amount of water evaporated, drying period, useful energy required for drying, and other related parameters by knowing the drying kinetics of a substance. Drying kinetics of the product is the critical data necessary for designing and simulating an air dryer (Fontaine & Ratti, 1999). Also, it adequately describes the transport characteristics involved in drying, which include moisture transfer coefficient, heat transfer, water diffusion, and so on. A kinetic model (first-order), as shown below, may explain the removal of moisture during drying (Vijayan et al., 2017).

$$\frac{dX}{dt} = k(M - M_e) \tag{8.39}$$

where X is the moisture content (dry basis) of the material during drying (kg water/kg dry product), M_e is the equilibrium moisture content of dried material (kg water/kg dry product), k is the drying rate (1/min), and t is the drying time (min).

A slope of the falling rate drying curve may be used to estimate the drying rate of any product.

At $t = 0$, the water content of the material is the same as the initial water content $(M = M_i)$.

For water content, the following expression at a time t and M_t can be obtained by integrating Eq. (8.39).

$$M_t = M_e - (M_e - M_i)e^{-kt} \tag{8.40}$$

The moisture ratio (MR) can be written from Eq. (8.40) as

$$\mathrm{MR} = \frac{M_t - M_e}{M_i - M_e} = e^{-kt} \tag{8.41}$$

8.6 Role of solar drying systems

There are many advantages of using a solar energy drying system, and the benefits such as economic and environmental have got paramount importance. For assessing the solar drying system's commercial sustainability, giving attention to these aspects is necessary for developing a dryer.

8.6.1 Economic aspects

The solar energy drying system usually needs substantial investment. In order to justify whether such investment is economically viable, one typically estimates the return on investment for a solar drying system and compares it with the options available locally. The aim of the economic assessment of the solar dryer is to determine the payback period. For example, Hollick (Hollick, 1999) carried out an analysis of the economic performance of the solar drying system for crops and computed a total cost (purchase and installation) of \$42,850 for the solar dryer. The fuel-saving was 0.5 L/m^2 panel each day, and fuel (oil) consumption saving was 168,000 L per year. The cost of the oil was \$0.12 per liter, and the payback period was about two years. Notably, the payback time in this study was calculated only on the expenses of the solar panel, not on the dryer system as a whole. For determining the overall cost for developing a new solar dryer, other cost benefits, including enhanced quality of the final product, higher yields, less land use, faster drying, and so on, should be incorporated into the analysis in achieving a more precise cost-benefit analysis. Sreekumar et al. (Singh, Singh, & Dhaliwal, 2006) analyzed the economic evolution of an indirect solar dryer and estimated the payback period of 3.26 years. In another study, the payback period of the mixed-mode forced convection solar dryer is calculated as 0.65 years (Lakshmi, Muthukumar, Layek, & Nayak, 2019). This payback period is very much less than the total lifetime of the dryer, which is approximately 20 years. A very shorter payback time infers that the dryer will dry products for almost its entire lifespan without any cost (Singh et al., 2006).

8.6.2 Economic evaluation using dynamic method

The payback time is defined as the time required for the accumulated savings to equal the total initial investment (Duffie, Beckman, & Blair, 2020). The effect of inflation is considered by the dynamic method of economic evaluation. This method is summarized below, following the reference (Böer, 1978).

Payback happens if the cumulative savings S become equal to the sum of investment capital I and annual interest as well as the cumulative expenses E.

$$S = I + E \tag{8.42}$$

For a natural solar dryer, the annual cumulative savings (S) can be estimated by subtracting the mass and quality losses of the product from the net income D. In contrast, the savings (S) of semiartificial as well as solar-powered artificial dryers may be determined based on the cost of the alternative conventional energy D. Note that an increase in the market price of the product can compensate for the losses incurred in the case of a natural solar dryer. The calculation of yearly cumulative savings can be performed using the following equation.

$$S = \frac{(1+r)^n - (1+e)^n}{r-e} D \tag{8.43}$$

where r is the annual interest rate, e is the annual inflation rate for the price of energy, and n is the number of years for calculating the cumulative savings.

The sum of investment (I) can be calculated considering the initial investment cost (C) with interest and the number of operating years (n).

$$I = C(1+r)^n \tag{8.44}$$

Considering the annual fixed charge rate mC as well as the inflation rate i for equipment, the cumulative annual cost is

$$E = \frac{mC(1+r)^n - mC(1+i)^n}{r-i} \tag{8.45}$$

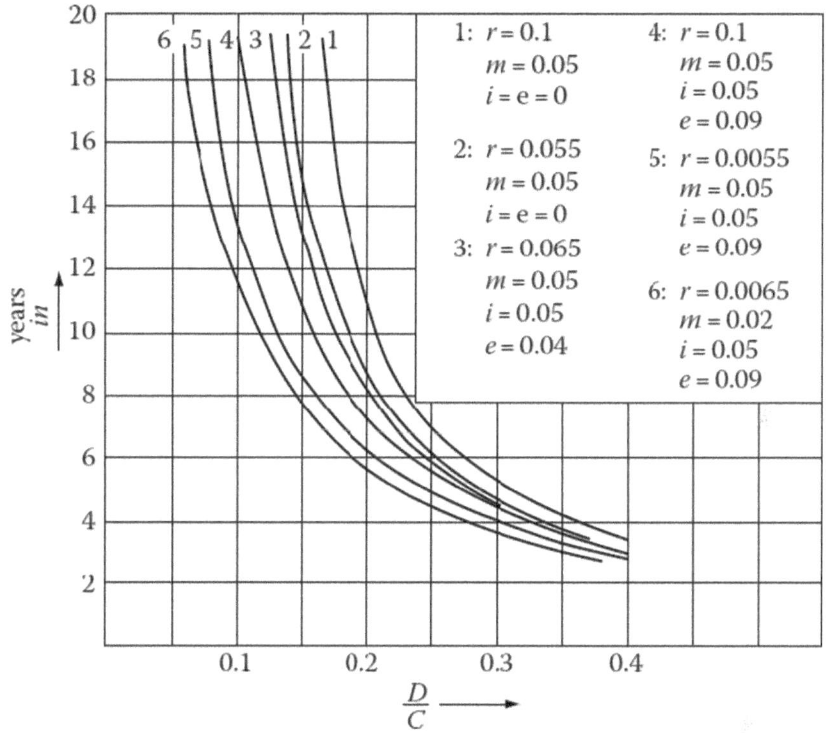

FIGURE 8.9 Variation in the payback period (n) in years with D/C, and various parameters r, m, i, and e (Böer, 1978).

Knowing the values of the required parameters such as C and D, a diagram, as shown in Fig. 8.9, can be made to calculate the payback period, referring to different known values of r, m, i, and e. However, if the values of the parameters are different from the values mentioned in the diagram, one then needs to estimate the payback time using Eqs. (8.42)–(8.45).

The payback period can also be calculated using Eq. (8.46), as mentioned by several authors in (Sreekumar, 2010; Sreekumar, Manikantan, & Vijayakumar, 2008).

$$n_p = \frac{\ln\left[1 - \frac{c}{s}(d - i)\right]}{\ln\left(\frac{1 + i}{1 + d}\right)} \tag{8.46}$$

where c is the capital cost ($), s is the saving ($) during the first year of the solar dryer, d is the interest rate on long-term investments, and i is the inflation rate.

Problem 8.2. The material and labor cost for the construction of the dryer is \$500, with an annual interest rate of 7%, an inflation rate of 4%, and the first-year saving of \$172. Calculate the payback period of the dryer.

Solution. Applying Eq. (8.46), the payback period is

$$n_p = \frac{\ln\left[1 - \frac{c}{s}(d - i)\right]}{\ln\left(\frac{1+i}{1+d}\right)}$$

$$n_p = \frac{\ln\left[1 - \frac{500\ \$}{172\ \$}(0.07 - 0.04)\right]}{\ln\left(\frac{1+0.04}{1+0.07}\right)}$$

$$n_p = 3.21 \text{ years}$$

8.6.3 Environmental aspects

In general, the environmental assessments of a solar system focus on the consumption and emission of energy and carbon dioxide (CO_2), respectively. Reducing both of them is the main advantage of a solar drying system. Also, this system saves energy, land area, and time, increases the quality of the product, ensures the drying process more efficient, and more importantly, it helps to protect the environment (Sharma, Chen, & Lan, 2009). An indirect solar drying system contributes to reducing energy consumption and preventing 34% of CO_2 emissions per year (Lamrani et al., 2019). In the case of solar crop drying, an aperture area of 1 m^2 of solar irradiation may prevent 463 kg of CO_2 from being produced during the calculated life cycle (Kumar & Kandpal, 2005). Nevertheless, environmental assessment data are not widely available, so it is essential to evaluate all phases of the system, including raw materials to construction, packaging, distribution, end-use, and disposal, based on comprehensive assessments, especially using life-cycle assessments (Berville, Croitoru, & Nastase, 2019).

8.7 Application of solar dryers

Thermal drying is a vital operation in various industrial, agricultural, and marine processes — ranging from vegetables and fruits to wood and textile manufacture and from polymer and ceramics to pharmaceutical and mineral processing. Solar drying in horticulture also plays an essential role in developing countries, especially in many rural locations where most areas do not have electricity connected to the grid and provisions of other nonrenewable energy sources or unable to afford heaters fuel to provide hot air for drying crops (Pirasteh et al., 2014).

8.7.1 Application of solar dryer in agriculture

A direct solar dryer can be used to dry small quantities of wheat, bananas, coffee, nuts, and vegetables. The low-cost natural convection solar dryers (indirect-type) can be used to dry cauliflower, cassava, grapes, bananas, green peas, rough rice, strawberries, and mangoes. In contrast, a mixed-mode dryer (driven by a fan) can be applied for agricultural products such as tomatoes, mango slices, and grains, which need to dry at ambient temperature in a short time (Mustayen, Mekhilef, & Saidur, 2014). The mixed-mode solar dryer with forced convection is generally useful, especially in humid tropical regions. Agricultural products such as tea, tobacco leaves, onion slices, turmeric rhizomes, pineapple, and so on can be dried applying the hybrid solar dryer.

8.7.2 Application of solar dryer in industry

In industry, solar drying may be applied to various materials, which include solid biomass, leather, polymers, textiles, pulp and paper, timber and wood, and various processes such as porous materials drying, processes in pharmaceutical manufacturing, and wastewater treatment (Sandali et al., 2019). Industrial applications also include the drying of cement, bricks, sewage sludge, tea, food, breweries, tobacco, and dairy products.

8.7.3 Application of solar dryer in marine

Marine products generally deteriorate very rapidly and can be processed through canning, freezing, salting, and dehydrating (Posomboon, 1998). The preservation of about 40% of fish landing in most developing countries is achieved either by solar drying or salt curing or drying (Visavale, 2012). Solar tunnel drying can be used as an effective process of drying silver jewfish (Bala & Mondol, 2001) and tilapia fish (Kituu et al., 2010), along with pretreatment with salting. The solar cabinet dryer can be applied to dry salted shark fillets. Other marine products such as prawns, Bombay duck, chela fish, and so on may also be dried using the solar dryer.

References

Aboul-Enein, S., El-Sebaii, A., Ramadan, M., & El-Gohary, H. (2000). Parametric study of a solar air heater with and without thermal storage for solar drying applications. *Renewable Energy, 21*(3–4), 505–522.

Amer, B. M., Gottschalk, K., & Hossain, M. (2018). Integrated hybrid solar drying system and its drying kinetics of chamomile. *Renewable Energy, 121*, 539–547.

ASHRAE. (1974). *ASHARE hand book and product dictionary: Application.* New York, USA: American Society of Heating, Refrigerating and Air-Conditioning Engineers (ASHRAE).

Babalis, S., Papanicolaou, E., & Belessiotis, V. (2017). Fundamental mathematical relations of solar drying systems. In *Solar drying technology* (pp. 89–175). Springer.

Bala, B., & Mondol, M. (2001). Experimental investigation on solar drying of fish using solar tunnel dryer. *Drying Technology, 19*(2), 427–436.

Belessiotis, V., & Delyannis, E. (2011). Solar drying. *Solar Energy, 85*(8), 1665–1691.

Berville, C., Croitoru, C.-V., & Nastase, I. (2019). Recent advances in solar drying technologies-A short review. In *Paper presented at the international conference on energy and environment.*

Böer, K. (1978). Payback of solar systems. *Solar Energy, 20*(3), 225–232.

Chauhan, P. S., Kumar, A., & Tekasakul, P. (2015). Applications of software in solar drying systems: A review. *Renewable and Sustainable Energy Reviews, 51*, 1326–1337.

Djebli, A., Hanini, S., Badaoui, O., Haddad, B., & Benhamou, A. (2020). Modeling and comparative analysis of solar drying behavior of potatoes. *Renewable Energy, 145*, 1494–1506.

Duffie, J. A., Beckman, W. A., & Blair, N. (2020). *Solar engineering of thermal processes, photovoltaics and wind.* John Wiley & Sons.

Ekechukwu, O. (1999). Review of solar-energy drying systems I: An overview of drying principles and theory. *Energy Conversion and Management, 40*(6), 593–613.

El-Sebaii, A., & Shalaby, S. (2012). Solar drying of agricultural products: A review. *Renewable and Sustainable Energy Reviews, 16*(1), 37–43.

Fontaine, J., & Ratti, C. (1999). Lumped-parameter approach for prediction of drying kinetics in foods. *Journal of Food Process Engineering, 22*(4), 287–305.

Fuller, R. (2000). Solar drying-a technology for sustainable agriculture and food production. *Solar Energy Conversion and Photoenergy Systems, 25.*

Green, M. G., & Schwarz, D. (2001). *Solar drying technology for food preservation.* Gate information Service Gtz.

Guiné, R.d. P. F., & Barroca, M. J. (2017). Drying kinetics in solar drying. In *Solar drying technology* (pp. 317–355). Springer.

Hollick, J. (1999). Commercial scale solar drying. *Renewable Energy, 16*(1–4), 714–719.

Imre, L. (2014). Solar drying. In *Handbook of industrial drying* (pp. 304–346). CRC Press.

Jain, D. (2005). Modeling the system performance of multi-tray crop drying using an inclined multi-pass solar air heater with in-built thermal storage. *Journal of Food Engineering, 71*(1), 44–54.

Kalogirou, S. A. (2014). Industrial process heat, chemistry applications, and solar dryers. *Solar Energy Engineering*, 379–429.

Kamarulzaman, A., Hasanuzzaman, M., & Rahim, N. A. (2021). Global advancement of solar drying technologies and its future prospects: A review. *Solar Energy, 221*, 559–582. https://doi.org/10.1016/j.solener.2021.04.056

Kandpal, T. C., & Garg, H. P. (2003). *Financial evaluation of renewable energy technologies.* MacMillam India Limited.

Khouya, A. (2020). Effect of regeneration heat and energy storage on thermal drying performance in a hardwood solar kiln. *Renewable Energy, 155.*

Khouya, A., & Draoui, A. (2019). Computational drying model for solar kiln with latent heat energy storage: Case studies of thermal application. *Renewable Energy, 130*, 796–813.

King, C. J. (1977). Heat and mass transfer fundamentals applied to food engineering. *Journal of Food Process Engineering, 1*(1), 3–14.

Kituu, G. M., Shitanda, D., Kanali, C., Mailutha, J., Njoroge, C., Wainaina, J., & Silayo, V. (2010). Thin layer drying model for simulating the drying of Tilapia fish (*Oreochromis niloticus*) in a solar tunnel dryer. *Journal of Food Engineering, 98*(3), 325–331.

Kumar, A., & Kandpal, T. C. (2005). Solar drying and CO_2 emissions mitigation: Potential for selected cash crops in India. *Solar Energy, 78*(2), 321–329.

Lakshmi, D., Muthukumar, P., Layek, A., & Nayak, P. K. (2019). Performance analyses of mixed mode forced convection solar dryer for drying of stevia leaves. *Solar Energy, 188*, 507–518.

Lamrani, B., Khouya, A., & Draoui, A. (2019). Energy and environmental analysis of an indirect hybrid solar dryer of wood using TRNSYS software. *Solar Energy, 183*, 132–145.

Mennouche, D., Bouchekima, B., Boubekri, A., Boughali, S., Bouguettaia, H., & Bechki, D. (2014). Valorization of rehydrated Deglet-Nour dates by an experimental investigation of solar drying processing method. *Energy Conversion and Management, 84*, 481–487.

Motevali, A., Minaei, S., & Khoshtagaza, M. H. (2011). Evaluation of energy consumption in different drying methods. *Energy Conversion and Management, 52*(2), 1192–1199.

Mujumdar, A. S. (2014). Principles, classification, and selection of dryers. In *Handbook of industrial drying* (pp. 4–28). CRC Press.

Mujumdar, A. S., & Beke, J. (2003). *Grain drying: Basic principles. Handbook of postharvest technology: Cereals, fruits, vegetables, tea, and spices.*

Mustayen, A., Mekhilef, S., & Saidur, R. (2014). Performance study of different solar dryers: A review. *Renewable and Sustainable Energy Reviews, 34*, 463–470.

Nair, P., Espinosa-Santana, A. L., Guerrero-Martínez, L., López-Ortiz, A., & Nair, M. (2020). Prospects toward UV-blue filtered solar drying of agricultural farm produce using chemically deposited copper chalcogenide thin films on cellular polycarbonate. *Solar Energy, 203*, 123–135.

Norton, B. (2017). Characteristics of different systems for the solar drying of crops. In *Solar drying technology* (pp. 69–88). Springer.

Pardhi, C. B., & Bhagoria, J. L. (2013). Development and performance evaluation of mixed-mode solar dryer with forced convection. *International Journal of Energy and Environmental Engineering, 4*(1), 23.

Pirasteh, G., Saidur, R., Rahman, S., & Rahim, N. (2014). A review on development of solar drying applications. *Renewable and Sustainable Energy Reviews, 31*, 133–148.

Posomboon, W. (1998). *Processing effect on quality of dried shrimp* (Master of Engineering). Thailand: Asian Institute of Technology.

Sandali, M., Boubekri, A., & Mennouche, D. (2019). Improvement of the thermal performance of solar drying systems using different techniques: A review. *Journal of Solar Energy Engineering, 141*(5).

Sandali, M., Boubekri, A., Mennouche, D., & Gherraf, N. (2019). Improvement of a direct solar dryer performance using a geothermal water heat exchanger as supplementary energetic supply. An experimental investigation and simulation study. *Renewable Energy, 135*, 186–196.

Schweiger, H., Mendes, J. F., Benz, N., Hennecke, K., Prieto, G., Cusí, M., & Gonçalves, H. (2000). *The potential of solar heat in industrial processes. A state of the art review for Spain and Portugal.*

Sharma, A., Chen, C., & Lan, N. V. (2009). Solar-energy drying systems: A review. *Renewable and Sustainable Energy Reviews, 13*(6–7), 1185–1210.

Simo-Tagne, M., Zoulalian, A., Rémond, R., & Rogaume, Y. (2018). Mathematical modelling and numerical simulation of a simple solar dryer for tropical wood using a collector. *Applied Thermal Engineering, 131*, 356–369.

Singh, P. P., Singh, S., & Dhaliwal, S. (2006). Multi-shelf domestic solar dryer. *Energy Conversion and Management, 47*(13–14), 1799–1815.

Sreekumar, A. (2010). Techno-economic analysis of a roof-integrated solar air heating system for drying fruit and vegetables. *Energy Conversion and Management, 51*(11), 2230–2238.

Sreekumar, A., Manikantan, P., & Vijayakumar, K. (2008). Performance of indirect solar cabinet dryer. *Energy Conversion and Management, 49*(6), 1388–1395.

Ssemwanga, M., Makule, E., & Kayondo, S. (2020). Performance analysis of an improved solar dryer integrated with multiple metallic solar concentrators for drying fruits. *Solar Energy, 204*, 419–428.

Tonui, K., Mutai, E., Mutuli, D., Mbuge, D., & Too, K. (2014). *Design and evaluation of solar grain dryer with a back-up heater.*

Vijayan, S., Arjunan, T., & Kumar, A. (2017). Fundamental concepts of drying. In *Solar drying technology* (pp. 3–38). Springer.

Visavale, G. (2012). Principles, classification and selection of solar dryers. In C. L. Hii, S. P. Ong, S. V. Jangam, & A. S. Mujumdar (Eds.), *Solar drying: Fundamentals, Applications and innovations* (pp. 1–50). Published in Singapore.

Weiss, W., & Buchinger, J. (2012). *Solar drying* (Vol. 19). Feldgasse: AEE Intec Publication, A-8200 Gleisdorf.

Xu, Z., Leong, S. Y., Farid, M., Silcock, P., Bremer, P., & Oey, I. (2020). Understanding the frying process of plant-based foods pretreated with pulsed electric fields using frying models. *Foods, 9*(7), 949.

Chapter 9

Solar water desalination system

M. Asaduzzaman Faysal[1], Nahin Tasmin[2], M.M. Rashid[1] and Md Hasanuzzaman[3]

[1]*Department of Mechatronics Engineering, International Islamic University of Malaysia, Kuala Lumpur, Malaysia;* [2]*Department of Mechanical Engineering, Rajshahi University of Engineering & Technology, Rajshahi, Bangladesh;* [3]*Higher Institution Centre of Excellence (HICoE), UM Power Energy Dedicated Advanced Centre (UMPEDAC), University of Malaya, Kuala Lumpur, Malaysia*

Chapter outline

Technologies for Solar Thermal Energy. https://doi.org/10.1016/B978-0-12-823959-9.00007-6

9.1 Basic concept of solar water desalination system

Desalination specifically refers to the removal of minerals and contaminated substances from saltwater. Although water occupies around 71% of the earth's surface, around 97% of the total water reservoir of the earth is ocean water, leaving us with only 3% of freshwater. Nevertheless, about 2.5% of the reserved freshwater is unavailable being in a frozen condition in glaciers, ice caps or is confined in soil, or atmosphere, or too deep inside the earth surface. Thus, only less than 1% of the reserved freshwater is drinkable which signifies the gravity of careful consumption of scarce freshwater supply as well as the purification of inconsumable saline water via desalination systems. The abundance of solar rays over the earth's atmosphere makes it highly judicial and beneficial to be implemented in saline water desalination systems. Although large-scale centralized desalination plants are the most typical systems, recently small-scale decentralized desalination plants are gaining more attention considering several facts. Large-scale centralized desalination plants require high maintenance and operating costs, however, such systems often fail to sustain the quality of water owing to adverse maintenance and functioning. Also, it is seldom unfeasible to ensure the supply of uncontaminated pure water to the remotest and poverty-stricken areas with adverse infrastructure. Here come the considerations of the decentralized desalination plants. The advantages of decentralized desalination plants are that such systems require less space thus reducing the construction cost, better maintenance, and operation, improved community relation, reduced chance of impurity, can more easily be sited directly adjacent to demand minimizing pipeline construction. In addition, an environmental gain of decentralized desalination is that the brine discharge is spread out over a giant area, and accordingly the environmental have an effect is considerably less than that related to giant-scale centralized desalination plants. Inflated cost of fossil fuels, environmental pollution, decreasing cost, and availability of renewable energy sources are factors propelling the implementation of small-scale decentralized solar-powered desalination plants. According to recent estimations, in favor of the highest, fossil fuels will last no longer than approximately a 100 years (Abed, Eleiwi, Hasanuzzaman, Islam, & Mohammed, 2020; Ritchie & Roser, 2021; Taylor, 2002). So, the spreading of solar-based technologies in all

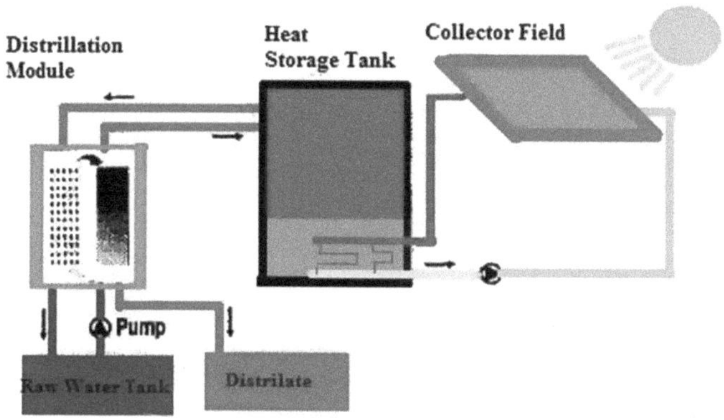

FIGURE 9.1 Schematic diagram of solar water desalination system (Müller-Holst et al., 1999).

technological respects is inevitable. It is possible to attain a high-level pure water production at a considerably low cost and with efficient energy utilization by competently combining renewable energy and renewable energy (Qiblawey & Banat, 2008). The schematic diagram of a solar water desalination system is presented in Fig. 9.1(Müller-Holst, Engelhardt, & Schölkopf, 1999).

There are many different types of water desalination technologies. The water desalination processes are primarily classified into two groups (Chandrashekar & Yadav, 2017):

- thermal desalination processes/phase change processes
- membrane desalination/single-phase processes.

Thermal desalination processes involve the application of heat sources either from fossil fuels, nuclear energy, or renewable energy, such as solar or geothermal energy to remove concentrated particles from saline/salty water. On the other hand, membrane desalination systems include the use of semipermeable membranes to make impure water pass through them.

9.2 Solar water desalination processes

The water desalination process by utilizing solar energy can be performed in two modes: direct and indirect. In direct mode, the saline water absorbs the solar energy via a solar collector whereas in indirect mode the solar energy is converted to thermal energy in one system that is further integrated into another subsystem for water distillation.

Examples of direct solar water desalination processes include solar stills that are further categorized as active and passive groups. In an active solar still, an external heating source like solar collectors, waste heat, boilers, or by using preheated water while in a passive system there is no requirement of additional

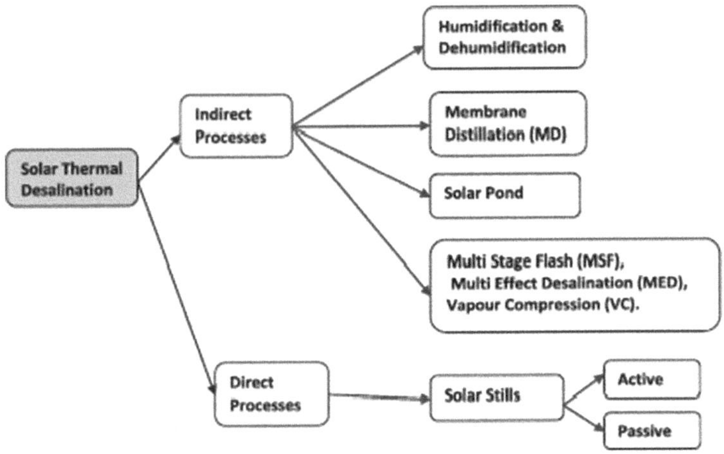

FIGURE 9.2 Possible configurations of solar energy resources with desalination technologies (Chandrashekara & Yadav, 2017).

heat sources (Li, Yuan, Wang, Li, & Xu, 2014; Sarkar, Sifat, Reza, & Sadique, 2017). Examples of indirect solar water desalination processes are multistage flash (MSF), membrane distillation (MD), multieffect desalination (MED), and thermal vapor compression (TVC). Fig. 9.2 illustrates the broad classification of solar thermal desalination systems.

9.3 Indirect processes

Indirect solar desalination processes comprise two system components namely a solar collector and a desalination section. The solar collectors used in an indirect desalination process include a broad range.

Concentrating solar collectors (Kumar, Hasanuzzaman, & Rahim, 2019):

- Parallel trough collectors,
- Linear Fresnel collectors,
- Transpired solar collectors.

Nonconcentrating solar collectors (Kumar et al., 2019):

- Flat plate,
- Evacuated tube,
- Heat pipe collectors.

Indirect solar desalination systems can be powered mechanically, electrically, or thermally. Important types of indirect solar desalination systems include (Narayan et al., 2010):

- Vapor compression (VC),

- Membrane distillation (MD),
- Multistage flash (MSF), and
- Multiple effect distillation (MED)

9.3.1 Solar humidification and dehumidification desalination

Solar humidification and dehumidification system follows the very basic natural water purification system (Kalidasa Murugavel & Srithar, 2011; Narayan et al., 2010). The principle of the solar humidification and dehumidification desalination (SHDD) system is basically the evaporation of salty water into an airstream and then the successive condensation of the air stream to collect pure water. Air can hold water molecules and this capacity increases with increasing temperature. For an increase of about 20°C in the temperature, there is an increase of around 400 gr water vapor per kg of dry air.

Fig. 9.3 presents an SHDD system integrated with a solar collector. The major segments of an SHDD system are a humidifier, a dehumidifier, and a solar collector as presented in Fig. 9.3 (Narayan et al., 2010; Nassar, Yousif, & Salem, 2007). Air passes through a solar air collector and gets heated. The warm dry air then enters into the humidifier and passes through the sprayed salty water and grabs water molecules. The heated humid air then flows through the dehumidifier where the water molecules get condensed to provide potable water.

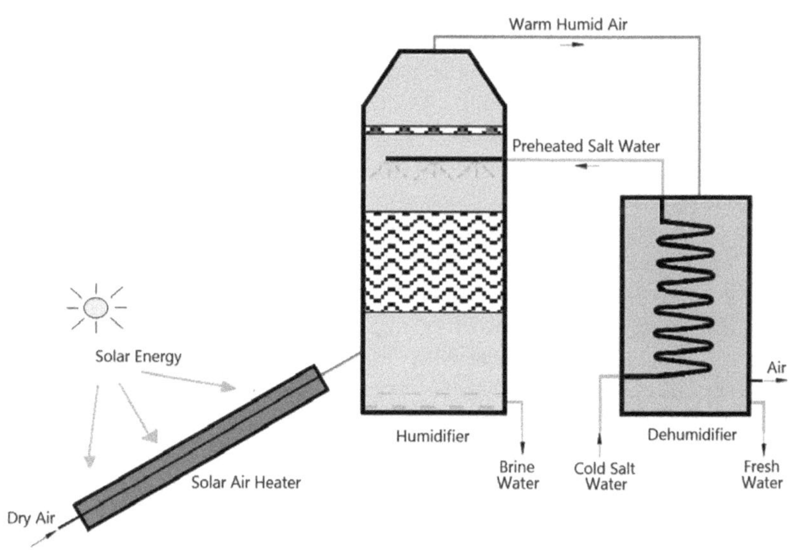

FIGURE 9.3 Humidification and dehumidification desalination unit coupled with solar collector (Narayan et al., 2010).

9.3.2 Membrane distillation

MD is a procedure to remove impurities (salt and minerals) from salty/brackish water as it passes through a semipermeable membrane. Usually, hydrophobic or nonwetting membranes are used for this purpose which is made of polypropylene (PP), polyvinylidene fluoride, polyethylene, or polytetrafluoroethylene (Sharon & Reddy, 2015). In the MD system, solar energy heats the feed water which then passes through the porous hydrophobic membrane, either in molecular/ionic form. The flow of water through the membrane happens due to the difference in vapor pressure or electric potential (Chen, Jiang, Ye, Yang, & Hou, 2019; Sarkar et al., 2017). Hydrophobic materials with excellent porosity and less thermal conductivity are ideal to be used as membranes. Large pore sizes with a decreased thickness of membrane enhance the flux. The continuing operation, flow rate, porosity, thermal conductivity, and membrane width have a considerable effect on the system operation.

9.3.3 Multistage flash

As the name suggests, in a MSF system seawater is made to flash to vapor in stages on its way through a number of adjoining chambers/vessels, each with gradually decreasing pressure. These chambers are also called stages where the seawater vapor suddenly evaporates and subsequently gets condensed in the contact with the bundle of heat exchanger tubes above (Condenser).

In a solar-powered MSF process, a solar collector is employed to provide additional heat to the cold seawater coming through the colder to hotter chambers of the system. Thus, the seawater is heated above its saturation point and is then made to flow through the vessels. In the vessels, the low pressure is maintained using a vacuum pump. As the seawater passes through the stages, by sudden low pressure it flashes to steam and then gets condensed by releasing its heat to the condenser. As the seawater moves from one step to the subsequent step, flashing takes place continually and at a gradually decreased pressure removing the need for additional heat.

The liquid condensates are collected in a freshwater collector and are taken out using control valves. The efficiency of an MSF plant increases with the number the stages but with an increasing installation cost. So, there are about 19–30 stages in a regular large MSF plant (Kalidasa Murugavel & Srithar, 2011; Narayan et al., 2010). Fig. 9.4 presents the schematic of a solar-based MSF.

9.3.4 Vapor compression

Solar energy heats saline water and then is flashed into vapor in a container in a vapor compression system, as presented in Fig. 9.5. The produced vapor is

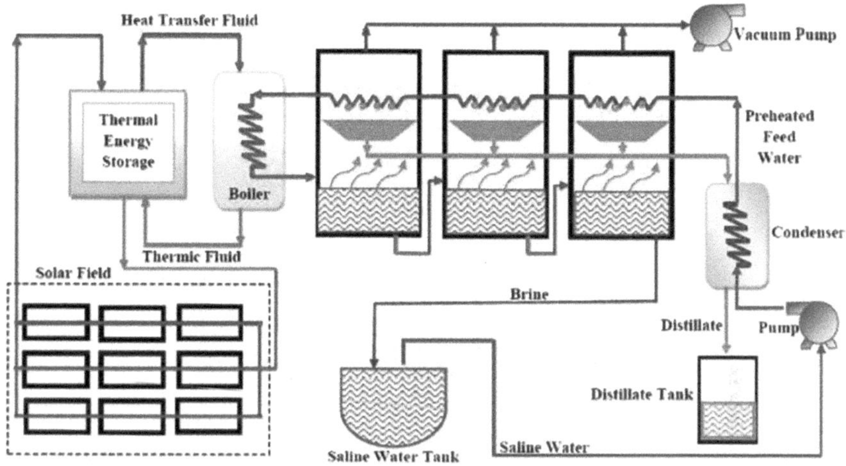

FIGURE 9.4 Solar-based multistage flash system (Kalidasa Murugavel & Srithar, 2011).

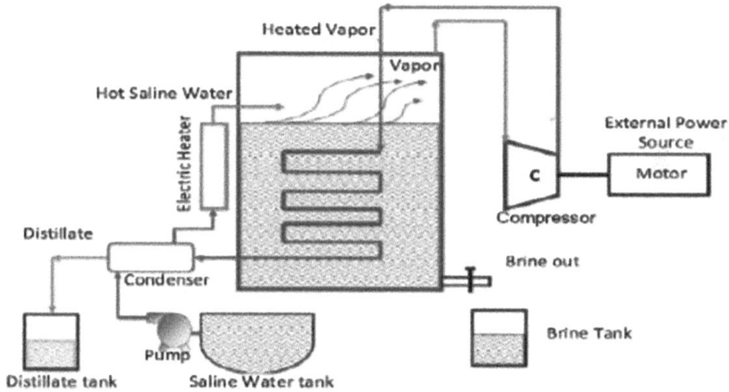

FIGURE 9.5 Mechanical vapor compression desalination system (Sharon & Reddy, 2015).

then compressed in a mechanical or thermal vapor compressor. As a result, both the pressure and temperature of the vapor increase. In turn, the saline water is preheated by utilizing the pressurized vapor.

9.3.5 Solar pond

Solar ponds are thermal devices like large-scale solar collectors that can absorb and store solar energy for later use. Solar ponds can even be used for constant and continuous water generation with thermal energy storage during nights or adverse weather conditions and are less expensive compared to other solar technologies. Solar ponds can also be used to power other conventional

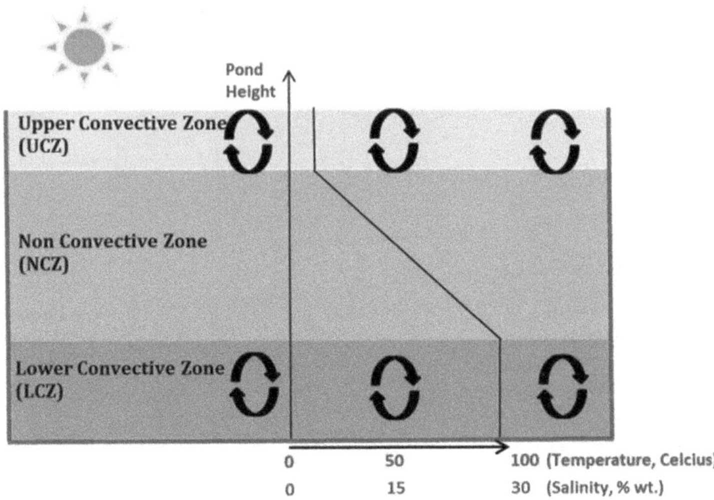

FIGURE 9.6 Schematic of a solar pond (Ding, Akbarzadeh, & Tan, 2018).

water desalination systems. Fig. 9.6 presents a solar pond with three layers namely the upper convective zone, nonconvective middle zone, and the lower convective zone. The salt concentration increases gradually with the depth of the pond and is the highest at the lowest level. The function of the upper zone is to provide heat insulation and to maintain the equilibrium of the solution in the pond. The middle zone prevents heat losses from the bottom zone. The bottom zone absorbs heat and acts as thermal storage. When the temperature of the upper layer is nearly ambient, the bottom layer could achieve the highest temperature of about 90°C. Small solar ponds are suitable for domestic appliances while large solar ponds are used for industrial purposes such as the generation of electricity, heating, and desalination (Esmaeilion, 2020). It has been found that based on obtained results for desalinating seawater, producing each cubic meter of seawater needs 1000–4000 m^2 surface of solar pond per day.

9.3.6 Multieffect distillation

In a MED system, steam is produced in several vessels arranged in series that is shown in Fig. 9.7. A solar-assisted MED system can use both thermal vapor compression (TVC) and mechanical vapor compression (MVC) system. In the MVC system, a mechanical compressor is included. The schematic of a TVC based MED system is illustrated in Fig. 9.7. The system comprises several chambers in series containing internal heat exchangers. Solar energy is used to produce steam and is made to pass through a thermocompressor. Some portions of the vapor from the last chamber in the series are consumed by the

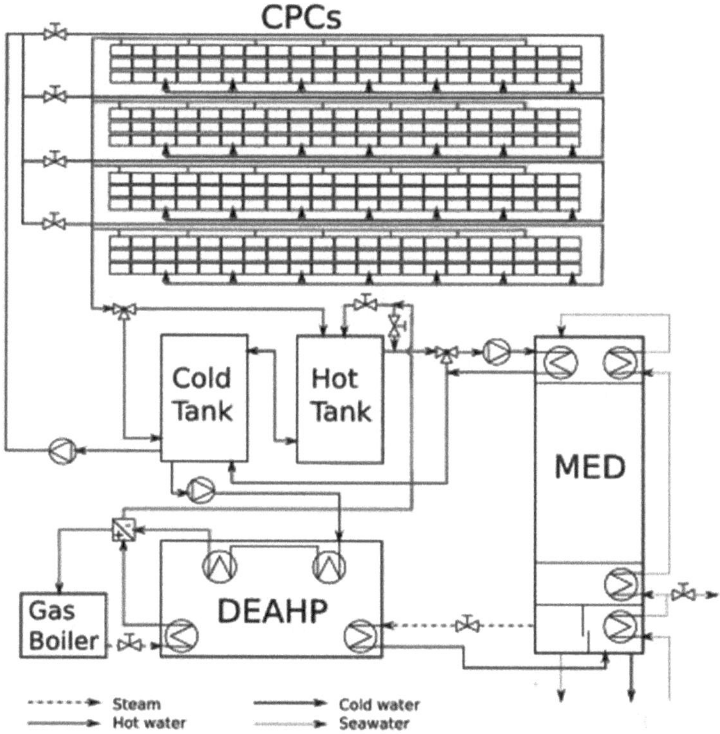

FIGURE 9.7 A solar-assisted multieffect distillation system (de la Calle, Bonilla, Roca, & Palenzuela, 2015).

thermocompressor. As the heated steam passes through the heat exchanger inside the chamber, saltwater is sprayed over the tube bundles and thus heat exchange takes place. The steam flowing through the tubes gets condensed in exchange for the latent heat of condensation to the saltwater spray. The vapor produced in one chamber passes to the next chamber until the last one and finally gets consumed by the thermocompressor to restart the cycle.

Note: CPC: Compound parabolic collector; DEAHP: Double effect absorption heat pump.

9.4 Direct processes

As the name suggests direct desalination system refers to the direct absorption of solar energy by saline water. In this system, both condensation and evaporation happen in the same apparatus. Direct desalination systems are mostly used for residential and small community purposes with a water generation capacity of up to 200 m^3/day. Examples include solar stills, water cones, water pyramids, etc. Some of the direct processes are discussed in the following sections.

9.4.1 Solar stills

The most basic and simplest forms of direct solar desalination system are solar stills excluding the requirement of high-grade energy and so are environmentally friendly. The construction and operation principles of solar stills are quite easy. As presented in Fig. 9.8, a basin with a translucent inclined top cover contains saline water up to a certain level. Generally, the basin is colored in black or dark color to enhance absorptivity. As the solar radiation enters through the cover, the basin absorbs heat causing the water to evaporate. The water vapor then rises up and gets condensed in contact with the transparent top cover. Because of the gravitational force, the vapor condensation gets accumulated at the bottom of the sloped glass cover (Chen et al., 2019; Sarkar et al., 2017).

9.4.2 Bucket defluoridator

Defluoridation is the process of reducing fluoride content in water as an excess amount of fluoride in water can lead to dental and bone problems. There are four types of defluoridation systems as given below (Chandrashekara & Yadav, 2017):

Adsorption methods: These methods include the use of bone charcoal, activated alumina, and clay as sorbents in a batch or column system.

Ion exchange methods: Mercantile ion exchange resins are used in these methods which are also costly.

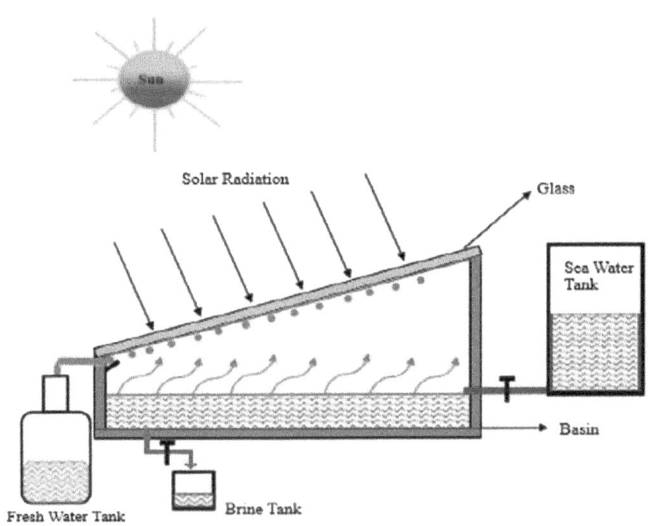

FIGURE 9.8 Simple solar still (Sharon & Reddy, 2015; Sharshir, Yang, Peng, & Kabeel, 2016).

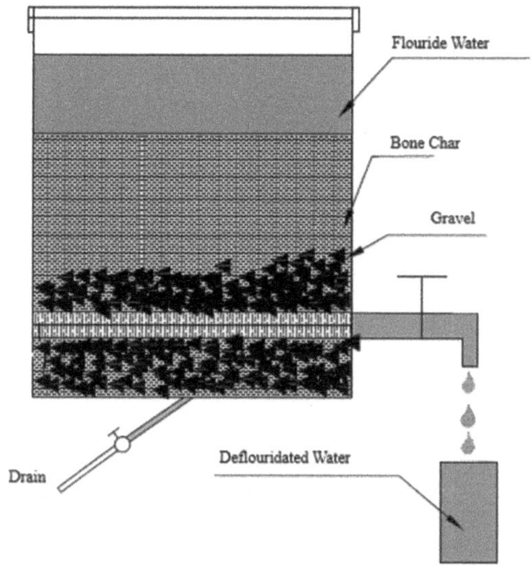

FIGURE 9.9 Bucket defluoridator (Chandrashekara & Yadav, 2017).

Coprecipitation and contact methods: These methods coprecipitate or precipitate fluoride ions with aluminum sulfate and lime or calcium and phosphate compounds, respectively.

Membrane processes: Such processes include electrodialysis, reverse osmosis, and nanofiltration methods.

Fig. 9.9 shows a bucket defluoridator system with bone char, gravel, and fluoride water arranged in a column in a plastic bucket (Chandrashekara & Yadav, 2017; Olkis, Brandani, & Santori, 2019). A PVC pipe has been used as a drain. A tap is placed about 3–4 cm above the bottom for directly collecting purified water. Such systems are one of the low-cost methods of removing fluoride and are suitable for residential and small-scale usage. The setup operates with high efficiency to reduce the original contents from the water.

9.4.3 Still with a separate section of vapor condensation and condensate collection

As the name refers to, such kind of system comprises two segments: an evaporator and a condenser section to reach the optimum condensation rate. Fig. 9.10A presents the schematic diagram of still with a separate section of vapor condensation and condensate collection. Fig. 9.10B presents an increase in heat transfer rate with rectangular cross-section channels.

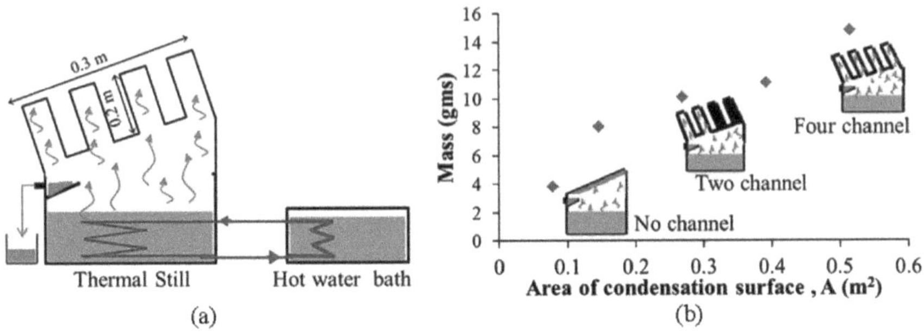

FIGURE 9.10 Schematic diagram, (A) Experimental setup and (B) Result with increasing condensation area (Chandrashekara & Yadav, 2017).

9.4.4 Solar still coupled with flat-plate collector

When combined with solar collectors, the performance of solar stills gets enhanced. Fig. 9.11 shows such a system. Usually, pump-driven water circulation than thermosyphon as well as a different condensation segment provides better functions. Also, wicking materials like a sponge can be added to increase the water production rate.

9.5 Characteristics of solar water desalination system

Conventional desalination systems include the following limitations (Goosen, Mahmoudi, & Ghaffour, 2010):

(a) Consumption of high-grade energy
(b) Release of contaminants
(c) Corrosion, salt, or scale accumulation over heat exchangers and blockage of membrane and mass exchangers with high maintenance cost.

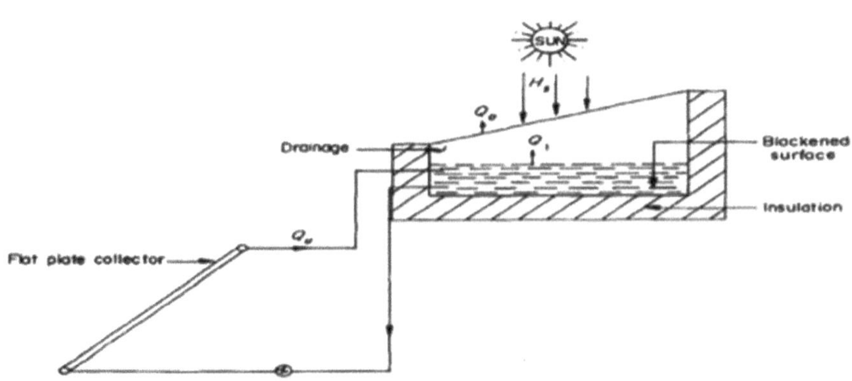

FIGURE 9.11 Schematic diagram of active solar still coupled with a flat-plate collector (Sathyamurthy et al., 2017).

9.5.1 Adsorption desalination system

To address the problems associated with conventional desalination systems, the adsorption desalination (AD) system has attained popularity. Such systems have several advantages:

- AD systems are low temperature (50°C) devices and so can be powered by renewable energy sources (solar, geothermal, etc);
- Due to an excellent adsorption ability, no vacuum pump is required;
- AD systems can provide high-quality drinking water by disinfecting it using porous adsorbent material.

9.5.2 Physical and schematic diagram of AD system

The physical and schematic diagram of a pilot AD System is presented in Figs. 9.12 and 9.13. As depicted, the system comprises two beds A and B, an evaporator, a condenser, a cooling tower, a water-heating tank, a seawater tank, vacuum pumps, water pumps, metal brackets, flowmeters, valves, pressure gauges, and flowmeters. The functions of the components are as follows.

Beds A and B: These are the most important parts of an AD system that contains the adsorbents.

Falling-film evaporator: This is used to heat and circulate salty water.

Condenser: This is used to condensate the distilled water vapor.

Metal brackets: These are mainly finned tube heat exchangers that carry the adsorbent particles in the fin clearance.

Steam conduits: For maintaining the even flow of adsorbent particles entering the beds.

Water pump: Used to circulate seawater to the evaporator.

Water-heating tank: Supplying heated water to heat the seawater.

Cooling tower: Providing cooling water to cool the water vapor.

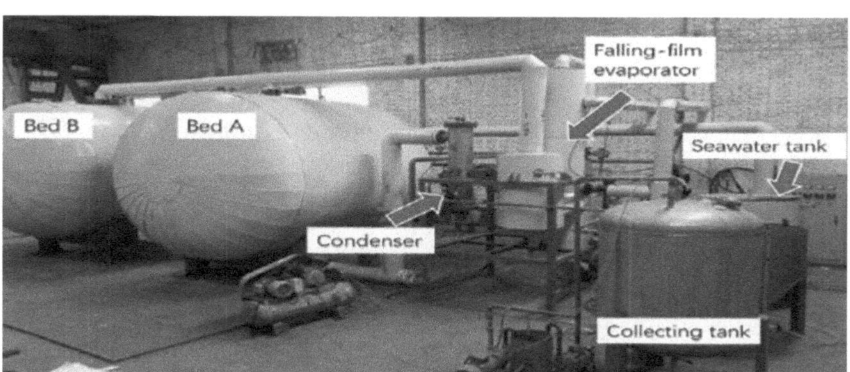

FIGURE 9.12 Physical diagram of the AD system (Zhang et al., 2020).

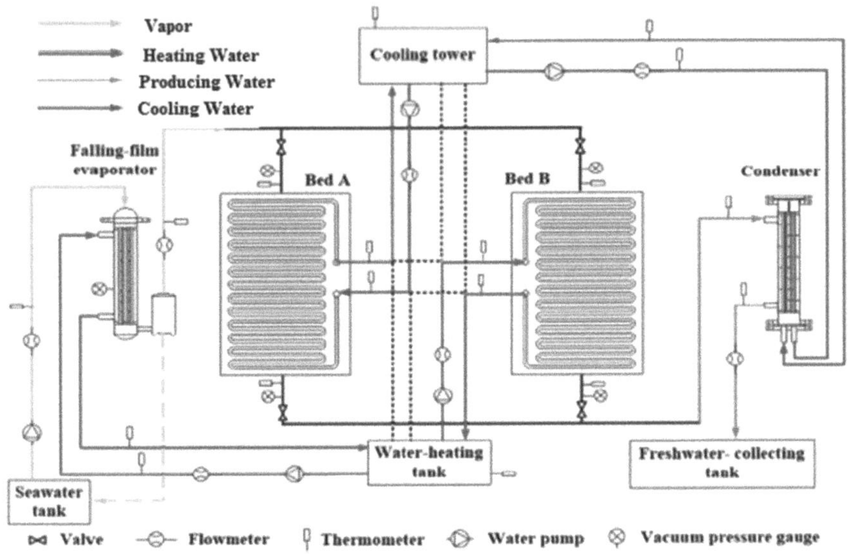

FIGURE 9.13 Schematic diagram and instruments location of the AD system (Zhang et al., 2020).

Working Principle:

A water pump continuously provides seawater to the falling-film evaporator. There the seawater is heated with the aid of hot water from the water-heating tank. Part of the heated water gets evaporated in the evaporator at a negative pressure attained by a vacuum pump and a low temperature. The resulting vapor gets cooled by cold water from the cooling tower and enters into the adsorber bed (Bed A) and gets absorbed by the adsorbents. Another part of the heated water passes to the desorber bed (Bed B) after being heated by another water-heating pump to release and heat the water from the adsorbents. The heated water is then gets cooled in the condenser and is collected in a tank and this completes one cycle. The accumulated brine is ejected from the system.

9.5.3 Design parameters and performance evaluation of the pilot-scale system

Design parameters

The design parameters of the pilot-scale system were deduced experimentally. The experimental period was 30 min with a salt concentration of 3.5%. The operating parameters were hot-water temperature at 55°C, ambient air pressure of 101.3 kPa, ambient air temperature of 25°C, and silica gel as adsorbent.

The primary design parameters of the system are presented below.

- Hot-water inlet to the evaporator
- Hot-water inlet to the desorber
- Cooling water inlet to the condenser
- Cooling water inlet to the adsorber
- Seawater supply

9.5.4 Equations for the required and released heat in the system

The needed heat for the AD system involves three quantities: latent heat of seawater evaporation (Q_{SW}), latent heat of water evaporation (Q_e), silica gel (Q_s), and the metal of the system (Q_m). Eq. (9.1) presents the essential equations to determine the required heat (Zhang et al., 2020).

$$Q = Q_{sw} + Q_e + Q_s + Q_m \tag{9.1}$$

9.5.5 Performance evaluation

Using the gain output ratio (GOR), it is possible to compare thermal desalination systems.

The GOR (Olkis et al., 2019) is the ratio of the latent heat of the water production to the input thermal energy, presented in the following equation.

$$\text{GOR} = \frac{m_w r_e}{Qs + Qe + Qm} \tag{9.2}$$

9.6 Design of an autonomous solar water desalination system

To address the problem associated with meeting the load demand to provide freshwater, the grid-independent hybrid renewable energy systems (GIHRESs) are gaining popularity. Examples of renewable energy-based systems are solar/battery or/hydrogen reverse osmosis desalination (ROD), wind/battery or/hydrogen/ROD, etc. It has been found that GIHRES-based battery energy storage systems are more economical than GIHRES-based hydrogen energy storage systems.

9.6.1 Mathematical modeling

Schematic of sustainable energy-based desalination systems is shown in Fig. 9.14. In combined conditions, the PV panels and wind turbines enhance

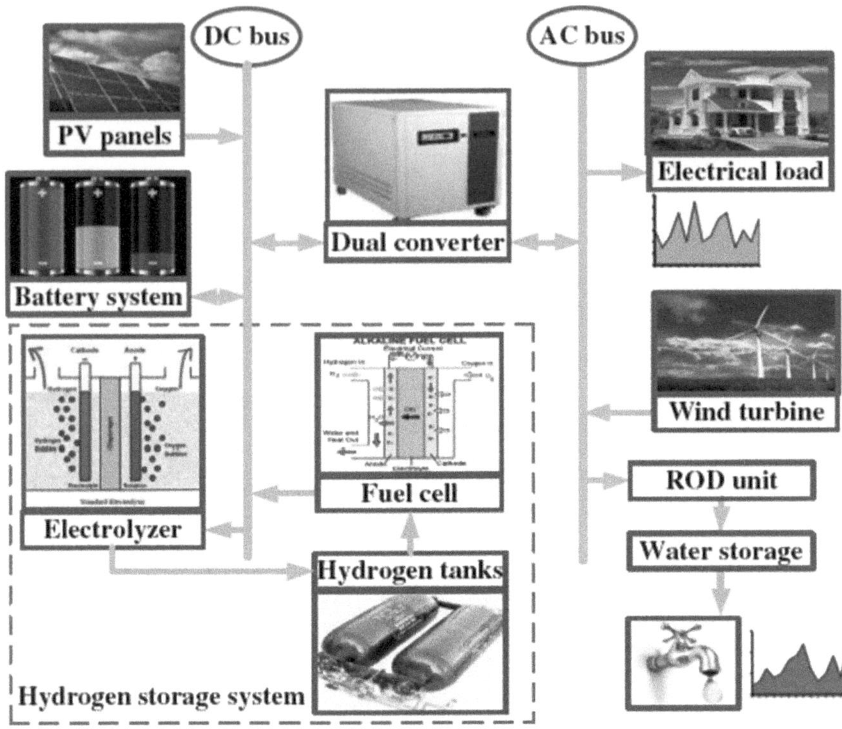

FIGURE 9.14 Schematic of the six hybrid renewable energy system (Maleki, 2018).

the availability of pure water as well as can meet the demand. Also, the generated electricity can be supplemented by energy storage systems when needed. Some of the energy storage systems are battery storage systems, hydrogen storage systems (fuel cells), etc. Water storage tanks are used to store excess water. From the mathematical modeling, the energy production of various components of the hybrid system can be deduced. In the following sections, the modeling of the photovoltaic system is analyzed.

9.6.2 Photovoltaic system: power and economic

The equation for photovoltaic power generation is as follows (Bakelli, Hadj Arab, & Azoui, 2011; Rahman, Hasanuzzaman, & Rahim, 2015, 2017):

$$P_{PV}(t) = \eta_{PV} R_t A_{PV} \tag{9.3}$$

where R_t is the solar insolation, kW/m^2, A_{PV} is the PV surface area (m^2), and η_{PV} is the photovoltaic efficiency which can be expressed as (Bakelli et al.,

2011; Kumar, Hasanuzzaman, & Rahim, 2021; Mamun, Islam, Hasanuzzaman, & Selvaraj, 2021; Markvart & Castañer, 2003):

$$\eta_{PV} = \eta_r \eta_{PC}\left[1 - N_T\left(\left(T_{air} + \left[\frac{\text{NOCT} - 20}{800}\right]R_t\right) - T_{ref}\right)\right] \tag{9.4}$$

Here η_{PC} = power conditioning efficiency
And, η_r = reference module efficiency
T_{air} = ambient air temperatures (°C)
T_{ref} = cell temperature at ideal conditions (°C)
NOCT = nominal cell operating temperature
N_T = temperature coefficient of the PV collector efficiency

9.6.3 Reverse osmosis desalination

The relation among volumetric water demand (H_{WD}), electric power (P_{DEM}), and average specific energy consumption (S_{DC}) is provided by the following equation (Maleki, 2018):

$$P_{DEM} = H_{WD} S_{DC} \tag{9.5}$$

The energy required by present ROD technology is 2–5 kWh/m^3 freshwater (Spyrou & Anagnostopoulos, 2010).

Water production capacity (D_{WC}) by desalination process on a daily basis can be written as (Maleki, 2018; Spyrou & Anagnostopoulos, 2010):

$$D_{WC} = 24\left(\frac{P_D}{S_{DC}}\right) \tag{9.6}$$

Here P_D = Power of the desalination process.

The total daily volumetric freshwater demand D_{WD} is proportional to the volumetric capacity of the freshwater tank (W_{Ta}) of the ROD system. For example, if the storage period is assumed two days, then the freshwater tank volumetric capacity V_{WTa} can be expressed as (Maleki, 2018):

$$V_{WTa} = 2D_{WD} \tag{9.7}$$

9.7 Design method of solar water desalination system

This section discusses a small-scale decentralized production system that is almost maintenance-free and does not require an electricity grid connection. The system applies the multieffect humidification (MEH) method which is mainly evaporation and condensation of water inside a thermally insulated box at ambient pressure. Water vapor is transported from the evaporator to the condenser by air medium and then condensed water is collected. Such systems with high heat recovery ratios require very careful fabrication of evaporator and condenser. Thermal storage might be required if the system runs on solar energy.

9.7.1 Distillation method

Seawater desalination by multieffect humidification

This desalination system follows the process of evaporating water and then condensing the steam from moist air. The humid air flows by natural convection between condenser and evaporator that are located in the same thermally insulated box. The flow of air is clockwise as presented in Fig. 9.15. In the evaporator, the hot seawater gets heated by solar energy and is distributed onto vertically hanging fleeces made of polypropylene and slowly trickles downwards. Air gets humidified as it moves over the brine through the evaporator. As a result, the brine becomes cool by partial evaporation and exits the evaporator with a high salt concentration and at about 45°C. The saturated air gets condensed in the condenser by exchanging heat with seawater. The distilled water is collected in a collecting basin. Hence, in the condenser, the brine temperature increases from 40°C to approximately 75°C. After that using solar energy or waste heat, the brine is heated to the evaporator inlet temperature (80–90°C). If recirculation is needed, the brine is made cool by using feed water from a cooler (Spirkl & Ries, 1995).

Process control

Under ideal operating conditions, the natural convection air loop can make self-adjustments. inlet temperature (evaporator), volume flow rate of saline water can maintain suitable thermal efficiency of the system whereas high-temperature thermal storage tank can maintain a definite temperature range and volume flowrate. Because of polypropylene, the temperature inside the

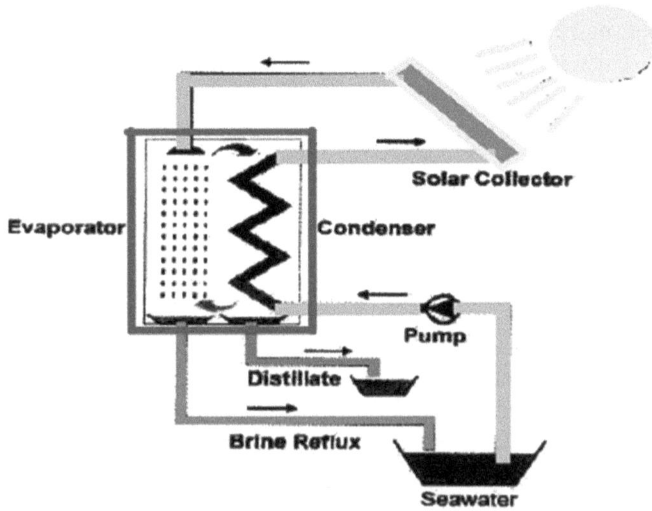

FIGURE 9.15 Distillation system (Müller-Holst et al., 1999).

distillation unit is up to 85°C with a set temperature range of 75–85°C for the evaporator. An optimized operation strategy for the whole system minimizes the energetic and economic specific demands per distillate unit fulfilling those boundary conditions (Spirkl & Ries, 1995).

Desalination system supplemented by a thermal storage tank

For 24-h operating conditions, thermal storage systems are required to the mentioned system in the previous section. Also, distillation cost reduces significantly if the distillation system is a 24 h working system. Thermal storage can be indirectly connected with heat exchangers which also precludes the need for pressure-proof storage tanks (Islam et al., 2021; Kumar et al., 2021). The configuration favored in the present SODESA pro

gram is one with a thermal storage tank and collector field being flown through by the seawater.

Simulation model of the distillation module

The model of the distillation unit is a core element for the simulation of the process. The exchange ratio between the evaporator and the condenser is influenced by the sum of four resistances as shown in Fig. 9.16.

(a) The latent and sensible heat transfer coefficient from the evaporator surface to the humid air
(b) Convective airflow friction force coefficient
(c) Convective airflow to the condenser surface transfer coefficient
(d) Heat flow through the condenser plate coefficient

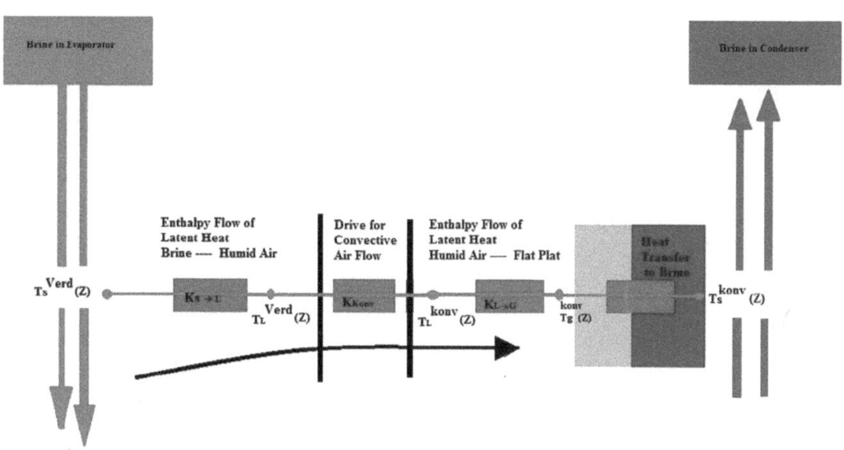

FIGURE 9.16 Physical model of the distillation unit: counter flow heat exchanger (Müller-Holst et al., 1999).

As the friction of the airflow is very small, it can be neglected. The water vapor capacity of humid air and the heat conductivity of the fiat plate heat exchanger can limit the heat transfer. The physical properties of the distillation system depend on the surface capacity of the connecting pipes of the heat exchangers.

9.8 Performance analysis of solar water desalination system

In this section, a noble combined system for freshwater generation and cooling purposes has been investigated. The presented system is based on evaporative cooling using wind towers that allow passive cooling and ventilation. Wind towers are environmentally friendly and can diminish the use of electrical load. Air is received by the wind towers at a higher elevation which is then made to pass through a wet mesh of porous clay and is cooled by evaporative cooling. The cold air is then directed to the selected building areas by forced suction. This humid air can be effectively used in the water desalination process by dehumidification since it is highly enriched with water vapor even after fulfilling the cooling requirements. So, humidifying the air for thermal comfort and dehumidifying for potable water is the proposed technique proposed in this section. The humidification-dehumidification desalination process has the following attractive characteristics:

(a) Low working temperature
(b) Can utilize renewable energy sources
(c) Technical feasibility and suitability for small-scale operation

9.8.1 Advanced humidification-dehumidification desalination system

A humidification-dehumidification desalination system comprises an evaporator to humidify incoming air with saline and a condenser to dehumidify the moist air for generating potable water. In a functional sense, the evaporators and condensers are actually heated exchangers made of resin and polyester to resist corrosion. Also because of these elements, heat exchangers attain many advantages like less weight, economical, easy maintenance, less pollution, and high resistance to corrosion as well as to hostile media. In addition, air heaters are more suitable compared to water heaters because of reduced thermal requirement as suggested by Li et al. (2014). A number of desalination systems such as solar-powered desalination systems, geothermal desalination systems, mechanical compression-based desalination systems were investigated (Bourouni, Chaibi, & Tadrist, 2001).

9.8.2 Proposed combined system

The proposed combined integrated system is shown in Fig. 9.17. The system is designed to provide fresh water and thermal comfort. The tower heat receives the ambient air which is then made to flow through the clay conduit column (CCC). Inside the column, water particles are applied to the air to cool it by evaporation. However, capturing the wind in a region with an inconsistent direction of wind flow can be difficult which can be solved by implementing a swivel tower head.

The CCC is highly effective to attain the required thermal comfort for residential purposes. It is commonly useful for both sensible cooling and evaporative cooling. The wide surface area increases the rate of sensible cooling while the applied water at the top for evaporative cooling is allowed to flow downwards. The sump at the base stores the remaining water after evaporation.

The sump acts as a reservoir for both evaporative cooling in CCC and water condensation in the dehumidifier. The cold air obtained from evaporation can be supplied from the bottom of the CCC column to the rooms in two ways as depicted in the figure.

(I) to floor 2 using the wall opening.

(II) to floor 1 and then to floor 2 through the roof opening of floor 1. In this method, the dust particles in room 1 are removed thus providing clean air

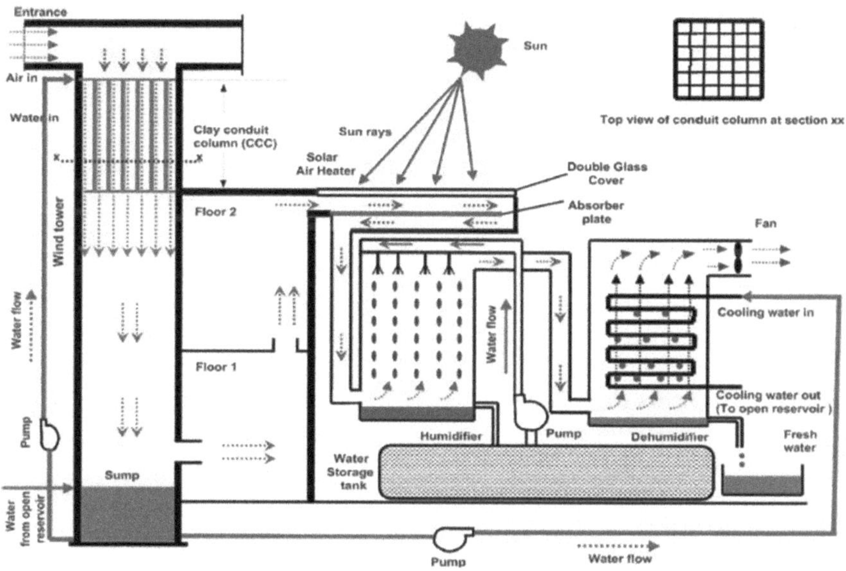

FIGURE 9.17 Proposed combined system (Sachdev, Gaba, & Tiwari, 2020).

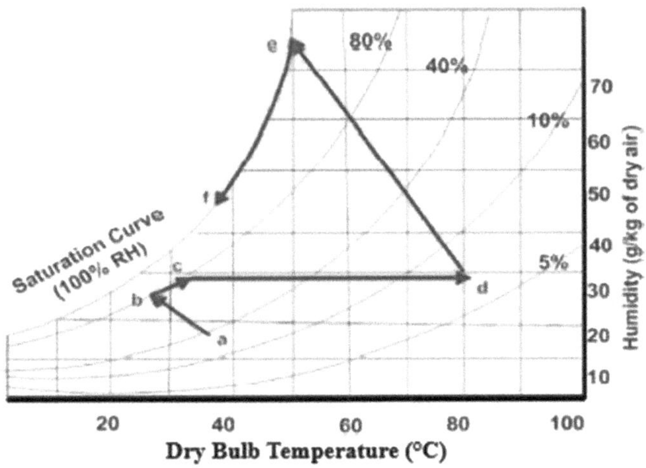

FIGURE 9.18 States of air during the flow through the proposed system (Sachdev et al., 2020).

toward room 2. The resulting moist air is then added to the HDH desalination system to generate freshwater as a byproduct of the arrangement after providing thermal comfort on floor 2.

To enhance the humidification rate of the air, a double pass solar air heater is employed in the proposed system. A storage tank is used to provide sprayed water from the top of the humidifier over the air entering the humidifier. As a result, the extra cost of a water heater can be diminished. Also, the high temperature of air and water toward the humidifier increases the vaporization rate. The water is not to be recycled to maintain a steady temperature of the storage tank which will otherwise increase to close the water cycle.

Lastly passing through the dehumidifier, the humid air gets condensed and generates pure water. An attached fan ensures the proper airflow through the dehumidifier and removes the losses. The conditions of the flowing air can be analyzed by the given psychrometric chart in Fig. 9.18.

Process a-b: The decrease in temperature of the air by evaporation as it flows through the columns of the clay conduits.

Process b-c: Heating of air and an increase in the moisture constituents while flowing through floor 1 and floor 2.

Process c-d: Heating of air by sensible heat in solar air heater.

Process d-e: Humidifying air to the saturation point using salty water within the humidifier.

Process e-f: Dehumidifying the saturated air obtained from the humidifier using cooling water to produce the freshwater.

Nomenclature

1 Copper
2 Aluminum
3 Aluminum alloy
ρ Density of water
τ Time
AD Adsorption desalination
c_1 Specific heat capacity of copper
c_2 Specific heat capacity of aluminum
c_3 Specific heat capacity of aluminum alloy
c_s Specific heat capacity of silica gel
e Evaporation
h Hot water
m Metal
m_1 Quantity of copper
m_2 Quantity of aluminum
m_3 Quantity of aluminum alloy
m_s Quantity of silica gel
m_w Quantity of water
p Pressure
Q Total heat required
Q_{sw} Heat for seawater evaporation
Q_e Heat for water evaporation
Q_s Heat for silica gel
Q_m Heat for metal of the system
r_e The average latent heat of seawater evaporation
r_{sw} Average latent heat for water evaporation
s Silica gel
s_w Seawater
Δx_i Uncertainty
t Temperature
v_h The flow rate of hot water
y Directly measured value
w Water

References

Abed, F. M., Eleiwi, M. A., Hasanuzzaman, M., Islam, M. M., & Mohammed, K. I. (2020). Design, development and effects of operational conditions on the performance of concentrated solar collector based desalination system operating in Iraq. *Sustainable Energy Technologies and Assessments, 42*, 100886. https://doi.org/10.1016/j.seta.2020.100886

Bakelli, Y., Hadj Arab, A., & Azoui, B. (2011). Optimal sizing of photovoltaic pumping system with water tank storage using LPSP concept. *Solar Energy, 85*. https://doi.org/10.1016/j.solener.2010.11.023

Bourouni, K., Chaibi, M. T., & Tadrist, L. (2001). Water desalination by humidification and dehumidification of air: State of the art. *Desalination, 137*(1), 167–176. https://doi.org/10.1016/S0011-9164(01)00215-6

de la Calle, A., Bonilla, J., Roca, L., & Palenzuela, P. (2015). Dynamic modeling and simulation of a solar-assisted multi-effect distillation plant. *Desalination, 357*, 65–76. https://doi.org/10.1016/j.desal.2014.11.008

Chandrashekara, M., & Yadav, A. (2017). Water desalination system using solar heat: A review. *Renewable and Sustainable Energy Reviews, 67*, 1308–1330. https://doi.org/10.1016/j.rser.2016.08.058

Chen, C., Jiang, Y., Ye, Z., Yang, Y., & Hou, L.a. (2019). Sustainably integrating desalination with solar power to overcome future freshwater scarcity in China. *Global Energy Interconnection, 2*(2), 98–113. https://doi.org/10.1016/j.gloei.2019.07.009

Ding, L. C., Akbarzadeh, A., & Tan, L. (2018). A review of power generation with thermoelectric system and its alternative with solar ponds. *Renewable and Sustainable Energy Reviews, 81*, 799–812. https://doi.org/10.1016/j.rser.2017.08.010

Esmaeilion, F. (2020). Hybrid renewable energy systems for desalination. *Applied Water Science, 10*(3), 84. https://doi.org/10.1007/s13201-020-1168-5

Goosen, M., Mahmoudi, H., & Ghaffour, N. (2010). Water desalination using geothermal energy. *Energies, 3*. https://doi.org/10.3390/en3081423

Islam, M. M., Hasanuzzaman, M., Rahim, N. A., Pandey, A. K., Rawa, M., & Kumar, L. (2021). Real time experimental performance investigation of a NePCM based photovoltaic thermal system: An energetic and exergetic approach. *Renewable Energy, 172*, 71–87. https://doi.org/10.1016/j.renene.2021.02.169

Kalidasa Murugavel, K., & Srithar, K. (2011). Performance study on basin type double slope solar still with different wick materials and minimum mass of water. *Renewable Energy, 36*, 612–620. https://doi.org/10.1016/j.renene.2010.08.009

Kumar, L., Hasanuzzaman, M., & Rahim, N. A. (2019). Global advancement of solar thermal energy technologies for industrial process heat and its future prospects: A review. *Energy Conversion and Management, 195*, 885–908. https://doi.org/10.1016/j.enconman.2019.05.081

Kumar, L., Hasanuzzaman, M., & Rahim, N. A. (2021). Real-Time experimental performance assessment of a photovoltaic thermal system cascaded with flat plate and heat pipe evacuated tube collector. *Journal of Solar Energy Engineering, 144*(1). https://doi.org/10.1115/1.4051861

Li, X., Yuan, G., Wang, Z., Li, H., & Xu, Z. (2014). Experimental study on a humidification and dehumidification desalination system of solar air heater with evacuated tubes. *Desalination, 351*, 1–8. https://doi.org/10.1016/j.desal.2014.07.008

Maleki, A. (2018). Design and optimization of autonomous solar-wind-reverse osmosis desalination systems coupling battery and hydrogen energy storage by an improved bee algorithm. *Desalination, 435*, 221–234. https://doi.org/10.1016/j.desal.2017.05.034

Mamun, M. A. A., Islam, M. M., Hasanuzzaman, M., & Selvaraj, J. (2021). Effect of tilt angle on the performance and electrical parameters of a PV module: Comparative indoor and outdoor experimental investigation. *Energy and Built Environment*. https://doi.org/10.1016/j.enbenv.2021.02.001

Markvart, T., & Castañer, L. (2003). *Practical handbook of photovoltaics*. Amsterdam: Elsevier Science.

Müller-Holst, H., Engelhardt, M., & Schölkopf, W. (1999). Small-scale thermal seawater desalination simulation and optimization of system design. *Desalination, 122*(2), 255–262. https://doi.org/10.1016/S0011-9164(99)00046-6

Narayan, G. P., Sharqawy, M. H., Summers, E. K., Lienhard, J. H., Zubair, S. M., & Antar, M. A. (2010). The potential of solar-driven humidification–dehumidification desalination for small-scale decentralized water production. *Renewable and Sustainable Energy Reviews, 14*(4), 1187–1201. https://doi.org/10.1016/j.rser.2009.11.014

Nassar, Y. F., Yousif, S. A., & Salem, A. A. (2007). The second generation of the solar desalination systems. *Desalination, 209*(1), 177–181. https://doi.org/10.1016/j.desal.2007.04.039

Olkis, C., Brandani, S., & Santori, G. (2019). Cycle and performance analysis of a small-scale adsorption heat transformer for desalination and cooling applications. *Chemical Engineering Journal, 378*, 122104. https://doi.org/10.1016/j.cej.2019.122104

Qiblawey, H. M., & Banat, F. (2008). Solar thermal desalination technologies. *Desalination, 220*(1), 633–644. https://doi.org/10.1016/j.desal.2007.01.059

Rahman, M. M., Hasanuzzaman, M., & Rahim, N. A. (2015). Effects of various parameters on PV-module power and efficiency. *Energy Conversion and Management, 103*(0), 348–358. https://doi.org/10.1016/j.enconman.2015.06.067

Rahman, M. M., Hasanuzzaman, M., & Rahim, N. A. (2017). Effects of operational conditions on the energy efficiency of photovoltaic modules operating in Malaysia. *Journal of Cleaner Production, 143*, 912–924. https://doi.org/10.1016/j.jclepro.2016.12.029

Ritchie, H., & Roser, M. (2021). *Energy - our world in data*. Retrieved from https://ourworldindata.org/energy. (Accessed 18 August 2021).

Sachdev, T., Gaba, V. K., & Tiwari, A. K. (2020). Performance analysis of desalination system working on humidification-dehumidification coupled with solar assisted air heater and wind tower: Closed and open water cycle. *Solar Energy, 205*, 254–262. https://doi.org/10.1016/j.solener.2020.04.083

Sarkar, M. N. I., Sifat, A. I., Reza, S. M. S., & Sadique, M. S. (2017). A review of optimum parameter values of a passive solar still and a design for southern Bangladesh. *Renewables: Wind, Water, and Solar, 4*(1), 1. https://doi.org/10.1186/s40807-017-0038-8

Sathyamurthy, R., El-Agouz, S. A., Nagarajan, P. K., Subramani, J., Arunkumar, T., Mageshbabu, D., … Prakash, N. (2017). A Review of integrating solar collectors to solar still. *Renewable and Sustainable Energy Reviews, 77*, 1069–1097. https://doi.org/10.1016/j.rser.2016.11.223

Sharon, H., & Reddy, K. S. (2015). A review of solar energy driven desalination technologies. *Renewable and Sustainable Energy Reviews, 41*, 1080–1118. https://doi.org/10.1016/j.rser.2014.09.002

Sharshir, S. W., Yang, N., Peng, G., & Kabeel, A. E. (2016). Factors affecting solar stills productivity and improvement techniques: A detailed review. *Applied Thermal Engineering, 100*, 267–284. https://doi.org/10.1016/j.applthermaleng.2015.11.041

Spirkl, W., & Ries, H. (1995). Optimal finite-time endoreversible processes. *Physical Review E, 52*(4), 3485–3489. https://doi.org/10.1103/PhysRevE.52.3485

Spyrou, I. D., & Anagnostopoulos, J. S. (2010). Design study of a stand-alone desalination system powered by renewable energy sources and a pumped storage unit. *Desalination, 257*(1), 137–149. https://doi.org/10.1016/j.desal.2010.02.033

Taylor, P. (2002). What happens when the fossil fuels run out? In P. Taylor (Ed.), *Alkenes and aromatics* (pp. 170–171). The Royal Society of Chemistry.

Zhang, H., Ma, H., Liu, S., Wang, H., Sun, Y., & Qi, D. (2020). Investigation on the operating characteristics of a pilot-scale adsorption desalination system. *Desalination, 473*, 114196. https://doi.org/10.1016/j.desal.2019.114196

Chapter 10

Economic assessment of solar thermal energy technologies

A.B.M. Abdul Malek[1], Maliha Kawsary[2] and Md Hasanuzzaman[3]

[1]*Department of Industrial and Production Engineering, Shah Jalal University of Science and Technology, Sylhet, Bangladesh;* [2]*Department of Enterprise Engineering, The University of Bordeaux, Talence Campus, Talence, France;* [3]*Higher Institution Centre of Excellence (HICoE), UM Power Energy Dedicated Advanced Centre (UMPEDAC), University of Malaya, Kuala Lumpur, Malaysia*

Chapter outline

Technologies for Solar Thermal Energy. https://doi.org/10.1016/B978-0-12-823959-9.00004-0

10.1 Introduction

The solar energy systems impart low environmental impact. One of the most important benefits of such systems is that they produce a very low level of CO_2 or other toxic gases or radioactive materials, unlike the ones that are produced by the nonrenewable energy systems. But we are used to these traditional systems for producing our final energy forms (Hasanuzzaman, Al-Amin, Khanam, & Hosenuzzaman, 2015; Islam, Hasanuzzaman, Rahim, Nahar, & Hosenuzzaman, 2014). Another important feature of the solar energy systems is that the cost of energy consists only of the construction and maintenance of the plant, as the source of energy is free and theoretically unlimited (Hasanuzzaman, Zubir, Ilham, & Che, 2017; Stoddard, Abiecunas, & O'Connell, 2006). One of the major limitations of the solar energy systems is that these systems can only be installed in the areas where solar insolation is longer during the days and also available for maximum days of the year. Another important factor that makes the use of these systems is the limiting efficiency as compared to fossil fuel systems. So, a combination of the solar energy system and conventional fossil fuel system can be a suitable solution foreseen as these blends can supply continuous supply of energy in the absence of solar irradiation.

In the arena of solar energy systems, solar thermal energy has an enormous potential as they are proved in maximum of the prototypes experimented on. The predictions about this type of technology show that the efficiency of these particular systems can be increased to a significant level. Different techniques of active solar heating and solar thermal power generation are technically feasible and shown to have cost effectiveness, and some of them are commercially available. But the limitation is that these energy systems are highly dependent on the local climatic conditions and may play a supplementary role in meeting energy demands which needs continuous operation (Kamarulzaman, Hasanuzzaman, & Rahim, 2021). So, certain regions of the world are privileged to have the solar thermal energy systems efficient enough to be implemented on a commercial basis.

Nowadays, people want to reduce the dependence on nonrenewable energy. In this regard, many projects are running on solar thermal energy (Guo, Zhai, Xu, & Li, 2017; Wang, Huang, Zheng, & Zhao, 2018). Energy produced from all the water-based solar thermal systems amounted 335 TWh in 2014, corresponding to an energy savings of 36.1 Mtoe and reduction of 116.4 million tons of CO_2 (Mauthner, Weiss, & Spörk-Dür, 2016).

China is in the leading position for solar thermal collector installation and has successfully employed for district water heating and space heating (Wang et al., 2018). Extensive use of solar thermal energy for water heating and industrial process heating purpose are also seen in the US, Canada, and many EU countries (Nikoofard, Ismet Ugursal, & Beausoleil-Morrison, 2014). European Solar Thermal Industry Federation estimated the installed capacity in EU countries to reach 1019 GW, and forecasted to contribute about 15% of the low temperature heat demand. The optimistic figure for 2050 is 2716 GW and that amount is roughly 47% of the overall heat demand in the EU. Solar thermal systems can provide up to 60% of households hot water demands and need very little maintenance cost over the life span. If compared to gas boiler, solar thermal systems are better options and the potential for reducing global warming and depletion of fossil fuel can be as lower as by 88% and 83%, respectively (Greening & Azapagic, 2014).

10.2 Basic concept of solar thermal project planning

Project planning is an orderly process of achieving targets by ensuring integration in different defined tasks like policies, procedures, and program within a forecasted environment. Project includes objective, quality criteria, manpower, funds, timeframe, and uncertainty.

Project planning must be disciplined in documentation, investigation, and maintenance and staffing. It should have the ability to accept multifunctional inputs.

A project manager directs and supervises a project through developing, making changes, and executing the plans. The way of project is not always smooth. Sometimes, it may have shortage of resources, conflicts, or political influences or threat from emerging new rivals. Top administration helps to eradicate these barriers.

Nowadays, People want to reduce the dependence on nonrenewable energy. In this regard, many projects are running on solar thermal energy (Guo et al., 2017; Wang et al., 2018). China is the leader in solar thermal collector installation and successfully employed for district water heating and space heating that is shown in Fig. 10.1 (Wang, Guo, Zhang, Yang, & Mei, 2017; Wang et al., 2018). Extensive use of solar thermal energy for water heating and industrial process heating purpose are also seen in the US, Canada, and many EU countries (Nikoofard et al., 2014).

A typical solar thermal energy system is comprised of a collection component, working fluid circulation system, storage component, and controls. All these components function together to convert solar radiation into thermal energy of low to moderate temperature. This energy can be used in various applications. Collector types include (Owens, 2002):

- Simple flat-plate collectors
- Flat-plate collectors with heat-pipe
- Evacuated-tube collectors
- Parabolic-trough collectors
- Paraboloidal-dish collectors
- Power-tower collector systems

Each type of solar thermal energy systems got its distinctive features and has differential parametric scenario. The specifications corresponding to a particular type set a specified design window. Comparisons among energy plants with different parametric scenarios are restricted to degree of freedom of the axes for their alignment, shapes of the concentrators, also in the influx of solar fields. The results of the experiments show that the optimum system performance can be ensured by the introduction of a thermal energy storage (TES) integrated with the design configurations. The capacity of the TES

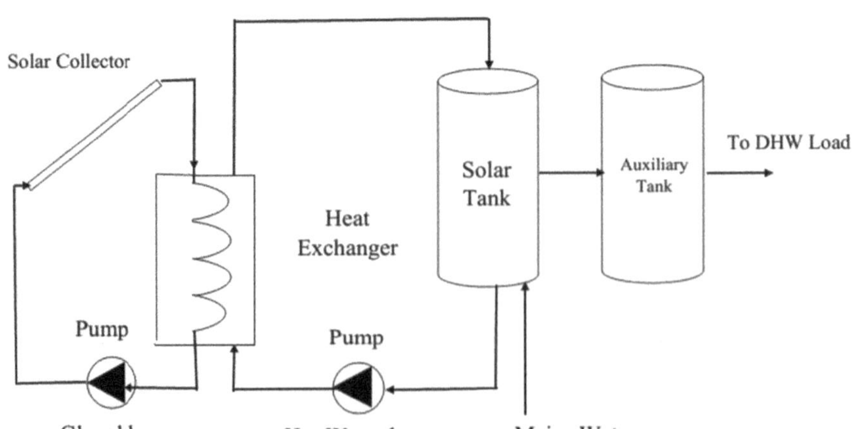

FIGURE 10.1 Simple block diagram of solar domestic hot water plant (Wang et al., 2017).

depends on the solar field and size of the concentrator used. Moreover, it is found that some parametric scenarios lead to an increase in the cost of energy when the energy storage capacity is broadened. On the contrary, parametric scenarios associated to cheaper solar fields yield a better economic performance when TES is integrated to the system (González-Portillo, Muñoz-Antón, & Martínez-Val, 2017).

10.2.1 Causes for project planning

The causes in the project planning of solar thermal energy include:

- Obtaining better comprehension of the targets
- Stamping out or lessening unreliability
- Enhancing the competence of function
- Giving a platform of monitoring and controlling work
- Making decisions regarding future and embody endeavor to carry out these decisions
- Finding out the systematic planning substitute in decision-making based on history

10.2.2 General planning

The general planning of solar thermal energy includes:

- An aim, intention which is to be accomplished within a time frame
- Instructions which are carried out in performing tasks
- Lists of activities to be implemented showing project completion time
- Estimating the spending to complete the project
- Pronging the future of a certain time period based on present and past
- Distributing duties and responsibilities among various positions
- Scheme for different organizational tasks and decision-making
- A detailed way to carry out an action
- Measurement of acceptance

A list of questions in solar thermal energy includes:

- To know about environmental impact
  a. What is the project's present situation?
  b. What are the reasons and conditions to get into that situation?
- Defining goals
  a. What are or is our project objective or objective?
  b. What is our expected position within a specific timeframe?
- Finding out strategies
  a. Is it okay to carry on with the present track?
  b. What path should be followed to achieve objectives?

- Finding out threats and opportunities
  a. What obstacles might create a barrier to obtain the desired output?
  b. How can a project overcome those hurdles?
- Making forecasts

  How much a project capable to make a successful completion?

- Choosing strategy portfolio

  What are the potential advantages and disadvantages of the project?

- Making working schemes
  a. What are the actions necessary to accomplish the project?
  b. Who will be the responsible person and how will he do it?
- Supervising and controlling
  a. Is the project on the right track?
  b. What measures should be taken to bring the project back on right track?

10.2.3 Life cycle phases

The life cycle phase is a framework that provides manageable phase consistency to carry out a project from initiation to the completion (Lam & Law, 2018).

Every organization outlines the activities regarding each phase.

A manager can also practice his own idea along with the outline. He/she asks for funding. Gradually, a project team gets the structure by identifying required activities, budget, risk potentiality, equipment, facility utilization, etc.

After each phase, top management, manager, sponsor, stakeholder, and in some cases, customer discuss the output of the respective phase and based on performance, get acceptance for the next phase.

The life cycle phases of solar thermal energy include:

- Formulation
- Feasibility
- Preliminary planning
- Detailed design
- Implementation
- Testing and commissioning

Formulation

In this phase, brainstorming and common sense play a vital role. It includes two factors:

- To recognize and specify the problem/objective/need
- To find out the possible solution options

Feasibility

Is it possible to solve the problem/objective/need?—is a great matter in feasibility analysis.

In feasibility analysis, it is necessary to define the approach, limitation, or opportunity of a project.

Technical and operating decisions must be analyzed properly. Improper analysis may result in faulty decisions. It may stop the project or cause error to the project during the continuation or termination. Budget estimation, improvement scheme are also very important in feasibility.

The purposes of the feasibility phase are:

- Identify the total expenditure, time, manpower, and machinery requirement
- Develop improvement scopes to get the desired results from the project

Summary level of feasibility analysis:

- Assess substitutes
- Assess the market prospective and costing
- Assess productivity and technical support
- Detailed level of feasibility analysis
- Finding out the exact problem
- Inspecting in-house technological competence
- Evaluating viability of substitute
- Estimating imperfections and undetermined defects
- Making trade-off analysis on time, cost, and performance

Preliminary planning

Preliminary planning includes:

- Opportunity of the task
- Measurement of performance
- Reference to the performed tasks
- Documentation
- Equipment requirement
- Resources, facilities, machinery, and aid
- Timetable of performance
- Presentation and addition

Detailed design

- All of the engineering evaluations are completed, when it is known what is to be designed
- Depiction and technological restrictions

Implementation

- Execution of project function
- Confirming sufficient construction of the function as designed
- Final product configuration plays an important role in this phase

Examining and completion

- When the implementation phase is completed, the final product is examined and commissioned
- Project completion and financial issues are documented

10.3 Basic of project assessment

The real outcomes of a solar thermal project can be evaluated mainly from two aspects. Number one is from energy aspects and number two is economic aspects. In case of energy generation concern, a combined analysis is taken seriously for the project assessment. The evaluation process follows some steps like (Kumar, 2015):

1. Evaluation of costs and benefits over the years of operation. Costs involved are investment costs and costs for maintenance, servicing and insurance against damage, etc. On the other hand, benefits may be direct revenue earning or savings for replaceable energy, i.e., the avoided bill costs, the incentives received from CER, or revenue earning from the selling of energy to the utility companies under certain tariff rates.
2. Analysis of cash inflows and outflows.
3. Evaluation of the economically effective space otherwise would have been left empty.
4. Estimation of the energy recovery factor pertaining to the analysis done.
5. Sensitivity analysis for the most significant parameters.

10.3.1 Economic analysis for project assessment and use of discounted cash flow

The discounted cash flow (DCF) concept can be presented in a simple equation. The total earning from the project in its life span is represented by, G. Overall return from the project activities is R with the cost incurred C. The simple relation then looks like the following equation:

$$G = R - C \tag{10.1}$$

As the initial investment and the subsequent cash flows occurs in different time frames, so a time value effect is imparted to this simple equation. This time value is included in the relation using some correlation coefficients which equalize the time value of the money or the future payments of receipts are discounted. So, DCF tries to figure out the value of an investment on the base year and highlights on how much money it will generate in the future. This applies to both financial investments for the stake holders of the venture and business partners looking to make changes in their businesses joining to the venture (suppliers of new equipment specific to the venture).

The formula for DCF is:

$$\mathrm{DCF} = \frac{CF_1^1}{1+r} + \frac{CF_2^2}{1+r} + \frac{CF_n^n}{1+r} \tag{10.2}$$

where:

CF=the cash flow for the given year

CF_1 is for year one, CF_2 is for year two

CF_n is for additional years

r = the discount rate

Each of the future cash flows must be "deflated" first to go back to base year. So, future cash flows must be multiplied by the discount factor:

$$\frac{1}{(1+r)^j} \tag{10.3}$$

where r is the cost of the capital given by the relation, $r = i{-}f$, i.e., difference between the estimated discount rate "i" and the rate of inflation "f."

Thus, DCF is used to get net present value (NPV) of an investment:

$$\mathrm{NPV} = \sum_{j=1}^{n} \frac{CF_j}{(1+r)^j} - Io \tag{10.4}$$

where:

Io = initial investment

n = years of duration of the investment

When the NPV results to a positive figure, it means at the end of life of investment, the DCFs produced throughout the entire life possess higher inflow than the cost of the initial investment, and other associated costs and therefore, the erection of a plant is justified from a financial point of view; vice versa when the NPV is negative.

Details of NPV and other indicators that uses discounted cashflow like life cycle cost (LCC), internal rate of return (IRR), and discounted pay back period will be discussed in details in later sections.

10.3.2 Economic feasibility indicators

First of all, the intension behind performing an economic analysis is to find out the feasibility of investment in a solar thermal energy system. Another issue is the profitability analysis so that the investment can be justified for the stake holders. Generally, the economic feasibility of such plants on unit basis is evaluated in terms of levelized cost of energy (LCOE). The LCOE evaluates the unit cost of energy produced throughout the life time of the energy system. Specifically, there are two terms available to assess the unit energy cost, one is real LCOE and another one is nominal LCOE. Real LCOE uses constant and inflation adjusted value of the dollar, and it is used for long-term analysis. In case of short-term analysis, nominal LCOE is determined by considering different factors including (Soomro et al., 2019):

- Total energy output from the system
- Direct capital invested for machineries and installation
- Indirect capital cost: cost for approvals, cost of signing power purchase agreement, cost of consultancy and engineering, and land cost
- Cost of operation and maintenance cost

10.4 Costs and cost estimating

10.4.1 Basic concept of cost

Cost is an important element in solar thermal energy project planning. In calculating cost, several factors like equipment, new construction, investment, manpower, intellectual asset, development course, maintenance, repairing need, inspection, attribute criteria, technical and general support.

Cost can be classified into:

- Fixed cost: It is a constant cost. It does not affect or vary regarding activity.
- Variable cost: Variable cost varies or changes depending on the quantity of yield or function.
- Average cost: Average cost = total cost for all units/total number of units.
- Marginal cost: Marginal cost is used to produce/purchase/enroll additional or extra unit.
- Sunk cost: A sunk cost is that expenditure that occurs due to the previous function.
- Opportunity cost: Opportunity cost is expensed in one function rather than the other function.

10.4.2 Fixed cost and variable cost

In a solar power plant, fixed cost remains the same. For example, the rent of solar thermal project floor space, cost of machineries like solar collector, insurance, property tax does not change depending on the solar power production volume.

On the other hand, in solar thermal energy plant, variable cost increases or decreases depending on the number of employees, solar power production quantity.

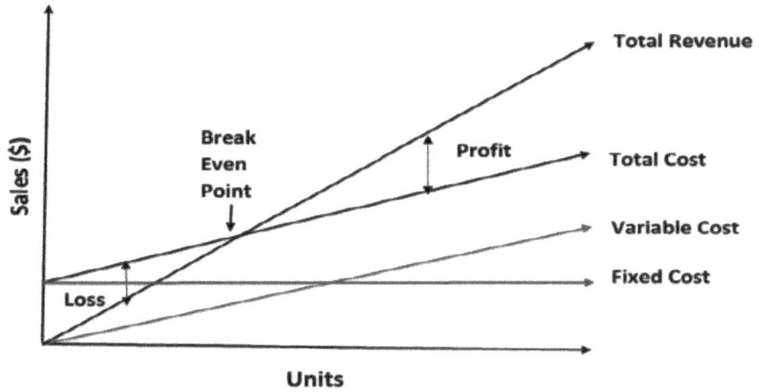

FIGURE 10.2 Break even point.

10.4.3 Breakeven point

In a solar thermal project, breakeven point is a point where total cost = total fixed cost + total variable cost which means there is no gain and no loss that is shown in Fig. 10.2.

Profit region

For making profit, the breakeven point is less than the production level of solar thermal energy.

Loss region

Loss occurs when the breakeven point is greater than the production level of solar thermal energy.

10.5 Feasibility analysis

10.5.1 Concept of feasibility analysis

The workability of a project through quality information is evaluated by feasibility analysis. It is an important tool in decision-making. When the list of possible alternatives are narrowed down to a few, it is the time to have an inner look on the feasibility of the project to have the ability to meet its customer needs, to have the ability to be operational, and after operating for the desired target parameters, it now must have the economic performance to return the stack holders for their valuable investment. So, a selected project must prove it for technical, operational, and economic feasibility. There are other inherent factors that need to be studied are the solar field, the power block, and the energy storage (González-Portillo, Muñoz-Antón, & Martínez-Val, 2017). Generally, the feasibility of the project is then justified by combining the results of the various subcomponents of the logistical drivers like demand and supply, unit price of supplied energy, investment cost, etc. This analysis is first done for the base case scenario. Of course, this base case information is not sufficient for the decision makers as there are many financial risks for project implementation. To assess this risk on the project level, David B. Hertz developed a method known as risk analysis (Hertz, 1964).

10.5.2 Technical feasibility (risk assessment and solutions)

Increasing quality but reducing time and costs always put pressure on the management of tasks in a project. Thus, risk management is particularly given emphasis in the projects based on new technology like solar thermal energy projects (Cagliano, Grimaldi, & Rafele, 2015). So, here risk is defined as an uncertain event or condition that, if it occurs, has either positive or negative effects on project objectives (Hillson & Management, 1997). Risk hinders the expected return from a project (Owens, 2002). So, risk management needs a benchmarking of current practices and continual learning for improvement (Yeo & Ren, 2009). The highest ranked factor for project failure was the poor planning in two distinct areas as reported by Yeo and Ren (2009). The next ranking goes for poorly addressed or inadequately addressed risks as part of the project planning process.

A project on solar thermal energy may confront with several risks, which are:

Construction

- If there is adequate resource supply or substitution opportunity in case of shortage?
- Is it possible to meet the specification, procedure requirement within the budget?
- If there is any probability of loss or accident that may hamper the project?

Operation/commercial risk

- Is there difficulty in operating the project efficiently or within the capacity?
- Challenges in experience, delivery, business entry?

Equipment

- Are there enough machineries in case of replacement?
- If they are suitable for the project works or if there is technological risk?

Economic

- Does the contract reliable?
- Is there unpredicted value?

- Is there money guarantee or problem with currency transformation?
- If there is possibility of variation in the tax policy?

Market availability

- Is it possible to access into local or international market?
- If the credit rating of buying contract is satisfactory?

Political

- If there is political denial in foreign exchange, agreement?
- If there is possibility of new tax imposition?
- If there is desperate nationalism?

Environmental

- If the condition of environment is suitable for project running?
- If there is coverage of insurance in case of environmental disaster?

It is practically not possible to foresee all the associated risks at the planning phase. A standard operating procedure must be maintained for risk identification and then plan out details of the recovery issues (Yeo & Ren, 2009).

Risk matrix is used to minimize risk in a project that includes (Yeo & Ren, 2009)

- Diagram of different types of risk
- Causes of risk
- Identification of most serious risk
- Risk minimization measures
- Visualize the risk situation for investor related to the project.

10.5.3 Operational feasibility (control, efficiency and service)

It is essential to identify the combination of criteria for assessing solar thermal energy systems (Kassem, Al-Haddad, Komljenovic, & Schiffauerova, 2016). Various researchers have used different parameters combinations for the evaluation of multiple criteria energy portfolio planning.

In solar thermal project, control helps to identify:

- Is it possible to run the project within the allowable limit?
- If it is possible to prevent divergence from standard?
- If the running working performance is satisfactory?

In order to calculate the efficiency of solar power system, "Hottel-Whillier-Bliss" equation is used (Zheng & Hatzell, 2020):

$$Q_u = AF_R h[I\tau\alpha - U(T_{in} - T_a)] = \eta_c AIh \tag{10.5}$$

A = Solar collector area
I = Average solar irradiance
F_R = Collector heat removal factor
τ = Transmissivity
α = Absorptivity
U = Heat dissipation coefficient
T_{in} = Temperature of the working fluid entering the collector
T_a = Ambient temperature
η_c = Average efficiency of the collector
h = Solar collector operational hours

$$\text{Average daily heat gains } = \eta_c HA \tag{10.6}$$

H(kWh/m^2/day) = Total solar irradiance value

$$\text{Annual water production, } M_e\left(\frac{\text{Kg}}{\text{year}}\right) = P_r D\frac{\eta_c HA + W}{h_{fg}} \tag{10.7}$$

P_r = Performance ratio
D = Days working per year
The daily water production m_e (kg/day) is relevant to system scale x (m^3/day):

$$m_e = \frac{M_e}{D} = \rho_w x \tag{10.8}$$

ρ_w = Density of water

Abundance of sunlight is also very important in solar thermal service. This type of renewable project is very suitable for areas having high emission of greenhouse gases, excessive dependence on nonrenewable energy, shortage of electricity, pure drinking water, etc. The performance coefficient of renewable energy project defines the service quality.

10.5.4 Schedule feasibility (timeline estimation and resources optimization)

Schedule feasibility refers to deadline and resource optimization of a project (Keshavarzzadeh, Ahmadi, & Rosen, 2020, p. 121760).

Timeline estimation includes:

- If it is possible to accomplish all the planned tasks of a project within the time limit.

Resource optimization determines the operating and design procedure regarding timespan and budget to make the best use of present resources. It identifies, which should be maximized and which should be minimized. It defines objective, parameters, and constraints.

10.6 Cost control

10.6.1 Basic concept of cost control

Cost control is used to identify and reduce cost in a project by monitoring and recording and helps to be proactive. Cost control in a project includes the task of reckoning cost; keeping the books of accounts; managing the cash inflows and outflows; and standardizing the costs. The main concern is to take corrective actions in direct costs and overheads. Other related cost control elements are providing incentives or penalties and maintaining the promotional programs continuously.

10.6.2 Importance of cost control

- Cost control secures the proper plan of budget.
- Every organization requires cost control.
- Failure of any or one project may put a company into money risk or other unwanted situations. Very often, small companies generally have to maintain a rigid budgetary control.
- Huge companies may have the scope to bear losses for several project but whereas the small companies may have few projects or not have any scope.

10.6.3 Purpose of cost control

Cost control helps management to determine the current project status.

- It helps to establish standard, policy, procedure, and techniques to be utilized in a project.
- It provides information that:
- Provides a pictorial view of actual project growth
- Connects cost and performance
- Determines possible risk
- Provides an estimation of cost to the responsible top management
- Represent that the goals are sound and convenient

10.6.4 Cost control planning

Cost control planning helps to improve planning, evaluate development, and control change.

The system must be able to:

- Organize and line up tasks
- Identify variables that are necessary for project evaluation
- Identify direct labor and overhead budgets
- Identify management fund capability

Almost all project planning and control systems have identifiable design requirements:

- To combine time, spending, and technological performance, a general layout is constructed
- Capability to find out the improvement of important variables
- Quick feedback
- Ability for end-value prediction
- Accurate and appropriate data for decision-making by each level of management
- To analyze problem with detail description
- Swift quantitative evaluation of substitute solutions

10.7 Tools for project economic performance analysis

Economic performance is better understood with the value a product or service provides to the willing customers. A higher value means a higher price customer willingly pays for the product or service. In economics, this value is called economic value, though there exist other values like moral value, aesthetic value, cultural value, etc. Economic value that a customer is willing to pay for tradable goods may be greater than the actual market price (thus creating an economic surplus), but it is not usually less (van de Poel, 2009). Otherwise, customer would not buy the product replacing available one.

Economic performance must be justified with proper tools, so that the user of the product put their preferences over other available alternatives.

The following tools are commonly used for economic assessment of solar thermal energy project:

- Life cycle cost (LCC)
- Net present value (NPV)
- Internal rate of return (IRR)
- Discounted payback (DPB)
- Levelized cost of energy (LCOE)
- Profitability index (PI)
- Sensitivity analysis

All of the above tools are based on the DCFs that occur throughout the useful life of a solar thermal project. The usual question is what do we mean by DCF in the project life?

10.7.1 Life cycle cost analysis

LCC gives a basis for comparing solar energy technologies to conventional energy technologies. This method accounts the total system cost during a specified time period (life of the project). It comprises the initial investment and operational cost during the useful life. LCC is the sum of total cost that includes not only initial investment but also costs directly related to repair, operation, maintenance, transportation to the site, and fuel used to run the system. All of these costs are discounted with a minimum attractive rate of return (MARR) to the present value (PV). An LCC analysis allows the designer to study the effect of using different components with different reliabilities and lifetimes. It is also helpful for comparing costs of different designs and/or determining whether a hybrid system would be a cost-effective option.

The equation of LCC analysis is (Foster, Ghassemi, & Cota, 2009):

$$\mathrm{LCC} = C + M_{wp} + E_{wp} + R_{wp} - S_{wp} \tag{10.9}$$

LCC = Life cycle cost.

C = initial cost of installation-the present value of the cost on capital resources.

M_{WP} = Aggregation of all yearly operation and maintenance costs-includes wage of the operators, site access, guarantees paid, and all other regular maintenance costs.

E_{WP} = Aggregation of all yearly energy cost including fuel cost and its transportation to the plant site.

R_{wp} = Aggregation of all yearly replacement costs.

S_{wp} = Salvage value.

10.7.2 Net present value analysis

The net future earnings are discounted to the base year with the rate selected to justify MARR. The investment is deducted from the present sum of benefits. This value is called NPV (Abdul Malek, Hasanuzzaman, Rahim, & Al−Turki, 2020).

$$\mathrm{NPV} = -S + \frac{CF_1}{(1+r)^1} + \frac{CF_2}{(1+r)^2} + \cdots + \frac{CF_T}{(1+r)^T} = -S + \sum_{j=1}^{T} \frac{CF_j}{(1+r)^j} \tag{10.10}$$

Net present value analysis sample problem

Problem: Two types of heat exchanger system are under consideration for a solar domestic hot water plant. The economic attributes are as follows (Table 10.1):

The opportunity cost of capital is 10%. Identify the better alternative on the basis of NPV.

TABLE 10.1 Economic performance heat exchanger A and heat exchanger B.

Type	A	B
Initial investment	\$50,000	\$70,000
Annual maintenance cost	\$2000	\$3000
Economic life	10 years	10 years
Salvage value	\$5000	\$7000

Solution: NPV for both type of heat exchanger system can be calculated using Eq. (10.10).

$$
\begin{aligned}
\text{NPV of type A: NPVA} &= -\$50,000 + \frac{-\$2000}{(1+0.1)^1} + \frac{-\$2000}{(1+0.1)^2} \\
&+ \frac{-\$2000}{(1+0.1)^3} + \frac{-\$2000}{(1+0.1)^4} + \frac{-\$2000}{(1+0.1)^5} \\
&+ \frac{-\$2000}{(1+0.1)^6} + \frac{-\$2000}{(1+0.1)^7} + \frac{-\$2000}{(1+0.1)^8} \\
&+ \frac{-\$2000}{(1+0.1)^9} + \frac{-\$2000}{(1+0.1)^{10}} + \frac{\$5000}{(1+0.1)^{10}} \\
&= -\$60,361.42
\end{aligned}
$$

$$
\begin{aligned}
\text{NPV of type B: NPVB} &= -\$70,000 + \frac{-\$3000}{(1+0.1)^1} + \frac{-\$3000}{(1+0.1)^2} \\
&+ \frac{-\$3000}{(1+0.1)^3} + \frac{-\$3000}{(1+0.1)^4} + \frac{-\$3000}{(1+0.1)^5} \\
&+ \frac{-\$3000}{(1+0.1)^6} + \frac{-\$3000}{(1+0.1)^7} + \frac{-\$3000}{(1+0.1)^8} \\
&+ \frac{-\$3000}{(1+0.1)^9} + \frac{-\$3000}{(1+0.1)^{10}} + \frac{\$7000}{(1+0.1)^{10}} \\
&= -\$85,734.90
\end{aligned}
$$

Since *Type A* heat exchanger system incurred least cost, it should be used for the solar domestic hot water plant.

NPV shows the following advantages for solar thermal project assessment

- Time value of money is included in NPV analysis.
- This method considers cash flows disregarding the accounting profits. All cash flows are considered, but noncash flow benefits are not taken into consideration.
- NPV method is consistent with the objective of profit maximization.

But NPV has the following limitations for evaluating a solar thermal project

- The calculation is complex and hence requires skill handling.
- It is particularly difficult to quantify the return on investment in an economy where inflation varies year to year and hence necessitates year to year adjustment.
- NPV method does not consider hidden costs or incomes not shown as cash flow.
- Misleading results are probable when the projects are mutually exclusive. In that case, profitability index (PI) is a more suitable method for summarizing the output.

10.7.3 Internal rate of return analysis

IRR discounts all the cash back, providing zero NPV throughout the investment life of the solar project (Abdul Malek et al., 2020).

$$NPV = -S + \sum_{J=1}^{T} \frac{CF_J}{(1 + IRR)^j} = 0 \tag{10.11}$$

This method uses a widely understood percentage rate as the decision variable to compare mutually exclusive investments or individual investments whether public or private. Incremental IRR analysis is preferred to individual analysis by analysts.

IRR method for solar thermal project analysis has the following merits:

- Time value of money is taken into account.
- Negative values can be used.
- No need to have a precise calculation of discount rate, only a guess is supplied for assessment.
- The output of IRR method is a rate, which can directly measure of project profitability.

In analyzing the economic performance of a solar thermal project, the use of IRR method got the following limitations:

- The calculation of IRR is to go through a trial-and-error method, so it is difficult to attain the final point if done manually.
- It is assumed that cash flows generated by the project can be reinvested at its IRR. This is quite unrealistic.
- IRR can have a negative value. Moreover, there is a possibility of having multiple internal rates of return to be produced for the same project.

There exists a huge theoretical preference for NPV analysis for project appraisal, and investigations suggest that corporate executives prefer IRR analysis over NPV analysis. Actually, managers like to compare projects of varying sizes in terms of predictive performance, using IRR as a decision metric put a summary value of the firm performance rather NPV returns a value of merit figure not a rate. IRR method is obviously a shortcut of assessing the economic viability of a project.

10.7.4 Discounted payback period

Discounted payback period (DPP) is a modified version of the payback period that accounts for the time value of money. This is the time period when the project cash inflows reach a "break even" or to get the point where the net cash flows generated is equal to the initial cost of the project. DPP can be used to evaluate the profitability and feasibility of a solar project (Alam, Singh, Suresh, & Redpath, 2017). DPP can be calculated by solving the following equation (Orioli & Di Gangi, 2015).

$$\sum_{1}^{DPP} \frac{CF_n}{(1+r)^n} = 0 \tag{10.12}$$

where

CF_n is the cash flow related to the n-th year and r is the discount rate.

Discounted payback period has the following limitations

- No track is kept for the cash flows in the project life after the recovery period is achieved.
- This method may not be consistent with the goals of profit maximizing for the business owners. And, the cash flows occur after the DPP attained are generally insignificant, but practically all the cash flows throughout the span of economic life contribute the project outcome.
- DPP method plays a minor role in mutually exclusive project selection.
- Discount rate is considered fix for the whole span of project. But in practical, the rate must be adjusted for inflation in a regular interval (Table 10.2).

TABLE 10.2 Sample microsoft excel worksheet to evaluate NPV, IRR, and PBP (Malek, Hasanuzzaman, Rahim, & Al Turki, 2017).

Year	Generation (kWh)	Revenue Earning (MYR)	Cost of sales (MYR)	Gross profit (MYR)	Operating Expenses (MYR)	CER/Tax (MYR)	EAT (MYR)	Cumulative Earning (MYR)
0							−54,576,000	−54,576,000
1	65,700,000	13,797,000	7,450,366	6,346,634	586,146	2,436,500	8,196,988	−46,379,012
2	70,080,000	14,716,800	7,972,960	6,743,840	606,827	2,601,500	8,738,513	−37,640,499
3	79,978,800	16,795,548	8,889,223	7,906,325	628,282	2,976,215	10,254,258	−27,386,241
4	79,978,800	16,795,548	9,172,407	7,623,141	650,544	2,976,215	9,948,812	−17,437,429
5	79,978,800	16,795,548	9,464,912	7,330,636	673,644	2,976,215	9,633,207	−7,804,222
6	79,978,800	16,795,548	9,767,059	7,028,489	697,617	2,976,215	9,307,087	1,502,865
7	79,978,800	16,795,548	10,079,180	6,716,368	722,497	2,976,215	8,970,086	10,472,951
8	79,978,800	16,795,548	10,401,620	6,393,928	748,321	2,976,215	8,621,822	19,094,773
9	79,978,800	16,795,548	10,734,737	6,060,811	775,127	2,976,215	8,261,899	27,356,672
10	79,978,800	16,795,548	10,734,737	6,060,811	802,955	2,976,215	8,234,071	35,590,743
11	79,978,800	16,795,548	11,800,419	4,995,129	831,847	1,040,821	3,122,461.5	38,713,205
12	79,978,800	16,795,548	11,538,820	5,256,728	861,845	1,098,721	3,296,162.25	42,009,367
13	79,978,800	16,795,548	11,910,593	4,884,955	892,996	997,990	2,993,969.25	45,003,336
14	79,978,800	16,795,548	12,294,800	4,500,748	925,346	893,851	2,681,551.5	47,684,888

Continued

TABLE 10.2 Sample microsoft excel worksheet to evaluate NPV, IRR, and PBP (Malek, Hasanuzzaman, Rahim, & Al Turki, 2017).—cont'd

Year	Generation (kWh)	Revenue Earning (MYR)	Cost of sales (MYR)	Gross profit (MYR)	Operating Expenses (MYR)	CER/Tax (MYR)	EAT (MYR)	Cumulative Earning (MYR)
15	79,978,800	16,795,548	12,691,877	4,103,671	958,944	786,182	2,358,545.25	50,043,433
16	79,978,800	16,795,548	13,102,279	3,693,269	993,842	674,857	2,024,570.25	52,068,003
17	79,978,800	16,795,548	13,526,474	3,269,074	1,030,094	559,745	1,679,235	53,747,238
18	79,978,800	16,795,548	13,964,952	2,830,596	1,067,755	440,710	1,322,130.75	55,069,369
19	79,978,800	16,795,548	14,418,218	2,377,330	1,106,884	317,612	952,835	56,022,203
20	79,978,800	16,795,548	14,886,798	1,908,750	1,147,541	190,302	570,907	56,593,110
21	79,978,800	16,795,548	15,333,402	1,462,146	1,181,967	70,045	210,134	56,803,244
						IRR	12.40%	
						PV	61,274,142.67	
						NPV	6,698,142.67	
						PBP	5.84	

10.7.5 Levelized cost of energy

LCOE is a uniform equivalent rate that is calculated from the revenue stream of an energy project. The revenue generated is discounted at IRR to yield an NPV. The calculated NPV is converted to annual payments and then divided by the project's annual energy output. The unit stands at \$/kWh. This LCOE is a first order parameter to evaluate projects attractiveness. The LCOE should be at a comparable level to defend the competitor's price. LCOE analysis of power generation plant is a price estimation based on specific assumptions. The assumptions are made for the simplification of calculations (Table 10.3). A standard form used by most of the industries worldwide is as below:

$$LCOE(T) = \frac{\sum_{n=0}^{T} \frac{TIC(n) + OM(n) + FC(n)}{(1+r)^n}}{\sum_{n=0}^{T} \frac{EP(n)}{(1+r)^n}} \tag{10.13}$$

where *TIC* is the total investment cost in the year, *OM* is the annual operation and maintenance cost, *FC* is the annual fuel cost, *EP* is the estimated annual generation, and *T* is.

10.7.6 Profitability index

PI is the ratio of the future cash flows to initial investment. If the value is 1, then the project is at breakeven point and greater than one means project is profitable. If mutually exclusive projects are ranked based on PI, then it eases the decision-making. If an individual project shows to have a PI ratio less than 1, then it is indicative that the future cash inflows cannot cover the expenditures.

The simple relation of PI in terms of NPV and I_0 can be written as:

$$PI = \frac{NPV}{I_0} + 1 \tag{10.14}$$

The present value of a single payment made in the future can be written as (Foster et al., 2009):

$$PV = FV(1+i)^{-n} \tag{10.15}$$

Profitability index (PI) is a relative parameter. It shows how much present value of cash inflows generated for each dollar invested. It is a ratio not having unit unlike NPV which is expressed in currency units.

Decisions for using the profitability index

Accept the investment project proposal if index is greater than 1.0.

Reject the project proposal if index is smaller than 1.0.

When the index equals 1.0, it makes it indifferent whether accept or reject.

TABLE 10.3 Sample microsoft excel worksheet for evaluating LCOE.

Item	Capital investment (MYR)	Total annual cost (MYR)	Energy output (kWh)	Discount factor (1 + d)^n	Discounted cost Cn/(1 + d)^n	Discounted generation Qn/(1 + d)^n
Year 0	54,576,000		1	1.00	54,576,000	1
Year 1		8,036,512	65,700,000	1.06	7,581,615	61,981,132
Year 2		8,579,787	70,080,000	1.12	7,635,980	62,370,951
Year 3		9,517,505	79,978,800	1.19	7,991,081	67,151,743
Year 4		9,822,951	79,978,800	1.26	7,780,697	63,350,701
Year 5		10,138,556	79,978,800	1.34	7,576,119	59,764,812
Year 6		10,464,676	79,978,800	1.42	7,377,184	56,381,898
Year 7		10,801,677	79,978,800	1.50	7,183,732	53,190,470
Year 8		11,149,941	79,978,800	1.59	6,995,611	50,179,689
Year 9		11,509,864	79,978,800	1.69	6,812,671	47,339,329
Year 10		11,537,692	79,978,800	1.79	6,442,587	44,659,744
Year 11		12,632,266	79,978,800	1.90	6,654,520	42,131,834
Year 12		12,400,665	79,978,800	2.01	6,162,751	39,747,013
Year 13		12,803,589	79,978,800	2.13	6,002,822	37,497,182
Year 14		13,220,146	79,978,800	2.26	5,847,283	35,374,700

Year 15		13,650,821	79,978,800	2.40	5,696,011	33,372,359
Year 16		14,096,121	79,978,800	2.54	5,548,886	31,483,357
Year 17		14,556,568	79,978,800	2.69	5,405,791	29,701,281
Year 18		15,032,707	79,978,800	2.85	5,266,616	28,020,076
Year 19		15,525,102	79,978,800	3.03	5,131,248	26,434,034
Year 20		16,034,339	79,978,800	3.21	4,999,583	24,937,768
			1,495,419,601		184,668,787	895,070,073
			LCOE	0.206318		

So, the investment alternatives should be ranked from highest index to lowest one.

Profitability analysis sample problem

The opportunity cost of capital is 10%. Identify the best alternative among those three using PI (Table 10.4).

Solution: Profitability index for the three mutually exclusive projects can be calculated using Eq. (10.14).

Profitability Index of Project A:

$$PI_A = \frac{\frac{\$12,500}{(1+0.1)} + \frac{\$12,000}{(1+0.1)^2} + \frac{\$11,500}{(1+0.1)^3} + \frac{\$11,000}{(1+0.1)^4} + \frac{\$7000}{(1+0.1)^4}}{\$40,000} = 1.05$$

Profitability Index of Project B:

$$PI_B = \frac{\frac{\$13,000}{(1+0.1)} + \frac{\$12,000}{(1+0.1)^2} + \frac{\$11,000}{(1+0.1)^3} + \frac{\$10,000}{(1+0.1)^4} + \frac{\$9,000}{(1+0.1)^5} + \frac{\$8,500}{(1+0.1)^5}}{\$42,000} = 1.14$$

Profitability Index of Project C:

$$PI_C = \frac{\frac{\$13,500}{(1+0.1)} + \frac{\$12,500}{(1+0.1)^2} + \frac{\$11,500}{(1+0.1)^3} + \frac{\$10,500}{(1+0.1)^4} + \frac{\$9,500}{(1+0.1)^5} + \frac{\$8,500}{(1+0.1)^6} + \frac{\$10,000}{(1+0.1)^6}}{\$55,000} = 0.99$$

Since project B has the highest profitability index, it should be chosen among the three alternatives.

TABLE 10.4 Three mutually exclusive projects of solar hot water plant are under consideration. The economic attributes are as follows.

Project	A	B	C
Initial investment	$40,000	$42,000	$55,000
Year 1 profit	$ 12,500	$ 13,000	$ 13,500
Year 2 profit	$ 12,000	$ 12,000	$ 12,500
Year 3 profit	$ 11,500	$ 11,000	$ 11,500
Year 4 profit	$ 11,000	$ 10,000	$ 10,500
Year 5 profit	-	$ 9000	$ 9500
Year 6 profit	-	-	$ 8500
Economic life	4 years	5 years	6 years
Salvage value	$7000	$8500	$10000

10.7.7 Sensitivity analysis

A sensitivity analysis illustrates how much the merit figures will change in response to a given change in an input variable. There always exist some critical parameters which have significant impact on the final sought parameters like NPV or IRR. For example, the estimate of energy produced from a solar energy project is often a major factor. Cost of the project, and estimated operation and maintenance cost are other factors generally considered to have greater impact in a sensitivity analysis.

A sensitivity analysis done for the operation of a power generation plant with revenue earning, costs of generation, and operational expenses as the parameters to have significant impact on IRR and payback period (PBP). These parameters are plotted with ±10% change from the business-as-usual scenario. Table 10.5 below gives the sensitivity analysis done in the three parameters, revenue earning; the cost of goods sold and operational cost, and the resulted changes in IRR and PBP. When all other parameters are fixed and revenue earning is declined by 10%, then the IRR becomes −3.13%. This negative IRR means the project cannot payback the investment in its lifetime and thus the PBP is not available in this condition. Similarly, the revenue earning increase by 10% causes IRR changes from 4.31% (base case) to 9.10%. Hence project turns to earn positive NPV and the corresponding PBP is 7.6 years only. Revenue earning (i.e., FiT rate) is the sensitive issue in the case of biomass-based power generation project.

As the operational cost of a plant run on biomass cannot be expected to decrease over the years, the first cost of project installation must be curtailed. These can happen if the government ensures the tax credit and subsidy in the import items of the equipment needed.

Fig. 10.3 shows the sensitivity of IRR with respect to revenue earning, the total cost of power generation and operational cost of generation. The IRR of a solar thermal project is highly sensitive to revenue earning and cost of investment. The operational cost shows a less sensitivity. Perhaps, the earning

TABLE 10.5 Sensitivity analysis of the project IRR and payback period.

	−10%		−5%		0%		+5%		+10%	
	IRR	PBP	IRR	PBP	IRR	PBP	IRR	PBP	IRR	PBP
Revenue	−3.1	NA	1.2	13.8	4.3	10.3	6.8	8.7	9.1	7.6
Costs	7.4	8.5	6.0	9.3	4.3	10.3	2.3	12.1	0.0	15.6
Operation and maintenance	4.5	10.1	4.4	10.2	4.3	10.3	4.2	10.5	4.0	10.6

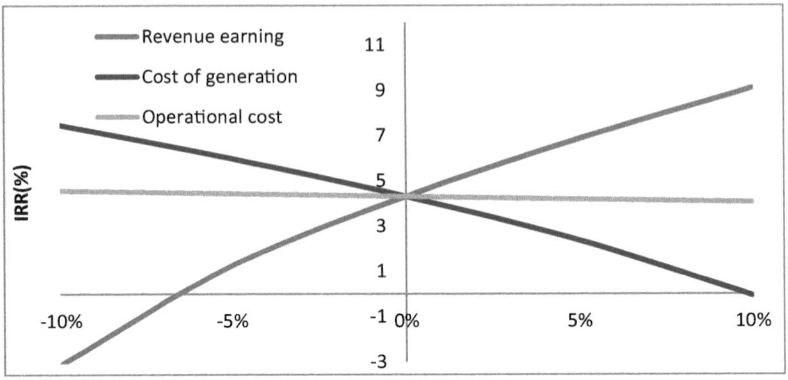

FIGURE 10.3 Sensitivity of IRR at the variation of revenue earning, generation cost, and operational cost.

is based on the selling to the utility company and the rate if low the IRR is low. The implication of the IRR sensitivity curve is that the pricing of the energy generated should be increased to make the plant operation to be competitive with traditional power generation units.

References

Abdul Malek, A. B. M., Hasanuzzaman, M., Rahim, N. A., & Al−Turki, Y. A. (2020). Energy, economic, and environmental analysis of 10-MW biomass gasification based power generation in Malaysia. *Energy & Environment*. https://doi.org/10.1177/0958305X20930386

Alam, M., Singh, H., Suresh, S., & Redpath, D. A. G. (2017). Energy and economic analysis of vacuum insulation panels (VIPs) used in non-domestic buildings. *Applied Energy, 188*, 1−8. https://doi.org/10.1016/j.apenergy.2016.11.115

Cagliano, A. C., Grimaldi, S., & Rafele, C. (2015). Choosing project risk management techniques. A theoretical framework. *Journal of Risk Research, 18*(2), 232−248. https://doi.org/10.1080/13669877.2014.896398

Foster, R., Ghassemi, M., & Cota, A. (2009). *Solar energy: Renewable energy and the environment*. CRC Press.

González-Portillo, L. F., Muñoz-Antón, J., & Martínez-Val, J. M. (2017). An analytical optimization of thermal energy storage for electricity cost reduction in solar thermal electric plants. *Applied Energy, 185*, 531−546. https://doi.org/10.1016/j.apenergy.2016.10.134

Greening, B., & Azapagic, A. (2014). Domestic solar thermal water heating: A sustainable option for the UK? *Renewable Energy, 63*, 23−36. https://doi.org/10.1016/j.renene.2013.07.048

Guo, P., Zhai, Y., Xu, X., & Li, J. (2017). Assessment of levelized cost of electricity for a 10-MW solar chimney power plant in Yinchuan China. *Energy Conversion and Management, 152*, 176−185. https://doi.org/10.1016/j.enconman.2017.09.055

Hasanuzzaman, M., Al-Amin, A. Q., Khanam, S., & Hosenuzzaman, M. (2015). Photovoltaic power generation and its economic and environmental future in Bangladesh. *Journal of Renewable and Sustainable Energy, 7*(1). https://doi.org/10.1063/1.4906910

Hasanuzzaman, M., Zubir, U. S., Ilham, N. I., & Che, H. S. (2017). Global electricity demand, generation, grid system, and renewable energy polices: A review. *Wiley Interdisciplinary Reviews: Energy and Environment, 6*(3). https://doi.org/10.1002/wene.222

Hertz, D. B. (1964). Risk analysis in capital investment. *Harvard Business Review, 42*, 95–106.

Hillson, D. A. J. T. I. J.o. P., & Management, B. R. (1997). Towards a risk maturity model. *The International Journal of Project and Business Risk Management, 1*(1), 35–45.

Islam, M. A., Hasanuzzaman, M., Rahim, N. A., Nahar, A., & Hosenuzzaman, M. (2014). Global renewable energy-based electricity generation and smart grid system for energy security. *Scienctific World Journal*, 2014. https://doi.org/10.1155/2014/197136

Kamarulzaman, A., Hasanuzzaman, M., & Rahim, N. A. (2021). Global advancement of solar drying technologies and its future prospects: A review. *Solar Energy, 221*, 559–582. https://doi.org/10.1016/j.solener.2021.04.056

Kassem, A., Al-Haddad, K., Komljenovic, D., & Schiffauerova, A. (2016). A value tree for identification of evaluation criteria for solar thermal power technologies in developing countries. *Sustainable Energy Technologies and Assessments, 16*, 18–32. https://doi.org/10.1016/j.seta.2016.02.003

Keshavarzzadeh, A. H., Ahmadi, P., & Rosen, M. A. J. J.o. C. P. (2020). *Technoeconomic and environmental optimization of a solar tower integrated energy system for freshwater production.*

Kumar, D. (2015). Economic assessment of photovoltaic energy production prospects in India. In *Paper presented at the global challenges, policy framework & sustainable development for mining of mineral and fossil energy resources.*

Lam, P. T. I., & Law, A. O. K. (2018). Financing for renewable energy projects: A decision guide by developmental stages with case studies. *Renewable and Sustainable Energy Reviews, 90*, 937–944. https://doi.org/10.1016/j.rser.2018.03.083

Malek, A. B. M. A., Hasanuzzaman, M., Rahim, N. A., & Al Turki, Y. A. (2017). Techno-economic analysis and environmental impact assessment of a 10 MW biomass-based power plant in Malaysia. *Journal of Cleaner Production, 141*(Suppl. C), 502–513. https://doi.org/10.1016/j.jclepro.2016.09.057

Mauthner, F., Weiss, W., & Spörk-Dür, M. J. I. S. H. C. P. (2016). *Solar heat worldwide: Markets and contribution to the energy supply 2014.*

Nikoofard, S., Ismet Ugursal, V., & Beausoleil-Morrison, I. (2014). An investigation of the technoeconomic feasibility of solar domestic hot water heating for the Canadian housing stock. *Solar Energy, 101*, 308–320. https://doi.org/10.1016/j.solener.2014.01.001

Orioli, A., & Di Gangi, A. (2015). The recent change in the Italian policies for photovoltaics: Effects on the payback period and levelized cost of electricity of grid-connected photovoltaic systems installed in urban contexts. *Energy, 93*, 1989–2005. https://doi.org/10.1016/j.energy.2015.10.089

Owens, G. (2002). *Best practices guide: Economic & financial evaluation of renewable energy projects*. Retrieved from Washington DC, USA https://pdf.usaid.gov/pdf_docs/PNADB613.pdf.

van de Poel, I. (2009). Values in engineering design. In A. Meijers (Ed.), *Philosophy of technology and engineering sciences* (pp. 973–1006). Amsterdam: North-Holland.

Soomro, M. I., Mengal, A., Memon, Y. A., Khan, M. W., Shafiq, Q. N., & Mirjat, N. H. (2019). Performance and economic analysis of concentrated solar power generation for Pakistan. *Processes, 7*(9). https://doi.org/10.3390/pr7090575

Stoddard, L., Abiecunas, J., & O'Connell, R. (2006). *Economic, energy, and environmental benefits of concentrating solar power in California*. Retrieved from Colorado, USA.

Wang, Z., Guo, P., Zhang, H., Yang, W., & Mei, S. (2017). Comprehensive review on the development of SAHP for domestic hot water. *Renewable and Sustainable Energy Reviews, 72*, 871–881. https://doi.org/10.1016/j.rser.2017.01.127

Wang, Z., Huang, Z., Zheng, S., & Zhao, X. (2018). 6 - solar water heaters. In T. M. Letcher, & V. M. Fthenakis (Eds.), *A comprehensive guide to solar energy systems* (pp. 111–125). Academic Press.

Yeo, K. T., & Ren, Y. (2009). Risk management capability maturity model for complex product systems (CoPS) projects. *Systems Engineering, 12*(4), 275–294. https://doi.org/10.1002/sys.20123

Zheng, Y., & Hatzell, K. B. (2020). Technoeconomic analysis of solar thermal desalination. *Desalination, 474*, 114168. https://doi.org/10.1016/j.desal.2019.114168

Chapter 11

Environmental impact assessment of solar thermal energy

Hossain Al Tanjil[1] **and Md Hasanuzzaman**[2]
[1]*Department of Environmental and Resource Management, Brandenburg University of Technology, Cottbus, Germany;* [2]*Higher Institution Centre of Excellence (HICoE), UM Power Energy Dedicated Advanced Centre (UMPEDAC), University of Malaya, Kuala Lumpur, Malaysia*

Chapter outline

Technologies for Solar Thermal Energy. https://doi.org/10.1016/B978-0-12-823959-9.00006-4

11.1 Introduction

Environmental impact is characterized as any change to the environment, regardless of whether detrimental or beneficial, resulting from a facility's activities, products, or services. Environmental impact assessment (EIA) is the examination of the environmental result of a scheme, strategy, program, or actual projects preceding the combined choice of pushing in front direction and the processed activity. While EIA identifies with explicit projects, environmental assessment is a generic term, which likewise consolidates strategic environmental assessment of approaches, plans, and programs, and different types of evaluation. The overview of impacts of solar thermal energy (STE) to the environment, its life cycle assessment (LCA), and global warming impacts are the main objectives of this chapter.

11.2 Environmental impact assessment

EIA is an official evaluation of probable impacts on the earth, which applied by European Council EIA Directive 85/337/EEC, provided lots of recognizable proof, depiction as well as evaluation of the positive and negative impacts of the program on: people, flora and fauna, water, air, soil, environmental resources, and on cultural traditions.

EIA first evolved in the USA, after the central government brought into impact, an open law "National Environmental Protection Act" (NEPA) on January first, 1970. NEPA made all the formative undertakings to experience an environmental investigation before execution. Consequently, this required the arrangement of rules and strategies in regard to the preparation, audit, assessment, and dispersion of environmental impact statement (EIS) with the essential target of assessing the environmental effects of any formative action. Thereafter, most of the nations established departments to execute EIA. Although an EIA can be led even by a private organization, yet a responsible authority does the decision-making.

Environmental assessments are significant environmental protection tools. By including specialists, authorities, and residents, and incorporating ecological reports, the potential environmental impacts of a planned project can be recognized at a beginning period and mulled over during the decision-making process. The objective of completing environmental assessment is to ensure human health protection and the natural environments from the predictable, hurtful impacts of planned modern industrial facilities and foundation measures. Likewise, by making straightforwardness and involving the public in decision-making processes, environmental assessment additionally helps to increase more extensive acknowledgment for the project in question. Another objective is to give candidates and specialists arranging security in regard to the project.

11.2.1 Purposes of EIA

Purposes of EIA includes:

- Determination of the principle natural manifestation and perspectives
- Understanding the general importance of the common ecological viewpoint
- Appraisement the environmental activity of the intended plan against proper significant prospects
- Distinguishing proof of indicative partial and impartial effects
- Appreciation of overall environmental effect of proposed plan in order to establish parallelism with other proposals
- Help a comprehensive methodology along with venture part

11.2.2 Procedures and stages of EIA

EIA was commenced under the US Environmental Policy Act in 1969 to recognize the potential result of a suggested construction scheme on the environment. A comparable methodology has been adopted by most developed and many developing nations, often embedded into their legislation program. How an EIA is done is not inflexible: it is a cycle containing a progression of steps. The fundamental strides in the EIA cycle are:

- Distinguishing proof of project needing EIA, at times known as screening
- Distinguishing proof of major issues to be tended to in an EIA, called scoping
- Effect distinguishing proof, assessment, and evaluation
- Arrangement of the environmental perspectives and their importance
- Impact alleviation and surveillance

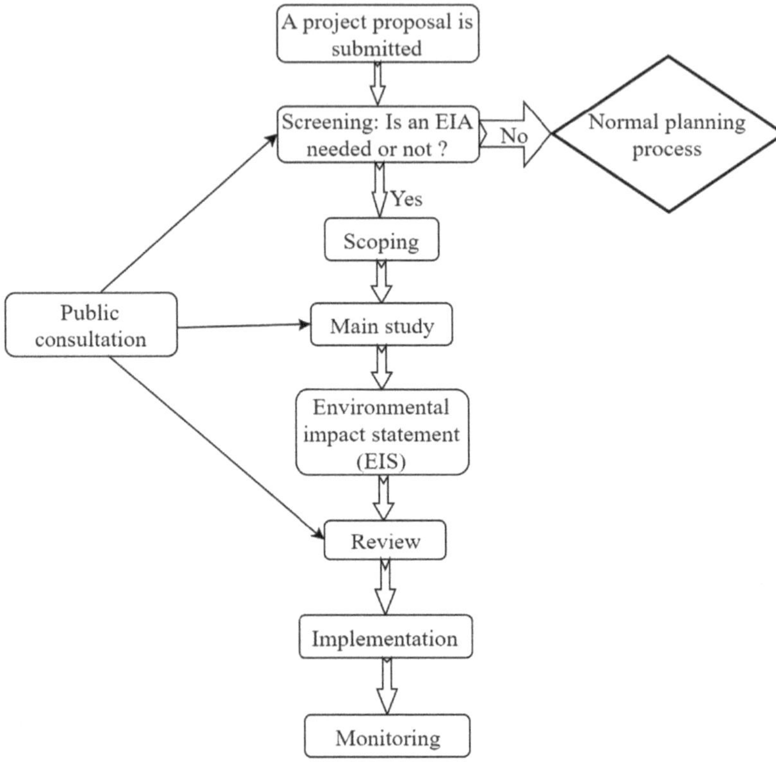

FIGURE 11.1 A standard modified flow diagram of EIA (Wang, Morgan, & Cashmore, 2003).

- Assembly of the aftereffect of an EIA in an archive known as an EIS that relates all the positive and negative impacts of a specific task on the environment
- Audit of the finalized EIS
- Communal involvement

The standard steps in a typical EIA process are shown in Fig. 11.1.

11.2.3 Methods for EIA

The major evaluating methods to assess the impacts of any formative action on the environment are enumerated in the following sections:

Adhoc method

Adhoc method demonstrates broad regions of potential effects by listing composite environmental parameters (viz., flora and fauna) liable to be affected by the proposed activity. These strategies include collecting a group of experts who recognize impacts in their subject matter. Moreover, every parameter is considered independently and the idea of impacts (long haul or short haul, reversible or irreversible) are thought of.

Checklist method

In checklist method, environmental factors are recorded in an organized pattern by giving significance weightings for factors and use of scaling methods for impacts of each alternatives. Checklists are solid markers of impact identification. They adequately earn the consideration and awareness with their audience. In checklist technique, the effects are classified as cells with data either in the graphic structure that gives data in regard to probability or potential presence of an impact while in the scaling or weighting methods, the magnitude or significance of impact is given. It is easy to comprehend, utilize, and use for site choice and need setting.

Matrix method

A framework of interaction of various exercises of a project with potential environmental effects is given by this method. A basic collaboration matrix is shaped when project activities are recorded on one axis (generally vertical) and environmental impacts are recorded along the other axis. This strategy was spearheaded by Leopold et al. (1971).The benefits of this matrix technique is that it links action to impact and an excellent technique for showing EIA results.

Network method

This technique uses the matrix approach and extends it to fuse both the primary and secondary effects. It is showed up as a tree called impact tree. This outline is in like manner called as reference or arrangement graph. Unmistakable confirmation of immediate, backhanded alongside short, long stretch effect is a dire and fundamental advance of having an effect tree. The effect tree is used to recognize cause-impact linkages and to show visual portrayal.

Overlay method

This strategy depends upon a lot of guides of a task domain's ecological attributes covering physical, social, natural, and stylish perspectives. It enables separate planning of basic natural features at a comparative scale as venture's site plan (e.g., wetlands, steep grades, soils, floodplains, bedrock outcrops, untamed life situations, vegetative systems, social resources, etc). In the old method, ecological highlights were planned on clear plastic in different hues. Present day strategy of a comparative activity is done using PC programming, equipment, information, and talented people. It is called geographic information system. The following methods are also used in some cases: environmental index using factor analysis, cost-benefit analysis, and predictive or simulation methods.

11.3 Life cycle assessment of industrial solar thermal systems

Industries have low energy requirements and low utilization of water temperature (40–80°C) that can be satisfied by solar thermal (ST) systems (Angeliki, Paris, Andreas, & Soteris, 2018). The environmental performance of any industrial ST systems is defined by LCA. Three separate assortments of solar energy transformation can be recognized: solar chemical energy, sunlight-based photovoltaic (PV) energy, and STE (light vitality to heat) (Nelson, Gambhir, & Ekins-Daukes, 2014, pp. 1–16). To evaluate the process of a STE, it should correlate with its industrial application and with the life cycle cost. So that, it will be easier to assess the life cycle. The whole procedures for the LCA can be presented under a framework in Fig. 11.2.

11.3.1 LCA scopes and inventory

For the duration of the existence pattern of a cycle, LCA permits the measurement of natural effects and the assessment of the improvement alternatives. In a few life cycles stages, the accompanying terms are utilized: measure choice, utilized materials, structure, end of life removal, and framework advancement. As expounded inside the ISO standard 14040 (ISO, 2006), LCA addresses the challenges of measuring ecological effects (e.g., the work of assets and furthermore the natural outcomes of releases) all through a product's life cycle from staple obtaining, through creation, use, end-of-life treatment, business, and last removal (e.g., cradle to-grave), in this way evading trouble moving. These qualities fabricate LCA an applicable and comprehensive philosophy that allows a precise ecoplan of products (Jaume et al., 2019). The products' life cycle boundaries conjointly illustrated in Fig. 11.3.

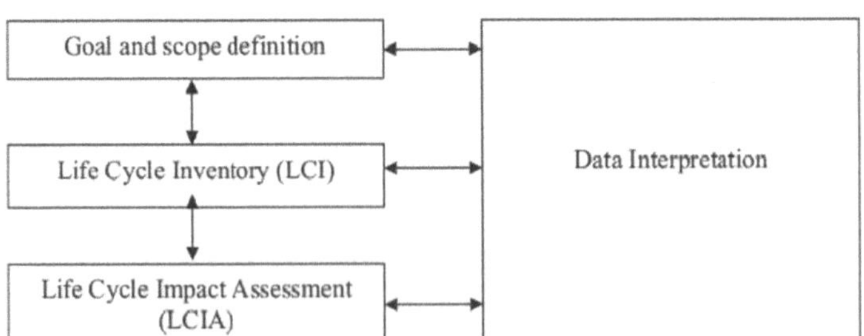

FIGURE 11.2 Framework of life cycle assessment (Norasikin et al., 2018).

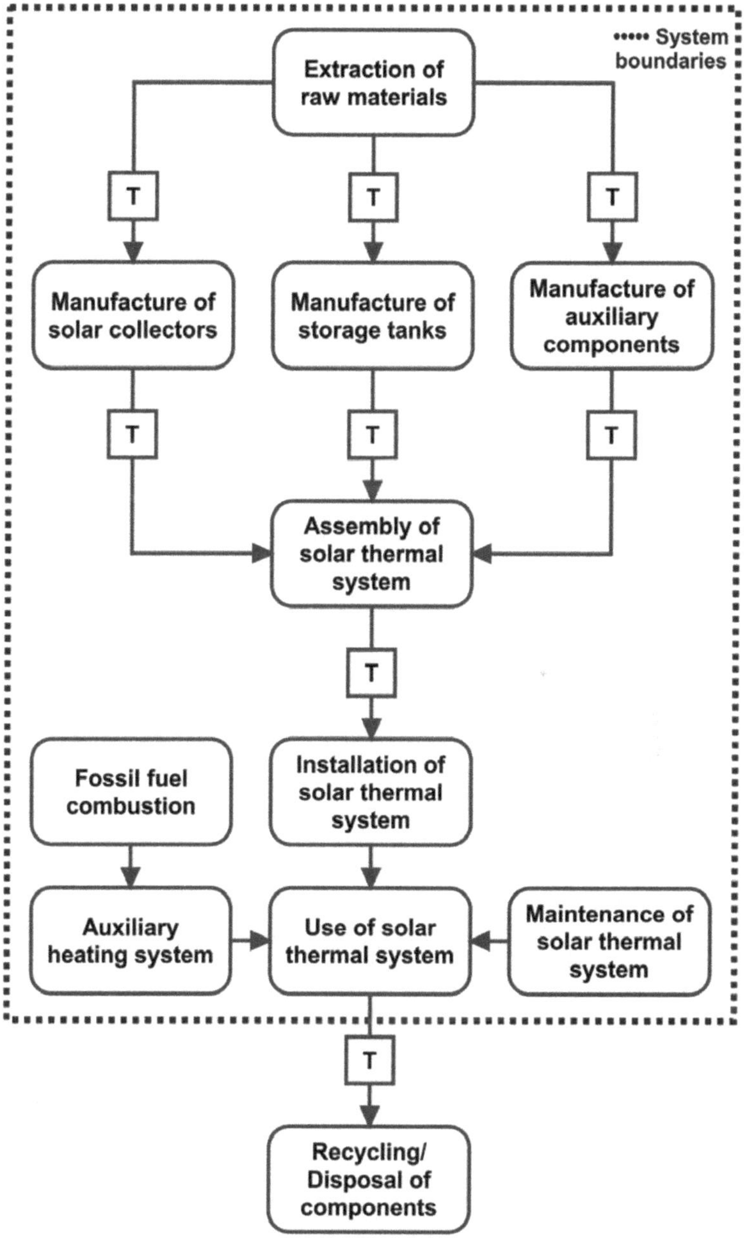

FIGURE 11.3 Solar thermal system boundaries (Angeliki et al., 2018).

11.3.2 Life cycle cost and support schemes

Economic life cycle analysis is a direct and advantageous innovative program (Colombo, Gilliaert, & Landabaso, 1992). It requires the summation of the energy prices of the instrumentality (or building) at the side of cyber web prices of purchase, installation, maintenance, operation, replacement, and everyone alternative prices attributed to the present investment. Salvage price is not thought of as value, however, as revenue. The investment that has the lower total life cycle value whereas meeting the investor's objective then selected. A support topic alludes to any instrument, topic, or system applied by a nature that advances the work of power from inexhaustible sources by decreasing the estimation of that energy, increasing the value of that energy and the energy's selling value, or increasing its sold value, or increasing the energy purchased degree. This could encapsulate venture aid, tax exceptions or decreases, tax returns, sustainable power source commitment uphold plots, and direct endowment plans (Angeliki et al., 2018).

The economic analysis is used in ELCCA-PC program is the yearly life cost approach using cash flow methodology.

The residual cash flow for a domestic ST system is (Colombo et al., 1992):

$$\begin{aligned}\text{Yearly cost with solar system} &= \text{Mortgage payment}\\ &+ \text{Auxiliary fuel costs} + \text{Property tax increase}\\ &+ \text{Insurance premium} + \text{Operating costs} + \text{Maintenance costs}\\ &- \text{Income tax saving for interests and taxes paid}\end{aligned} \tag{11.1}$$

11.3.3 Life cycle impact assessment

The life cycle impact assessment (LCIA) can be created dependent on the Eco-Indicator 99 (EI99) sway evaluation technique, extracted from the information base Eco-Invent 2009. The ecological loads into three separate injury classes: human well-being, conspire quality, and assets. The examination of each impact classification is given by Eduard, Antoni, Alvaro, Dieter and Luisa (2012):

$$\text{IMP}_j = \sum_k \left(d_{kj}.LCI_k\right) \tag{11.2}$$

where IMP_j = the j impact class (human well-being, conspire quality, and assets), d_{kj} = the constant of injury (will be extracted from database) related to the part k and impact j, LCI_k = the entry of Life Cycle Inventory. The constant of injury of the assets sway class is communicated in MJ of surplus energy required for future extraction. Life cycle boundary is the step in which the practitioner uses impact catagories, catagory indicators, characterization

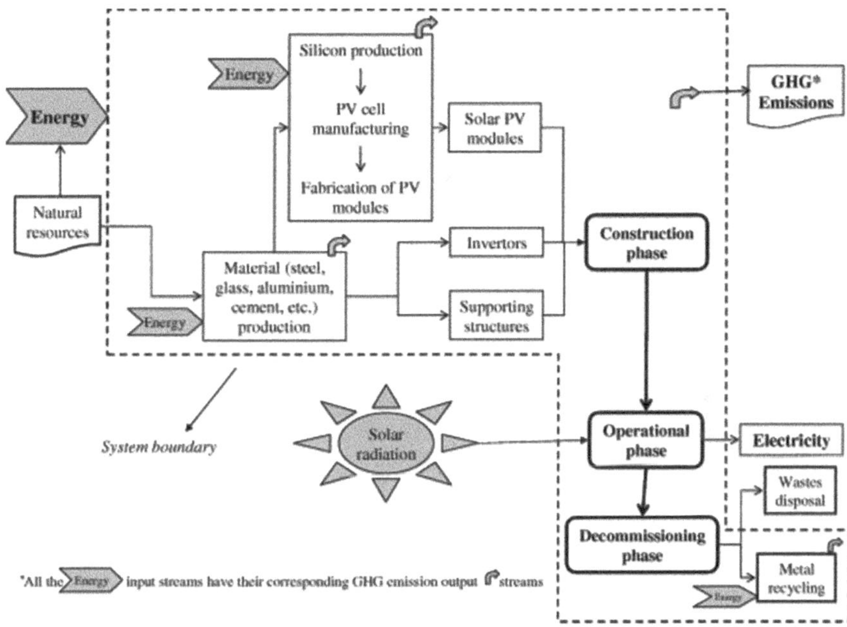

FIGURE 11.4 LCA boundary of a solar PV system (Kannan, Leong, Osman, Ho, & Tso, 2006).

models, equivalency factor and weighting values to realize the potential impact of data that are obtained in LCI analysis as shown in Fig. 11.4.

The life cycle of a ST force to be reckoned with is finished utilizing the three natural load profiles. To begin with, the energy balance factor speaks to the quantitative connection of energy output to input, that is utilized to confirm whether the framework is conceivable as related degree energy creation framework. Second, the energy payback time speaks to the record's list, by energy creation, for years expected to recuperate the general energy contribution to the framework over an entire life cycle. Third, the CO_2 emanation factor, that is, the CO_2 outflow per unit of electricity creation.

The fundamental hypothetical conditions for the three ecological load profiles laid out as:

$$\mathrm{EBF} = \frac{\mathrm{LCEO}}{\mathrm{LCOE} + \sum_i \mathrm{LCEE}_i} \tag{11.3}$$

$$\mathrm{EPT} = \frac{\sum_i \mathrm{LCEE}_i}{\mathrm{AEO} - \mathrm{AOE}} \tag{11.4}$$

$$\mathrm{CEF} = \frac{\sum_j \left[\left(\mathrm{LCOEE}_j + \sum_i \mathrm{LCEEE}_{ij}\right) \cdot \mathrm{CEE}_j\right]}{\mathrm{LCEO}} \tag{11.5}$$

where LCEO = the energy yield from related in nursing energy transformation plant over its whole cycle; LCEE_i and LCOE territory unit the "gear" and "activity" energies over the existence pattern of each strategy for the framework, severally (i demonstrates each technique, i.e., handling material and instrumentality, transportation, and development); AEO = the yearly energy yield; AOE = the yearly energy vitality of plant; $LCEEE_{ij}$ and $LCOEE_j$ zone unit the instrumentality and activity energies of each sensibly energy assets, e.g., power, coal, and oil, over the existence pattern of each technique, severally (j speaks to each sensible energy assets); and CEE_j = the CO_2 emission per unit energy of each energy asset.

The accompanying condition is utilized to ascertain the instrumentality energy over the entire life cycle of the framework:

$$\sum_i LCEE_i = \sum_i (ME_i + PE_i + TE_i + CE_i) \tag{11.6}$$

where ME_i = fabric energy, PE_i = production energy, TE_i = transportation energy, and CE_i = construction energy of each method, severally.

On the contrary hand, activity energy is plot on the grounds that the energy is expected to work a framework and comprises of the fuel utilization of warming turbines in startup and accordingly the power expended in useable instrumentality. The activity energy over the total life cycle of the framework is determined on the possibility of the resulting condition:

$$LCOE = FC_{fuel} + FC_{electricity} \tag{11.7}$$

where FC_{fuel} = the fuel utilization for warming turbines inside the startup of electricity generation sets, and $FC_{electricity}$ = electricity requested useable that is given via autogenic power.

Global warming potential and acid potential examination are needed to perform to recognize the ecological effect attributes of the ST force to be reckoned with. Upheld the ecological effect qualities of the STP plants, warming emanations like greenhouse discharge, NOx, CO, and regular activity outflows like SO_2, and Roman divinity inside the existence pattern of the capacity plant zone unit evaluated and broke down. The equivalency factor approach is required for the GWP investigation. At long last, the all out of the proportional qualities is accomplished on the grounds that the absolute impact potential (NARNRC, 2010).

11.4 LCA analysis of the fuel mix power generation

By 2050, the overall carbon dioxide emanations would be lessened up to 75th analyzed thereto of carbon dioxide content enrolled in the year of 1985 (Johanson, Kelly, Reddy, & Williams, 1993). A combined cycle may be proposed to help a substitution design of crossover biomass solar-powered energies composed of biomass segment, solar segment, and turbine segment (Anvari, Khalilarya, & Zare, 2019; Ilham, Hasanuzzaman, & Hosenuzzaman, 2017; Malek, Hasanuzzaman, & Rahim, 2020, Malek, Hasanuzzaman, Rahim, & Al Turki, 2017, Malek, Hasanuzzaman, Rahim, & Al Turki, 2021), which can be denoted as fuel mix power generation system.

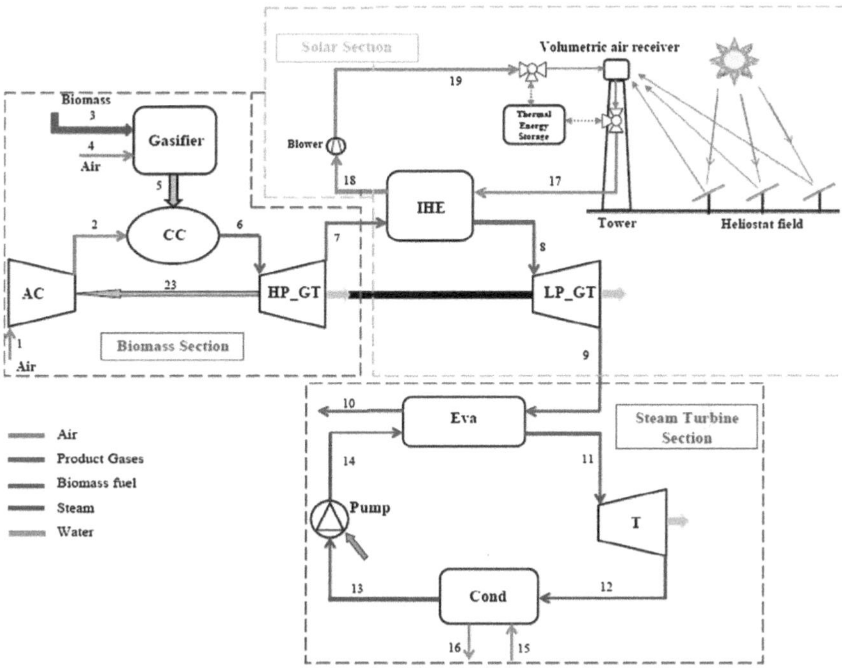

FIGURE 11.5 Simplified diagram of a proposed combined cycle (Anvari et al., 2019).

A combined fuel mix power generation cycle is indicated below in Fig. 11.5 to make it easy to explain. The cycle in three segments along with biomass (in that the chemical change of biomass is actualized), sunlight heat segment (included heliostat field with thermal energy storing tank), and turbine area created by steam (with squander heat healing from the nonforceful turbine by the evaporator).

11.4.1 System description and assumptions

A schematic diagram of a combined cycle is presented in Fig. 11.5. The framework comprises of three subframeworks: biomass area, star segment, and therefore the turbine segment. The ecological air is demonstrated in the figure on the grounds that the compound change agent enters to the gasifier wherever the biomass is taken care of for chemical change. It is significant that inside the biomass segment for assembling highest achievable power, associate in nursing inside jobless turbine is utilized. Accordingly, the biofuel made inside the gasifier streams into the combustion chamber any place it is responded with the compacted gas from the mechanical device. The high temperature burning product region unit swollen inside the aggressive turbine (HPGT) to concoct power, when that will endure the intermediate device (IHE) any place it is extra by the solar power. The solar power is gotten by the heliostat field so reflected into the collector. Inside the receiver, the warmth is moved to the passing air and warms it up, all together that it will warm the ignition stock inside the IHE as referenced. The ignition stock leaving the IHE endures the unaggressive turbine (LPGT) to concoct the power before coming into the evaporator of the temperature unit cycle. Accordingly, with sick the waste warmth of the aggressive turbine and utilization of that inside the unaggressive turbine, power is produced inside the unaggressive turbine also. For building up a fuel mix power generator, the accompanying suppositions are expected to consider for proposing a model. The system is operated under steady-state conditions. The basic parameters and assumed values for the purpose of proposing a combined cycle are given in Table 11.1.

TABLE 11.1 The basic parameters and assumed values for the purpose of proposing a combined cycle are given below (Gholamian et al., 2016; Soltani et al., 2013; Taghavifar et al., 2015; Zare & Hassanzadeh, 2016).

Parameter	Assumed value
P_o(bar)	1.013
T_o(K)	298.15
$\eta_{AC} = \eta_{HP\_GT} = \eta_{LP_{GT}}(\%)$	0.86
$\eta_{ST}(\%)$	0.88
$\eta_P(\%)$	0.92
$r_{P,AC}(-)$	10
MC (%)	20
$T_5 = T_g$(K)	1073
$p_5 = p_g$(bar)	1.013
T_6(K)	1370
HHV of wood as the biomass (kJ/kg)	449,568
$DNI(W/m^2)$	850
η_{cos}	0.8267
$\eta_{s\&b}$	0.9698
η_{int}	0.971
η_{att}	0.9383
η_{ref}	0.88

Continued

TABLE 11.1 The basic parameters and assumed values for the purpose of proposing a combined cycle are given below (Gholamian et al., 2016; Soltani et al., 2013; Taghavifar et al., 2015; Zare & Hassanzadeh, 2016).—cont'd

Parameter	Assumed value
N_{hel}	350
$A_{hel}(m^2)$	121.3
$\Delta T_{PP,\ Eva}$	15
$T_{Eva}s(K)$	753.2
$p_{Eva}(bar)$	140
T_{Cond} (K)	313.2

11.4.2 Designing and investigation

To demonstrate and recreate the thermodynamically execution of an extended framework, pc programs are manufactured exploitation of the engineering equation solver (EES) code. Inside the simulation strategy, all aspects of the idea of framework is treated as an effect volume that the standards of mass conservation and accordingly the first and second laws of thermodynamics region unit applied. The demonstrating strategy for proposing a consolidated cycle is required for investigating devices made of energy, exergy, environmental, and economic examination. The different equations of energy investigating instruments will be brought inside the content per totally various segments. Then, the total equations beside thought of assumptions governing the cycle will be resolved at the same period for the entire purposes of the cycle by EES and thus, the yield boundaries like performance parameters just as energy and exergy efficiencies, power creation, carbon dioxide outflow, and exergy obliteration of grouped portions will be inferred.

Energy analysis (first law)

The mass and energy change equation for the control volume under steady-state conditions could be written as:

$$\sum (\dot{m})_i - \sum (\dot{m})_o = 0 \tag{11.8}$$

$$\sum \left(\dot{m}h\right)_i - \sum \left(\dot{m}h\right)_o + \left(\sum \dot{Q}_i - \sum \dot{Q}_o\right) + W = 0 \tag{11.9}$$

Biomass segment

In the underlying, law examination applied for proposing a joined cycle. A colossal measure of consideration is paid to the demonstrating of biomass gasifier and burning chamber rather than the contrary parts, and subtleties of the simulation models are outline here. The applied energy balance and elective overseeing conditions to display the framework execution will be outline for each part as follows:

- **Air compressor**

Physical property potency and therefore the absorbed energy in AC are as follows (Anvari et al., 2015):

$$\eta_{AC} = \frac{h_{2s} - h_1}{h_2 - h_1} \tag{11.10}$$

$$\dot{W}_{AC} = \dot{m}_1 (h_2 - h_1) \tag{11.11}$$

- **Biomass gasifier**

Generally, wood is taken into account because of the biomass fuel. The material science balance model is probably going to simulate the chemical change measure. The equilibrium model accepts that everyone the responses square measure inside thermodynamic balance and furthermore the change item burns and arrives at the balance state in the reduction zone before deed the gasifier (Sahoo, Kumar, Pant, & Chaudhury, 2015). The responses inside the reduction zone square measure follows (Gholamian, Mahmoudi, & Zare, 2016; Grange, Dalet , Falcoz, Siros, & Ferriere, 2014):

$$C + CO_2 \leftrightarrow 2CO \tag{11.12}$$

$$C + H_2O \leftrightarrow CO + H_2 \tag{11.13}$$

$$C + 2H_2 \leftrightarrow CH_4 \tag{11.14}$$

Eqs. (11.11) and (11.12) can be jointed to provide the shift reaction as below:

$$CO + H_2O \leftrightarrow CO_2 + H_2 \tag{11.15}$$

The equilibrium constants for methane series formation reaction (Eq. 11.13) and for the shift reaction square measure shown by Eqs. (11.15) and (11.16), severally:

$$K_1 = \frac{n_5}{n_1^2} \left(\frac{\frac{P}{P_{ref}}}{n_{tot}} \right)^{-1} \tag{11.16}$$

$$K_2 = \frac{n_1 n_3}{n_2 n_4}\left(\frac{\frac{P}{P_{ref}}}{n_{tot}}\right)^0 \tag{11.17}$$

where n_1 to n_5 area unit the mole numbers within the chemical change merchandise in combining weight. In Eq. (11.21), K_1 and Godwin Austen area unit the equilibrium constants, which may be associated with the amendment within the chemist operate as follows:

$$\frac{-\Delta G_1^o}{RT_g} = \ln K_1 \tag{11.18}$$

$$\frac{-\Delta G_2^0}{RT_g} = \ln K_2 \tag{11.19}$$

wherc

$$-\Delta G_1^0 = \left(\underline{h}_{CH_4} - T_g \underline{s}_{CH_4}^0\right) - 2\left(\underline{h}_{H_2} - T_g \underline{s}_{H_2}^0\right) \tag{11.20}$$

$$\begin{aligned} -\Delta G_2^0 = &\left(\underline{h}_{CO_2} - T_g \underline{s}_{CO_2}^0\right) - 2\left(\underline{h}_{H_2} - T_g \underline{s}_{H_2}^0\right) \\ &- \left(\underline{h}_{CO} - T_g \underline{s}_{CO}^0\right) - \left(\underline{h}_{H_2O} - T_g \underline{s}_{H_2O}^0\right) \end{aligned} \tag{11.21}$$

Now, in the case for a biomass, the global gasification reaction can be written like (Anvari, Taghavifar, Khoshbakhti, Khalilarya, & Jafarmadar, 2015):

$$\begin{aligned} CH_aO_bN_c + wH_2O + m(O_2 + 3.76N_2) \rightarrow n_1H_2 + n_2CO \\ + n_3CO_2 + n_4H_2O + n_5CH_4 + n_6N_2 \end{aligned} \tag{11.22}$$

where $CH_aO_bN_c$ indicates the biomass, $w =$ biomass moisture content, and $m =$ the kilomoles of oxygen/kilomoles of biomass. As indicated by references (Gholamian et al., 2016; Soltani, Mahmoudi, Yari, & Rosen, 2013), the chemical formula of wood is $CH_{1.44}O_{0.66}$. The biomass moisture content, w, can be determined as:

$$w = \frac{M_{Biomass}MC}{18(1 - MC)} \tag{11.23}$$

where MC = moisture content/mole of wood as written:

$$MC = (\text{mass of water} / \text{mass of wet biomass}) \times 100 \tag{11.24}$$

The seven unknown coefficients of n_1–n_6 and m can be calculated by applying the mass conservation principle for H, O, C, and N. The energy balance equation for the gasifier, assuming an adiabatic gasification at a given temperature, can be expressed as (Gholamian et al., 2016):

$$\underline{h}^0_{\text{f-wood}} + w \times \underline{h}^0_{f\text{-H}_2\text{O}} + n_{\text{air},4} \times \underline{h}^0_{\text{air},4} = n_1\left(\underline{h}^0_{f\text{-H}_2} + \Delta\underline{h}_{\text{H}_2}\right) + n_2\left(\underline{h}^0_{f\text{-CO}} + \Delta\underline{h}_{\text{CO}}\right) + n_3\left(\underline{h}^0_{f\text{-CO}_2} + \Delta\underline{h}_{\text{CO}_2}\right) + n_4\left(\underline{h}^0_{f\text{-H}_2\text{O}} + \Delta\underline{h}_{\text{H}_2\text{O}}\right) + n_5\left(\underline{h}^0_{f\text{-CH}_4} + \Delta\underline{h}_{\text{CH}_4}\right) + n_6\left(\underline{h}^0_{f\text{-N}_2} + \Delta\underline{h}_{\text{N}_2}\right) \tag{11.25}$$

- **Combustion chamber**

The syngas leaving from the gasifier goes inside the burning chamber and responds with the compacted gas returning from the mechanical device.

For a complete combustion, the equation will be:

$$n_1\text{H}_2 + n_2\text{CO} + n_3\text{CO}_2 + n_4\text{H}_2\text{O} + n_5\text{CH}_4 + n_6\text{N}_2 + n_{\text{air},2}(\text{O}_2 + 3.76\text{N}_2) \rightarrow n_7\text{CO}_2 + n_8\text{H}_2\text{O} + n_9\text{O}_2 + \left(n_6 + 3.76n_{\text{air},2}\right)\text{N}_2 \tag{11.26}$$

For an adiabatic burning, the energy balance condition in the ignition chamber is written as:

$$\sum_j X_j\left(\underline{h}^0_{fj} + \Delta\underline{h}\right)_{\text{produced gas}} + \sum_j X_j\left(\underline{h}^0_{fj} + \Delta\underline{h}\right)_{\text{air},2} = \sum_j X_j\left(\underline{h}^0_{fj} + \Delta\underline{h}\right)_{\text{products}} \tag{11.27}$$

As to essential law investigation, the whole thermal energy contribution to the biomass area, related with the coming into biomass to the gasifier, are frequently written as:

$$\dot{Q}_{\text{Biomass}} = \dot{m}_{\text{biomass}} \times LHV_{\text{biomass}} \tag{11.28}$$

where $\dot{m}_{\text{biomass}}$ = mass flow rate

$\text{LHV}_{\text{biomass}}$ = lower heating value

- **High pressure gas turbine**

The isentropic productivity and the created power in HPGT are communicated as (Anvari et al., 2015):

$$\eta_{GT,HP} = \frac{h_6 - h_7}{h_6 - h_{7s}} \tag{11.29}$$

$$\dot{W}_{GT,HP} = \dot{m}_6(h_6 - h_7) \tag{11.30}$$

Solar segment

In the extended combined cycle, the star segment is supplemental to warm up the leaving gas from the HPGT. The star area comprises of the heliostat field (he), collector (rec), TE storage (TES), middle tool, and unaggressive turbine (LPGT). The principal suppositions and information record application for the reproduction of the star area square measure are given in Table 11.1. For each component of this area, the energy balance and distinctive administering conditions square measure measurable as follows:

- **Heliostat field**

The entire heat got by the heliostat field can be determined by Zare and Hassanzadeh (2016):

$$\dot{Q}_{\text{sun}} = \text{DNI} \times A_{\text{hel}} \times N_{\text{hel}} \tag{11.31}$$

where A_{hel} = realm of every heliostat, N_{hel} = the total variety of heliostat and DNI (direct traditional Irradiance) = quantity of star TE acquired by the heliostats. The potency of heliostat field, η_{field}, is expressed by Zare and Hassanzadeh (2016):

$$\eta_{\text{field}} = \eta_{\text{cos}} \times \eta_{\text{s\&b}} \times \eta_{\text{int}} \times \eta_{\text{att}} \times \eta_{\text{ref}} \tag{11.32}$$

where η_{cos} is that the trigonometric function result potency, $\eta_{\text{s\&b}}$ = shading and blocking efficiency, η_{int} = interception efficiency, η_{att} = attenuation efficiency, and η_{ref} = reflectivity of the heliostats.

- **Receiver**

Receiver sock up ($\dot{Q}_{\text{rec,in}}$) from the heliostat field, transferring a portion of it to the IHE ($\dot{Q}_{\text{rec,net}}$) while the rest of the aspect of the retained energy is lost to the ecology by convective, conductive, and reflective heat misfortunes ($\dot{Q}_{\text{rec,loss}}$). Here, energy balance for central receiver can be expressed as:

$$\dot{Q}_{\text{rec,net}} = \dot{Q}_{\text{rec,in}} - \dot{Q}_{\text{rec,loss}} \tag{11.33}$$

Also, the following equations are needed to apply:

$$\dot{Q}_{\text{rec,net}} = \dot{m}_{17}(h_{17} - h_{19}) \tag{11.34}$$

$$\dot{Q}_{\text{rec,in}} = \eta_{\text{field}} \times \dot{Q}_{\text{sun}} \tag{11.35}$$

- **Thermal energy storage**

The alternative energy vessel is being charged for the duration of the day and inside the hours of the stature sunlight-based illumination so the TE are frequently utilized once the radiation is declined or once at night there is no alternative power. In other words, the vessel is utilized to get the steady power for the duration of the day and night (Kamarulzaman, Hasanuzzaman, & Rahim, 2021; Pantaleo et al., 2017; Saha, Ruslan, Morshed, & Hasanuzzaman, 2020). The intentional temperature shift of the alternative energy vessel for the capacity pinnacle of the capacity plants is 300–1000°C (Alva, Lin, & Fang, 2018).

Among totally various materials which will be applied since the alternative energy storing substance, salts square measure certifiable to possess a high point of freeze thus proper for putting away high-temperature energy storage. The salts square proportion of grouped assortments like nitrites, hydroxides, chlorides, carbonates, sulfates, and fluorides. From these, nitrites have the low freezing focuses and hydroxides have the melting point somewhere in the range of 250 and 600°C, while chloride, carbonate, sulfate, and halide have the point of freezing of 600°C (Alva et al., 2018). During this examination, LiF-NaF-KF (46.5–11.5–42) is utilized as TE storage inside the TES (Forcberg, Peterson, & Zhao, 2007). TE amount which is charged by the tank can be calculated by (Forcberg et al., 2007):

$$Q_{\mathrm{TES}} = m \cdot C_p \cdot T \tag{11.36}$$

Here, m = mass of the salt (kg), C_p = salt specific heat (KJ/kg), and ΔT = temperature change.

- **Intermediate heat exchanger (IHE)**

Heat transfer between the hot air and combustion gases are leaving from high pressure gas turbine. Now, the equation stated as:

$$\dot{Q}_{\mathrm{IHE}} = \dot{m}_{17}(h_{17} - h_{18}) = \dot{m}_6(h_8 - h_7) \tag{11.37}$$

- **Low pressure gas turbine (LPGT)**

Isentropic efficiency power creation by low pressure gas turbine can be written by:

$$H_{\mathrm{GT,LP}} = \frac{h_8 - h_9}{h_8 - h_{9s}} \tag{11.38}$$

$$\dot{W}_{\mathrm{GT,LP}} = \dot{m}_6(h_8 - h_9) \tag{11.39}$$

Steam turbine segment

Concerning temperature of leaving gas from the low-pressure gas turbine, a conventional steam Rankine cycle would be a legit alternative to be joined with the idea about hybrid sun-based biomass framework to recuperate the waste heat and improve the overall framework strength. The demonstration of a simple Rankine cycle is simple and furthermore the required overseeing conditions for each component of the framework square measure are offered in Table 11.2.

11.5 Emission factor and analysis

Around 40% of global greenhouse gas (GHG) discharges are given by power generation. The renewable energy certificate (REC) which is a power following and accreditation structure can be an incredible arrangement apparatus for advancing low-carbon power acquisitions by clients of Scope 2 of the Kyoto Protocol. The Kyoto Protocol set a phase for worldwide coordinated effort for carbon discharge decrease, just as methodized a system to top worldwide carbon emanations and collected the estimation of carbon outflow by freeing carbon as a tradable product (Chuang, Lien, Den, Iskandar, & Liao, 2018).

The difficulties of sustainable power source confirmation frameworks composed in every nation are daintily extraordinary. Solar renewable energy certificates (SRECs) have become a huge propeller for advancing solar PV framework, yet SREC value instability projects a significant interference to advocating the establishment of sun-oriented PV framework detailed by a study in the United States (Lee et al., 2017). Government approaches, sustainable power source testament framework, and assembled consumption all can influence the usage of REC scheme (Chuang et al., 2018).

TABLE 11.2 Basic parts and essential energy-based equations for the steam turbine section modeling.

Basic parts	Essential energy-based equations
Evaporator	$\dot{m}_9(h_9 - h_{10}) = \dot{m}_{11}(h_{11} - h_{14})$
Steam turbine	$\dot{W}_T = \dot{m}_{11}(h_{11} - h_{12}),\ \eta_{ST} = \frac{W_a}{W_s} = (h_{11} - h_{12})/(h_{11} - h_{12s})$
Condenser	$\dot{m}_w(h_{16} - h_{15}) = \dot{m}_{11}(h_{12} - h_{13})$
Pump	$\dot{W}_P = \dot{m}_{11}(h_{14} - h_{13}),\ \eta_P = \frac{W_s}{W_a} = (v_{14}(P_{14} - P_{13})/(h_{14} - h_{13}))$

11.5.1 Alliance between renewable energy certificate and emission factor

In this subsection, the overall importance of sustainable energies such as hydro, solar, wind, water energy, etc., for the purpose of CO_2 emission reduction is not going to be discussed. The problems calculation system and their relationship with the emission factor will be indicated here.

- **Issuance of RECs**

To settle on decisions in the utilization of vitality REC power is a significant device for enabling customers RECs present energy properties or data identified with how energy is created, however, the vitality itself is not encapsulated in the RECs. At the point when sustainable energy plants satisfy REC generation guidelines, they can give 1 REC per 1 MW of power produced.

With RECs, power purchasers are competent to guarantee that they are tolerating sustainable energy in this manner forestalling the discharge of GHGs. Along these lines, RECs are given could be dealed to buyers who need to make claims for utilizing sustainable power source. At the point when the vitality use market inspires appeal for sustainable power source, sustainable power source generators can decide to benefit from REC exchanges under an organized convention, delivering the REC structure a successful device for emission counting (Sotos, 2015).

- **Greenhouse gas emission factor by power creation**

A combination of power sources produced and spread in a network framework through which obtained enterprises can fluctuate enormously. GHG emission factor must be evaluated through a lot of qualities and straightforward rules for explicit network framework. For instance, public ozone-depleting substance investigation rules gave by intergovernmental boards on environmental change were commonly received universally to figure local or public lattice emission factor (EF), which shifts every year relying upon the energy portfolio has been utilized by the energy maker. Power sources, fuel types, and power generation efficiency have a great place in making the value of power grid emission factor (Chuang et al., 2018).

- **Assumptions and simplifications**

Fig. 11.6 schematically illustrates the stake holder relationships. In order to illustrate the subtleties and potential consequences of the REC system, the six initial assumptions are:

1. Atomic energy is not countable as a renewable energy source as of preventing radioactive hazards (e.g., Fukushima nuclear disaster). In Taiwan, the national policy will eliminate their nuclear facilities by the year of 2025. As a result, the total power generation capacity will arise only from thermal energy and renewable energy.

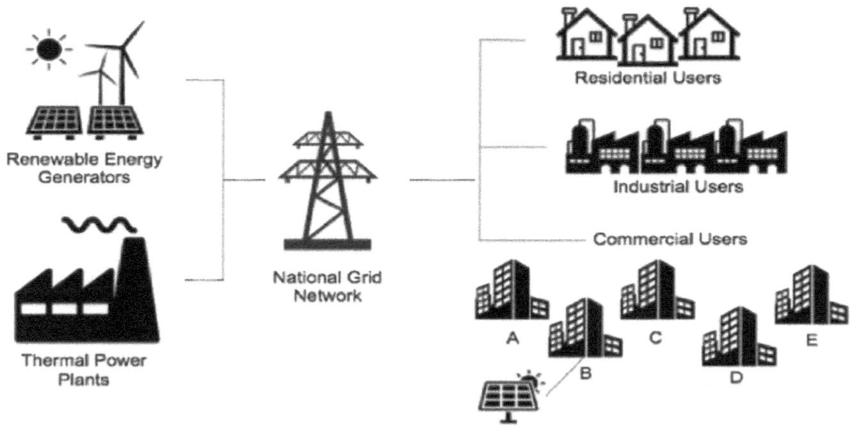

FIGURE 11.6 Schematic delineation of the circumstance examined in this area. The national power grid embraces 80% TE and 20% sustainable power source. Five theoretical business clients have been recognized as Entities A, B, C, D, and E. Substance B additionally acquires off-lattice sustainable force set inside its own limit (Chuang et al., 2018).

2. In the new energy portfolio, theoretically 20% is renewable energy and 80% is thermal energy.
3. From renewable energy power generation, there will be no carbon dioxide emission. In the case of thermal energy, it is considerable 1 CO_2 e M/Wh of the electricity produced.
4. In one megawatt/h renewable energy power there will be on REC.
5. Theoretical grid power generation is 100 megawatts/hour
6. The consumer pattern will be followed as by Fig. 11.6. Both REC distribution and REC generation are independent.

- **Evolution of framework**

Expanded scenes from EF's calculations are shown in Fig. 11.7. Basic and the modified EFs equations (later described here as "a mixture of residual EF") are listed below (Chuang et al., 2018):

$$\text{Electricity EF} = \frac{\text{Total amount of } CO_2 \text{ emission}}{\text{Total amount of electricity generation}} \tag{11.40}$$

$$\text{Residual Mix EF} = \frac{\text{Total amount of } CO_2 \text{ emission}}{\text{Total amount of electricity generation} - \text{REC}} \tag{11.41}$$

11.5.2 Achieving carbon emissions subsidence using solar energy

- **Ingredients/factors affecting carbon emissions mitigation**

The measure of CO_2 emissions that can be decreased by sustainable power source advances, for example, solar energy, over some undefined time frame

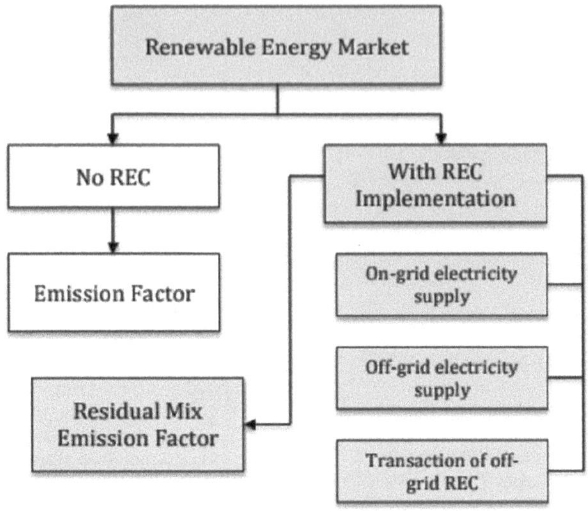

FIGURE 11.7 Logic flow in the situation (Chuang et al., 2018).

(typically, one year) relies upon various components. The emission magnitude is set by the measure of valuable power created by sun-based innovation in the separate time frame. The genuine degree of refinement is then found out by the distinction between the carbon density per (kWh) energy source dislodged by the sun-based and the carbon power of the subbed sun-based innovation. To compute the carbon power of a given sun-oriented innovation is not as direct as it might appear, on the grounds that it depends on where it is created as well as on how and where it is delivered.

The mitigation intensity of solar energy will be most noteworthy when the most effective sun-oriented power is introduced, when the carbon discharged per unit of electrical power delivered is reduced, and when sun powered confiscate the most carbon concentrated generation advances.

- Carbon intensity of solar electrical energy

To balance solar panels and systems are used carbon, through the energy used in production and transportation and the energy implanted in the materials used. This prompts a heightening of carbon for every unit limit between 600 kg CO_2/KWP for functional CdTe frameworks and around 1500–2500 kg CO_2/kWp for reasonable c-Si frameworks, mirroring a more power-escalated creation measure for silicon modules. This number relies upon the carbon power of the energy utilized underway, consequently on the production area. The intensity of carbon per kilowatt hour of electricity is acquired by partitioning the power of carbon per unit of energy by the energy created by the lifetime peak watts of the framework.

Relative amount is the carbon intensity of sun power which is utilized to fulfill the need instead of the power created. The disparity between asset changeability and flexibly and request implies that a portion of the measure of extra sun-oriented vitality expected to create uprooted power must be put away, dumped, or, maybe, enhanced by extra enough energy. These variables improve the carbon intensity and cost of valuable sunlight-based force. Dumping the power produced because of restricted storage capacity of the grid could double the force of solar carbon (Sandwell et al., 2016).

- Carbon intensity of displaced power

The intensity of carbon in grid power differs between areas, from around 900g CO_2 eq/kWh in India to around 500g CO_2 eq/kg every hour in the UK and less in nations like Norway and France. Sun-based assets additionally fluctuate between areas; the subsidence potential can differ by the situating of the PV arrangement. At the point when solar energy is utilized in off-framework circumstances, at that point the carbon intensity of the dis-positioned fuel and any significant plant ought to be utilized (Alstone, Mills, Jacobson, & Lai, 2014; Emmott, Jaoude, Urbina, Ekins-Dawkes, & Nelson, 2012; Hasanuzzaman, Zubir, Ilham, & Che, 2017).

Generally, the shipping amount was restricted by the significant expense of modules that brought about higher solar energy utilization than grid power, however since 2012 costs have ascended with grid-associated PV power matrices where retail power costs are higher, for example, Denmark, Germany, Italy, and other European nations are soon to follow (EPIA, 2011).

11.5.3 Carbon emissions subsidence costing

The cost adequacy of different carbon decrease techniques can be estimated by assessing the expense of executing a carbon saving measure for every unit CO_2, for example, against a reference situation. Most companies have outlined a marginal depreciation cost curve in the McKinsey report, highlighting CSP (18 dollar per ton carbon dioxide) and PV (19 dollar per ton carbon dioxide) as capacity solutions before the 2030 fossil solution (Barnhart, & Benson, 2013; Poyry, 2011; Sollmann, 2008).

For correlation, the Climate Change Committee's released estimates of the cost reduction of solar PV's 260/ton CO_2 deployed in existing residential buildings in the UK in 2020 show their strength, regional differences, and their affectability to previous studies, for this situation higher reconstruction costs and less sun radiation.

Despite modern possibilities, the high efficiency of modern gas boilers is expected to result in a significantly higher cost of 600/ton CO_2 compared to PV systems of solar water heating systems (Ardente, Beccali, Cellura, & Brano, 2005; Pineda, 2010). In Fig. 11.8, it is shown the GHG emissions for producing electricity generation for different renewable and nonrenewable energy sources.

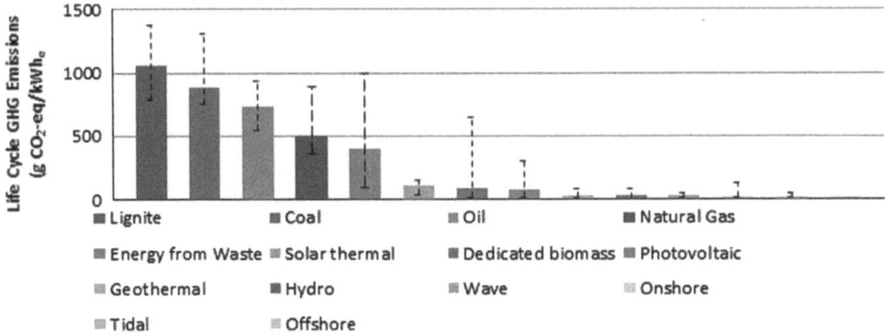

FIGURE 11.8 Life cycle GHG emission estimates of electricity generation methods (Amponsah, Troldborg, Kington, Aalders, Hough, 2014).

11.5.4 Essential emission calculation systems for proposing a combined system

In a conference on June 1992 at Rio de Janeiro, Brazil, the United Nations Conference on Environment and Development, expressed analytical concerns about sustainable global development. Accordingly, Agenda 21 (UNCED-endorsed) strategy has been included as "Appropriate new approaches or projects to upgrade the commitment of earth safe and savvy frameworks, particularly new and sustainable ones, through contaminating and more proficient energy creation, transmission, dissemination and use" (Kalogirou, 2013).

Recently, most natural experts and legitimate administrative devices share immersed for all intents and dangerous emanations. Nonetheless, natural difficulties have been exacerbated by the treatment of harmful air contaminations. Carbon dioxide is a noteworthy pollutant overall (Alauddin, 1996), which influences the ozone layer, the essential ozone-depleting substance.

For the purpose of proposing a fuel mix power generation system (Fig. 11.5), biomass-based systems are being added to solar segment to reduce CO_2 emissions. For biomass single gas turbine cycles, carbon dioxide emissions equation can be written as (Anvari et al., 2019):

$$\varepsilon_{CO_2-\text{Biomass Cycle}} = \frac{\left(\dot{m}_{CO_2}\right)_{\text{Biomass Cycle}}}{\dot{W}_{\text{Biomass Cycle}}} \tag{11.42}$$

Now, by integrating solar segment into biomass segment, numbers of carbon dioxide emissions can be written as:

$$\varepsilon_{CO_2-\text{Biomass Cycle Integrated With Solar Section}} = \frac{\left(\dot{m}_{CO_2}\right)_{\text{Biomass Cycle Integrated With Solar Section}}}{\dot{W}_{\text{Biomass Cycle Integrated With Solar Section}}} \tag{11.43}$$

Again, integrating straight turbine segment with Brighton cycle, carbon dioxide emissions for the projected technique can be written as:

$$\varepsilon_{CO_2-\text{Combined Cycle}} = \frac{\left(\dot{m}_{CO_2}\right)_{\text{Combined Cycle}}}{\dot{W}_{\text{Combined Cycle}}} \tag{11.44}$$

11.6 Global warming potential

Energy is the essential input and need for the advancement, modernization, and monetary development of any country in the mechanical division. The mechanical division plays a critical effect on monetary development for any country (Sharma, Sharma, Mullick, & Kandpal, 2017). Nonetheless, because of the fast development in worldwide energy request, the worry is the way to meet the future thermal power needs of the mechanical division (Abdelaziz, Saidur, & Mekhilef, 2011). Petroleum derivatives are singed to heat the cycle of modern applications that offer ascent to GHG outflows and lead to environmental change (Hasanuzzaman et al., 2012). Environmental change is a complex natural issue that prompts coastal soil degradation, ocean level ascent, and biological change. Financial development is straightforwardly relative to energy use (Ang, 2008; Saidur, Hasanuzzaman, & Rahim, 2012). Power utilization and outflows are legitimately joined, thus increment in energy utilization delivers a ton of emanations (Kumar, Hasanuzzaman, & Rahim, 2019; Mohamed, & Lee, 2006). The gathering of sun-based force in urban communities is unmistakably away to decrease our reliance on petroleum derivatives and might be an incredible method to relieve heating by bringing down the emanation of GHG (Gautam & Saini, 2020).

The large-scale energy storage can be defined as global warming. A worldwide temperature alteration is clarified here by disposing of heat from the worldwide utilization of nonsustainable power sources (petroleum products and atomic energy). Thus, net heat is delivered into the air. A little bit of this heat is produced into space as long wave radiation, while the rest is heated by the earth. A portion of this heat is put away as sensitively by the heat of air, land, and water. The rest is likewise put away as dormant heat, for example, vapor in the air and melting in the planet's enormous ice fields (Nordell, 2007).

Solar heating and cooling technologies are gaining ground in a number of countries with solar energy ground heating and cooling, although the primary application remains as the manufacture of hot water for sanitary purpose (ESTIF, 2015).

11.6.1 Global temperature

The research about an unnatural weather change is built on month to month temperature information from 1880 until today worldwide (Kundzewicz, Pińskwar, & Koutsoyiannis, 2020; Nordell, 2007). These temperatures are partitioned into three temperatures: land area temperature, sea surface temperature, and global average temperature. Ordinarily, MEAN is determined by the weight of the area relative to the global sea area (71%) and land territory (29%). In Fig. 11.9, it is shown the last 140 years global mean temperature change in land and ocean. Preceding that, the world average temperature was practically steady from 1856 to 1880 (Jones, Parker, Osborn, & Briffa, 2016, pp. 1850–2015; Nordell, 2007).

11.6.2 Net heat sources

There are two principal net heat sources on Earth: geothermal heat stream and net heat produced by human activities. The expanding utilization of non-sustainable power sources, predominantly petroleum products and atomic force has brought about extreme net heating on Earth. The utilization of sustainable power source scatters heat yet does not create any excess heat. In addition, a lot less unbroken yet regular net heat sources, for example, heat delivered from volcanoes, seismic tremors, and meteors. Besides, some ethnographic sources make up nuclear bomb tests, traditional bombs, and explosives (Nordell, 2007).

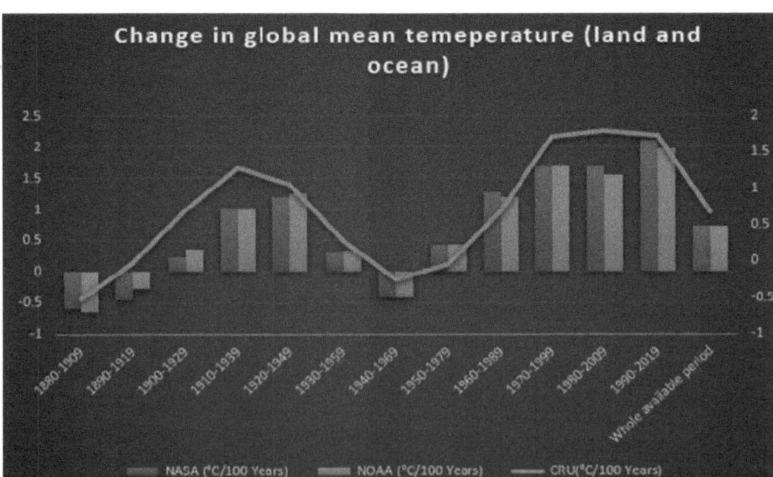

FIGURE 11.9 Change in global mean temperature in °C/100 years, for shifted 30-year intervals and for the whole examined interval Kundzewicz et al., 2020.

Geothermal heat flow

The ocean's geothermal heat flow varies from 0.065 to 0.101 W/m^2, with a mean estimation of 0.087 W/m^2. Estimations of mainland geothermal heat stream are made in boreholes typically penetrated to a profundity of a few hundred meters. Estimations on the ocean bottom are made in sediments (IHFC, 2001). Pollock and Chapman (1977) done the primary broad appraisal, which included 5500 estimations, 70% of which contained maritime information, bringing about a worldwide normal of around 0.060 W/m^2. Parasnis (1985, pp. 69–82) it was later indicated that high qualities happening along the Mediterranean locales give a maritime normal heat stream of 0.0798 W/m^2. By making up for the sea heat stream emerging from submarine magma ejections, the geothermal heat stream is 0.045–0.065 W/m^2 for both mainland and maritime estimations. The most recent plan of warmth stream data (Pollak, Hurter, & Johnson, 1993) numbers directly around 25,000 assessments.

Thermal pollution

The worldwide yearly utility of fossil power, including atomic force (BP-AMOCO, 2002; IEA, 1999), is near 9000 million tons of oil proportional. This power, proportionate to 11.63 T Wh/Mtoe, will radiate heat. At the point when the natural issue is deteriorated, the heat of the not completely consumed fuel will likewise be excluded. Along these lines, the aggregate sum of heat created by nonrenewable energy sources is 1014 kW h. By conveying this energy over the whole zone of the Earth, an extra 0.02 W/m^2 is warming the planet.

Additional net heat sources

A study revealed that, despite the power of such bombs, nuclear tests have indeed made a great contribution. The war studied (World War II and the Kuwait War (1990)) also depicts that explosions contributed less net heat than the secondary effects of war for burning the oil fields in Kuwait. Principal sources of discontinuity in net heat are volcanic eruptions, earthquakes, and falling meteorites.

11.6.3 Earth's radiation balance

The driving force of the Earth's air is engrossing sun-oriented power on its surface. The long wave radiation (OLR) heats the air while it emanates layer by layer through the atmosphere. At the top of the atmosphere, the long wave radiation, q (W/m^2) shot out from Earth to space as given by the Stefan–Boltzmann law:

$$q = \sigma T_s^4 \tag{11.45}$$

where T_s (K) is the effective mean temperature of the earth, and the Stefan–Boltzmann consistent $\sigma = 5.6697 \times 10{-}8$ (W/m^2 k^4). Estimations show that the average long wave radiation from the earth is 237 W/m^2 (Salby, 1996). Condition (12.45) shows that our planet is in thermal balance at an influenced temperature of 254.2 K.

Net long wave radiation

In the event that earth become observed from space, the yearly suggest qualities could show an amazing temperature of 254.2 K, where the approaching short-wave radiation is adjusted by utilizing the long wave radiation, i.e., a zero net warmness float. The overall mean temperature at floor surface is 33.1 K hotter than the effective Earth temperature, showing the thermally insulating highlights of nature. The suggested approach implies that the geothermal warmness accepts circumstances with the flow < internet OLR < (geothermal warmth skim + thermal pollutants) till the overall harmony temperature is accomplished. The worldwide mean temperature angle through the environmental factors is around linear between the ground surface of temperature Ts (K) and the Earth's blackbody layer of temperature Te (K). Mists typically spread around 1/2 of the planet's territory, which implies that the overall imply day is somewhat cloudy with light rain. During the day, short-wave radiation warms the ground surface that is later cooled off the guide of a similar measure of long wave radiation. The regular heat flow is emanated layer by layer through the environment. As indicated by the Stefan–Boltzmann law, between the N atmospheric layers the net OLR, Δq (W/m^2), becomes:

$$\Delta q = \frac{\sigma}{N}\sum_{1}^{N}\left(T_n^4 - T_{n+1}^4\right) = \frac{\sigma\left(T_1^4 - T_{n+1}^4\right)}{N} = \frac{\sigma\left(T_s^4 - T_e^4\right)}{N} \tag{11.46}$$

For net long wave radiation rates <0.1 W/m^2, Eq. (11.45) can be written as:

$$\Delta q = \frac{\sigma\left(T_s^4 - T_e^4\right)}{N} \approx \sigma\left(T_s - T_e\right)^4$$

The errors of this estimation is under 0.01% for applicable worldwide mean temperatures. The thickness of the environmental layers that is developing with the rise is connected to the optical profundity of the ecosystem.

By figuring the net long wave radiation from 1880 dependent on the worldwide normal month to month temperature (isolated by land area temperature and sea surface temperature), it very well may be seen that the worldwide temperature increment caused the net long wave radiation to ascend from 0.068 to 0.074 W/m^2 from the year 1880 to 2000. This heat flow rate is reliable with the prior heat flow assessments of Parasnis (1985, pp. 69–82), however lower than the most recent heat flow evaluations of Chapman (1998).

11.6.4 Global warming in large-scale thermal energy storage

At present the Earth's temperature is certainly not a thermal equilibrium, implying that the net long wave radiation is no higher than the net heat despite created. Common cold sinks in water, ground, and condition slow down the impact of thermal poisons. The chief assets of common cold are in water and ice. The general amount of world water is $1.4 \times 10^{18}\ m^3$, of which 94% is seawater simultaneously as $3 \times 10^{16}\ m^3$ (2%) of the water is ice inside the type of ice sheets and ice fields (Singh & Sing, 2001). The softening of 3×10^{16} ton of ice calls for around 3×10^{16} kWh of intensity. Since the once every year generally heat dissemination is 10^{14} kWh, the ice fields and ice glaciers would last 30,000 years, with blessing utilization of nonsustainable power source and the same cold sink.

The measure of resistance from long wave radiation will likewise build which demonstrates a developing temperature gradient. On the contrary hand, expanding humidity inside the environmental factors will mirror a huge bit of approaching sun-based radiation. Therefore, less long wave radiation is transmitted which must decrease Earth's effective temperature, T_e. A precise of global heat accumulation equation can be stated as:

$$
\begin{aligned}
&\text{Global net heat balance}\\
&= \frac{\text{Global Heat Accumulation due to Global Warming 1880} - 2000}{\text{Thermal Pollution(Global energy consumption} + \text{other net heat sources)}}\\
&\quad -\text{OLR}\\
&= \frac{\sum \text{Globally accumulated net heat(in air, ground and water)}}{\text{Preliminary estimations of accumulated heat}}\\
&\approx 40\%\ \text{in air}\\
&\approx 15\%\ \text{in ground}\\
&\approx 45\%\ \text{in water}\\
&= 100\%
\end{aligned}
\tag{11.48}
$$

11.6.5 Steady-state global warming

The future predictable-country worldwide temperature was foreseen by means of expecting that ceaseless warming follows the equivalent example as during the only last century. This implies the former LAT/SST increment proportion was thought to be 2.4 constant. In this manner, the land district temperature blast became accepted stuff 2.4 more noteworthy than the ocean surface temperature increment. The present utilization of noninexhaustible power changed into also expected unchanged. In a predictable state situation, all net warmness created is discharged to zone. Thus, by means of including the determined geothermal recommend warmth drift (0.068 W/m^2) and the measure of thermal contamination (0.20 W/m^2), the absolute net long wave radiation transforms into 0.088 W/m^2.

Around 1880, preceding overall warming, Earth was in thermal equilibrium at mean temperature of 13.6°C. The happening net long wave radiation was then proportionate to the geothermal warmth drift. Expanding groupings of gases, vaporizers, and stickiness into the environment will raise the Earth's albedo.

The present guidelines are targeting reducing the CO_2 discharges. There are a few thoughts accessible for putting away CO_2 in profound stores. Some global areas intend to intensify their atomic power industry. The main economical route is to utilize our own inexhaustible quality. With sustainable quality frameworks, it is not even achievable to pester the power strength of Earth (Nordell, 2007).

11.7 Energy and environmental impact

Risks from warm viscous liquids, pointed insolation and contamination of water sources are related with sun warm electric vegetation. Typically, air contamination is insignificant. There is little scattering to be discarded, anyway incidental release of hydrocarbons, oils, consumption inhibitors, bactericides, and glycols ought to undermine health (by means of ingestion or pores and skin contact) and water resources. The land place, around 0.4 ha/MW, required for a sun warm electric plant is comparative like conventionally fueled power create on because of the gigantic region identified with collector structures and related producing plant, subordinate and places of business, through landscaping and siting a long way from neighborhoods may likewise mitigate special visualization.

11.7.1 Principles of environmental impact

Biological effects can incorporate loss of or potentially changes to normal territories and, for exceptionally huge frameworks, microclimatic changes as indicated by nearby heat balance changes (OECD, 1988). Most locales considered as of now for sun-powered heat electric frameworks are in dry deserts. In spite of the fact that there are little human homes, there exist delicate ecologies which expect appraisal to affirm siting limits sway. Arranging, development and activity must be perceptive of the high disintegrating and environment misfortune potential.

Interruption as per development traffic is a small amount than for fuel or atomic plants as development is quicker. As light inadvertently engaged offhub may cause vision problems, insolation must not be reflected at houses, workplaces, and streets and individuals close by must wear suitable eye assurance. Auxiliary uses for land, notwithstanding, like to cow, sheep, or goats munching where grass can develop between the authorities stays practicable. Noise identified from ST facilities from fans, siphons, turbines, and cooling towers is similar to it of a steam producing plant. Sunlight-based warm electrical generation decisions do not cause radioactive material disposal issues, nor the most significant natural and social effects of the valley-flooding identified with hydropower.

11.7.2 Ecological impact

Sustainable development, futurity, equity, participation of normal civilized human beings and nature could be more rapidly gained through solar-based innovations than technologies using nuclear or fossil fuels (Mitchell, May, & McDonald, 1995). Conventional energy production fuel cycles (e.g., coal, nuclear) create noticeable environmental impacts in the place of CO_2 emission. For instance, a 1000 MW nuclear plant devours just 36 tons of prepared and improved uranium fuel, yet this needs the most minimal of 85.5×10 J huge amounts of mineral every year which produces poisonous tailings containing arsenic, cadmium, and mercury what is more as radionuclides. Transformation to enhanced uranium fuel requires preparing with fluorides, again producing huge amounts of harmful materials (Masters, 1991). For coal-controlled plant roughly multiple times the greatest measure of coal (94.5×105 tons every year) is needed to give the indistinguishable amount of power. Unmistakably, with sunlight-based force advances such fuel cycles are wiped out comparatively in light of the fact that the related natural effects (Norton, Eames, & Lo, 1998).

11.7.3 Environmental impacts of solar thermal electricity

Restricted implementation of ST power up to now suggests that there is little genuine encounters of the ecological effects that such a scheme may have. Thus, to different SESs, ST power frameworks present the basic ecological positive thing about the removal or the evasion of emissions identified with conventional power generation (Tsoutsos, Gekas, & Frantzeskaki, 2005).

ST electric frameworks are among the preeminent proficient SESs when it includes to land use (they produce every year around 4–5 GWh/ha). Until this point in time, most locales utilized or considered for ST frameworks are in parched desert territories, which as a rule have delicate soil and plant networks (OECD/IEA, 1998). Central concentrator power frameworks could represent a threat to flying animals, yet operational experience shows that flying animals maintain a strategic distance from any risk territories (OECD/IEA, 1998; Norton et al., 1998).

The coincidental release of warmth moves liquids (water and oil) from parabolic trough and focal beneficiary frameworks could shape a risk. The danger might be significant in some central tower frameworks, which utilize fluid sodium or liquid salts as a warmth move medium. Under ordinary working conditions, this could not represent any peril to operators, yet disappointment of the global positioning frameworks could end in straying beams which may represent an occupational safety and risk on location (Kaygusuz, 2009).

11.7.4 Energy and environmental impact analysis

Specialists have assessed the ecological effect of sunlight-based warming and cooling frameworks, chiefly by computing the amount of ozone harming substance outflows from the substituted petroleum derivatives that are disregarded. Moreover, scientists have reasoned that a commonplace sun-oriented thermal framework for respectable water production can decrease carbon dioxide produced nearby by 28% and while the deployment of a combined space and water utility can lessen emanations by very 60%. In Greece, one in all the pioneers inside the utilization of sunlight-based thermal frameworks for warm water creation, an examination that secured the commitment of sun-based thermal frameworks to the decrease of conventional energy and GHGs and other air toxin emanations for a 30-year time span demonstrated that the alternate fuel power ran from 21 GWh (0.1% of the domestic power use) to 1513 GWh (2.4%) from the year of 1978 to 2007, prompting the reduction of 1.67 Mt of carbon dioxide. The emissions of CO_2, NO_X, and SO_2 avoided, contrasted with those emitted yearly inside the nation were about 1.5% for CO_2, under 1% for NO_X, and around 2.5% for SO_2 of the nations in total (Tsilingiridis & Martinopoulos, 2010).

LCA could be a system that permits evaluation of ecological weights and their expected effects over the full life pattern of an item, cycle, or action. LCA is new and developing, with its foundations in research related with power prerequisites inside the 1960s and contamination avoidance, which was started inside the 1970s (Rebitez, 2004).

LCA technique was at first evolved by the Society of Environmental Toxicology and Chemistry and was later streamlined by the arrangement for standardization (ISO), yet its genuine discovery into the business world happened distinctly during the 1990s. The results of the appraisal could be utilized for the development of a natural naming guideline (Tsilingiridis, Martinopoulos, & Kyriakis, 2004). This idea is spoken to by utilizing a schematic outline in Fig. 11.10. ISO 14040, principles and framework; ISO 14041, goal and scope definition and inventory analysis; ISO 14042, life cycle impact assessment; ISO 14043, interpretation; are the four ISO standards which are particularly constructed for LCA application. In contrast to the 14000 standards, it is additionally unattainable to urge a political candidate enfranchisement stating that Associate in Nursing LCA, Associate in Nursing LCA methodology, or Associate in Nursing LCA software package like those

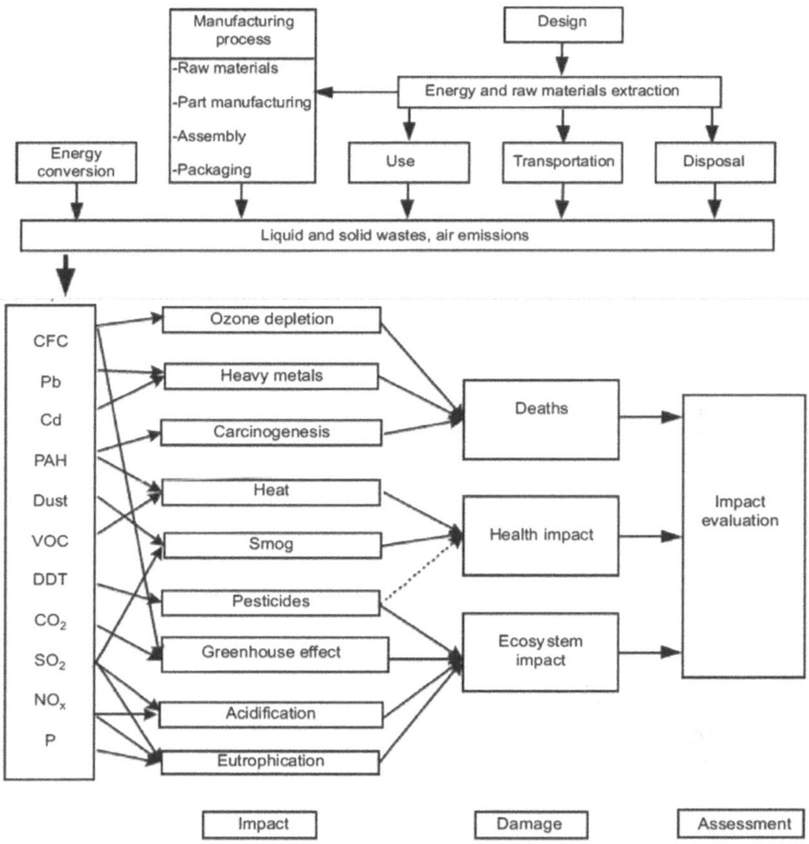

FIGURE 11.10 Flowchart of a product's life cycle and its environmental impacts (Martinopoulos, 2016).

employed in this work are created consistent with the quality. Best evaluates are interesting in LCA, with the, commonly unnoted, must be constrained to represent vulnerabilities once making differentiations among the outcomes (Tsilingiridis et al., 2004).

Probably the best renowned and most popular instruments among life cycle specialists are the ECO-it, GEMIS, SimaPro, and GaBi PC code. A small uninformed of materials and cycles is encased with the environmental impact markers beginning from the Eco-Indicator 99 technique in Eco-Indicator focuses. Eco-Indicator 99 incorporates ordinary qualities for (Goedkoop & Spriensma, 2001): materials, creation measures, transportation measures, energy generation cycles, removal circumstances, and so on. The quality Eco-Indicator 99 qualities is believed to be dimensionless figures, known as Eco-Indicator points (Pts). One respectable metal is sketched out as speaking to thousandth of the yearly natural heap of a mean European individual (Goedkoop & Spriensma, 2001).

The Eco-Indicator 99 approach examines the resulting five ecological effect classes: worldwide environmental change, acidification, ecopoisonousness, petroleum derivatives, and metabolism inorganics. The effect evaluation includes three primary advances: portrayal, standardization, and terminal weighted scores. Three injury classes of square measure for the ultimate weighted scores are public health, system quality, and assets (Rebitzer, 2004).

Besides, the software bundle has the adaptability to figure the amount of emanations that were released for the duration of the existence pattern of the framework. GaBi models every segment of an item or framework from a day-to-day existence cycle point of view, combat hardware organizations to make the best-educated decisions on the assembling and life pattern of any item though giving a lucid data to appraise the natural effect of sourcing and decontamination each stuff or handled aspect of a plant made thing. It can even bless different decisions for creating, appropriation, recyclability, contamination, and property.

References

Abdelaziz, E., Saidur, R., & Mekhilef, S. (2011). A review on energy saving strategies in industrial sector. *Renewable and Sustainable Energy Reviews, 15*(1), 150–168. https://doi.org/10.1016/j.rser.2010.09.003

Alauddin, Z. A. Z. (1996). *Performance and characteristics of a biomass gasifier system* (Doctoral dissertation). Cardiff: University of Wales.

Alstone, P., Mills, E., Jacobson, A., & Lai, P. (2014). High life-cycle efficacy explains fast energy payback for improved off-grid lighting systems. *Journal of Industrial Ecology.* https://doi.org/10.1111/jiec.12117

Alva, G., Lin, Y., & Fang, G. (2018). An overview of thermal energy storage systems. *Energy, 144*, 341–378. https://doi.org/10.1016/j.energy.2017.12.037

Amponsah, N. Y., Troldborg, M., Kington, B., Aalders, I., & Hough, R. L. (2014). Greenhouse gas emissions from renewable energy sources: A review of lifecycle considerations. *Renewable and Sustainable Energy Reviews, 39*, 461–475. https://doi.org/10.1016/j.rser.2014.07.087

Ang, J. B. (2008). Economic development, pollutant emissions and energy consumption in Malaysia. *Journal of Policy Modeling, 30*(2), 271–278. https://doi.org/10.1016/j.jpolmod.2007.04.010

Angeliki, K., Paris, F. A., Andreas, I., & Soteris, K. (2018). Environmental assessment of solar thermal systems for the industrial sector. *Journal of Cleaner Production, 176*, 99–109. https://doi.org/10.1016/j.jclepro.2017.12.150

Anvari, S., Khalilarya, S., & Zare, V. (2019). Power generation enhancement in a biomass-based combined cycle using solar energy : Thermodynamic and environmental analysis. *Applied Thermal Engineering, 153*, 128–141. https://doi.org/10.1016/j.applthermaleng.2019.02.112

Anvari, S., Taghavifar, H., Khoshbakhti, R., Khalilarya, S., & Jafarmadar, S. (2015). Implementation of ANN on CCHP system to predict trigeneration performance with consideration of various operative factors. *Energy Conversion and Management, 101*, 503–514. https://doi.org/10.1016/j.enconman.2015.05.045

Ardente, F., Beccali, G., Cellura, M., & Brano, V. L. (2005). Life cycle assessment of a solar thermal collector. *Renewable Energy, 30*, 1031–1054. https://doi.org/10.1016/j.renene.2004.09.009

Barnhart, C., & Benson, S. M. (2013). On the importance of reducing the energetic and material demands of electrical energy storage. *Energy & Environmental Science, 6*, 1083. https://doi.org/10.1039/C3EE24040A

BP-AMOCO. (2002). *BP statistical review of world energy 2002.*

Chapman, D. S. (1998). Global warming—just hot air' (1998). In *Gould distinguished lecture on technology and the quality of life, seventh annual address*. USA: University of Utah.

Chuang, J., Lien, H. L., Den, W., Iskandar, L., & Liao, P. H. (2018). The relationship between electricity emission factor and renewable energy certificate: The free rider and outsider effect. *Sustainable Environment Research, 28*(6), 422–429. https://doi.org/10.1016/j.serj.2018.05.004

Colombo, R., Gilliaert, D., & Landabaso, A. (1992). *Life cycle cost analysis of solar heating and DHW systems in residential buildings (Report No. EUR 13581 EN)*. Commission of the European Communities (CEC). ISSN 1018-5593. Retrieved from http://publications.europa.eu/resource/cellar/22ab5517-876f-451b-8b1c-5ba6db01a78a.0001.04/DOC_1.

Eduard, O., Antoni, G., Alvaro, G., Dieter, B., & Luisa, C. F. (2012). Comparative life cycle assessment of thermal energy storage systems for solar power plants. *Renewable Energy, 44*, 166–173. https://doi.org/10.1016/j.renene.2012.01.008

Emmott, C. J. M., Jaoude, A. A., Urbina, A., Ekins-Dawkes, N. J., & Nelson, J. (2012). *Life cycle analysis of an off-grid solar charging kiosk*. https://doi.org/10.1109/ICSET.2012.6357368

EPIA. (2011). *Solar photovoltaics competing in the energy sector – on the road to competitiveness.*

ESTIF. (2015). *Solar thermal markets in Europe trends and market statistics 2014*. European Solar Thermal Industry Federation.

Forsberg, C. W., Peterson, P. F., & Zhao, H. (2007). High-temperature liquid-fluoride-salt closed-Brayton-cycle solar power towers. *Journal of Solar Energy Engineering, 129*, 141–146. https://doi.org/10.1115/1.2710245

Gautam, A., & Saini, R. P. (2020). A review on technical, applications and economic aspect of packed bed solar thermal energy storage system. *Journal of Energy Storage, 27*, 101046. https://doi.org/10.1016/j.est.2019.101046

Gholamian, E., Mahmoudi, S. M. S., & Zare, V. (2016). Proposal, exergy analysis and optimization of a new biomass-based cogeneration system. *Applied Thermal Engineering, 93*, 223–235. https://doi.org/10.1016/j.applthermaleng.2015.09.095

Goedkoop, M., & Spriensma, R. (2001). *The eco-indicator 99: A damage oriented method for life cycle impact assessment: Methodology report, 2001.*

Grange, B., Dalet, C., Falcoz, Q., Siros, F., & Ferriere, A. (2014). Simulation of a hybrid solar gas-turbine cycle with storage integration. *Energy Procedia, 49*, 1147–1156. https://doi.org/10.1016/j.egypro.2014.03.124

Hasanuzzaman, M., Rahim, N. A., Hosenuzzaman, M., Saidur, R., Mahbubul, I. M., & Rashid, M. M. (2012). Energy savings in the combustion-based process heating in industrial sector. *Renewable and Sustainable Energy Reviews, 16*(7), 4527–4536. https://doi.org/10.1016/j.rser.2012.05.027

Hasanuzzaman, M., Zubir, U. S., Ilham, N. I., & Che, H. S. (2017). Global electricity demand, generation, grid system, and renewable energy polices: A review. *Wiley Interdisciplinary Reviews: Energy and Environment, 6*(3). https://doi.org/10.1002/wene.222

IEA. (1999). *Key world energy statistics*. International Energy Agency.

IHFC. (2001). International Heat Flow Committee.

Ilham, N. I., Hasanuzzaman, M., & Hosenuzzaman, M. (2017). European smart grid prospects, policies, and challenges. *Renewable and Sustainable Energy Reviews, 67*, 776–790.

ISO 14040:2006, Environmental management — Life cycle assessment — Principles and framework. Available at: https://www.iso.org/standard/37456.html

Jaume, A., Juliana, R., Marco, R., Rafael, A., Joan, T. R., Núria, S. G., … Pere, P. F. (2019). Life Cycle Assessment of a solar thermal system in Spain, eco-design alternatives and derived climate change scenarios at Spanish and Chinese National levels. *Sustainable Cities and Society, 47*, 101467. https://doi.org/10.1016/j.scs.2019.101467

Johanson, T. B., Kelly, H., Reddy, A. K. N., & Williams, R. H. (1993). *Renewable fuels and electricity for a growing world economy-defining and achieving the potential, Renewable energy- sources of fuels and electricity.* Washington DC, USA: Island Press. https://doi.org/10.15173/esr.v4i3.284

Jones, P. D., Parker, D. E., Osborn, T. J., & Briffa, K. R. (2016). *Global and hemispheric temperature anomalies: Land and marine instrumental records.* https://doi.org/10.3334/CDIAC/cli.002

Kalogirou, S. A. (2013). *Solar energy engineering: Processes and system.* Academic Press.

Kamarulzaman, A., Hasanuzzaman, M., & Rahim, N. A. (2021). Global advancement of solar drying technologies and its future prospects: A review. *Solar Energy, 221*, 559–582. https://doi.org/10.1016/j.solener.2021.04.056

Kannan, R., Leong, K. C., Osman, R., Ho, H. K., & Tso, C. P. (2006). Life cycle assessment study of solar PV systems: An example of a 2.7kWp distributed solar PV system in Singapore. *Solar Energy, 80*(5), 555–563. https://doi.org/10.1016/j.solener.2005.04.008

Kaygusuz, K. (2009). Environmental impacts of the solar energy systems. *Energy Sources, Part A: Recovery, Utilization, and Environmental Effects, 31*(15), 1376–1386. https://doi.org/10.1080/15567030802089664

Kumar, L., Hasanuzzaman, M., & Rahim, N. A. (2019). Global advancement of solar thermal energy technologies for industrial process heat and its future prospects: A review. *Energy Conversion and Management, 195*, 885–908. https://doi.org/10.1016/j.enconman.2019.05.081

Kundzewicz, Z. W., Pińskwar, I., & Koutsoyiannis, D. (2020). Variability of global mean annual temperature is significantly influenced by the rhythm of ocean-atmosphere oscillations. *The Science of the Total Environment, 747*, 141256. https://doi.org/10.1016/j.scitotenv.2020.141256

Lee, M., Hong, T., Yoo, H., Koo, C., Kim, J., & Jeong, K. (2017). Establishment of a base price for the Solar Renewable Energy Credit (SREC) from the perspective of residents and state governments in the United States. *Renewable and Sustainable Energy Reviews, 75*, 1066–1080. https://doi.org/10.1016/j.rser.2016.11.086

Leopold L.B., Clarke F.E., Hanshow B.B., Balsley J.E., et al. (1971). A procedure for evaluating environmental impact. *U.S. Geological Survey Circular 645, Washington,D.C.*

Malek, A. B. M. A., Hasanuzzaman, M., & Rahim, N. A. (2020). Prospects, progress, challenges and policies for clean power generation from biomass resources. *Clean Technologies and Environmental Policy, 22*(6), 1229–1253. https://doi.org/10.1007/s10098-020-01873-4

Malek, A. B. M. A., Hasanuzzaman, M., Rahim, N. A., & Al Turki, Y. A. (2017). Techno-economic analysis and environmental impact assessment of a 10 MW biomass-based power plant in Malaysia. *Journal of Cleaner Production, 141*, 502–513. https://doi.org/10.1016/j.jclepro.2016.09.057

Malek, A. B. M. A., Hasanuzzaman, M., Rahim, N. A., & Al–Turki, Y. A. (2021). Energy, economic, and environmental analysis of 10-MW biomass gasification based power generation in Malaysia. *Energy & Environment, 32*(2), 295–337. https://doi.org/10.1177/0958305x20930386

Martinopoulos, G. (2016). Energy efficiency and environmental impact of solar heating and cooling systems. *Advances in Solar Heating and Cooling*. https://doi.org/10.1016/C2014-0-03661-1. ISBN: 978-0-08-100301-5.

Masters, G. M. (1991). *Introduction to environmental engineering and science*. N.J: Prentice Hall.

Mitchell, G., May, A., & McDonald, A. (1995). PICABUE: A methodological framework for the development of indicators of sustainable development. *The International Journal of Sustainable Development and World Ecology, 2*, 104–123. https://doi.org/10.1080/135045095 09469893

Mohamed, A. R., & Lee, K. T. (2006). Energy for sustainable development in Malaysia: Energy policy and alternative energy. *Energy Policy, 34*(15), 2388–2397. https://doi.org/10.1016/j.enpol.2005.04.003

Nelson, J., Gambhir, A., & Ekins-Daukes, N. (2014). *Briefing Paper no 11 - solar power for CO_2 mitigation*. Grantham Institute for Climate Change. Retrieved from https://www.imperial.ac.uk/media/imperial-college/grantham-institute/public/publications/briefing-papers/Solar-power-for-CO2-mitigation—Grantham-BP-11.pdf.

Norasikin, L. A., Nur, M. I., Marlia, H. M., Mohd, I. A., Mohd, T. M. A., Suhaila, S., … Kamaruzzaman, S. (2018). Prospects of life cycle assessment of renewable energy from solar photovoltaic technologies: A review. *Renewable and Sustainable Energy Reviews, 96*, 11–28. https://doi.org/10.1016/j.rser.2018.07.048

Nordell, B. (2007). Global warming is large scale thermal energy storage. *Thermal Energy Storage for Sustainable Energy Consumption*, 75–86. https://doi.org/10.1007/978-1-4020-5290-3_4

Norton, B., Eames, P. C., & Lo, S. N. G. (1998). Full-energy-chain analysis of greenhouse gas emissions for solar thermal electric power generation systems. *Renewable Energy, 15*(1–4), 131–136. https://doi.org/10.1016/S0960-1481(98)00158-X

OECD. (1988). *Environmental impacts of renewable energy*. Paris.

OECD/IEA. (1998). *Benign energy: The environmental implications of renewables*. International Energy Agency.

Pantaleo, A. M., Camporeale, S. M., Miliozzi, A., Russo, V., Shah, N., & Markides, C. N. (2017). Novel hybrid CSP-biomass CHP for flexible generation: Thermo-economic analysis and profitability assessment. *Applied Energy, 204*, 994–1006. https://doi.org/10.1016/j.apenergy.2017.05.019

Parasnis, D. S. (1985). *Kosmos 1985. Yearbook of the Swedish physics association. (Svenska Fysikersamfundet)*. Uppsala: Swedish Science Press (in Swedish).

Pineda, I. (2010). *Carbon abatement cost of solar thermal technology* (M.Sc. thesis). Imperial College London.

Pollak, H. N., & Chapman, D. S. (1977). The flow of heat from the Earth's interior. *Scientific American, 8*, 60–76. www.jstor.org/stable/24954003.

Pollak, H. N., Hurter, S. J., & Johnson, J. R. (1993). Heat flow from the Earth's interior: Analysis of global data set. *Reviews of Geophysics, 31*(3), 267–280. https://doi.org/10.1029/93RG01249

Poyry. (March 2011). *The challenges of intermittency in North West European power markets*. Poyry Management Consulting. https://www.poyry.com/sites/default/files/intermittency_-_march_2011_-_energy.pdf.

Rebitzer, G. (2004). Life cycle assessment: Part 1: Framework, goal and scope definition, inventory analysis, and applications. *Environment International, 30*(5), 701–720. https://doi.org/10.1016/j.envint.2003.11.005

Saha, S., Ruslan, A. R. M., Morshed, A. K. M. M., & Hasanuzzaman, M. (2020). Global prospects and challenges of latent heat thermal energy storage: A review. *Clean Technologies and Environmental Policy*. https://doi.org/10.1007/s10098-020-01997-7

Sahoo, U., Kumar, R., Pant, P. C., & Chaudhury, R. (2015). Scope and sustainability of hybrid solar–biomass power plant with cooling, desalination in poly-generation process in India. *Renewable and Sustainable Energy Reviews, 51*, 304–316. https://doi.org/10.1016/j.rser.2015.06.004

Saidur, R., Hasanuzzaman, M., & Rahim, N. (2012). Energy, economic, and environmental analysis of the Malaysian industrial compressed-air systems. *Clean Technologies and Environmental Policy, 14*(2), 195–210. https://doi.org/10.1007/s10098-011-0386-9

Salby, M. L. (1996). *Fundamentals of atmospheric physics* (pp. 44–45). New York: Academic Press.

Sandwell, P., Chan, N. L. A., Foster, S., Nagpal, D., Emmott, C. J. M., Candelise, C., … Nelson, J. (2016). *Potential of solar photovoltaic technologies for rural electrification and carbon emissions mitigation: The case of India, report*. Grantham Institute for Climate Change.

Sharma, A. K., Sharma, C., Mullick, S. C., & Kandpal, T. C. (2017). Solar industrial process heating: A review. *Renewable and Sustainable Energy Reviews, 78*, 124–137. https://doi.org/10.1016/j.rser.2017.04.079

Singh, P., & Sing, V. P. (2001). *Snow and glacier hydrology. Water science and technology library* (Vol. 37, p. 742). Netherlands: Kluwer Academic Publishers, Softcover. ISBN: 978-90-481-5635-1.

Sollmann, D. (2008). Nocturnal solar power. *Photon International, 88*.

Soltani, S., Mahmoudi, S. M. S., Yari, M., & Rosen, M. A. (2013). Thermodynamic analyses of an externally fired gas turbine combined cycle integrated with a biomass gasification plant. *Energy Conversion and Management, 70*, 107–115. https://doi.org/10.1016/j.enconman.2013.03.002

Sotos, M. G. H. G. (2015). *Protocol scope 2 guidance: An amendment to the corporate standard*. Washington, DC: World Resources Institute.

Taghavifar, H., Anvari, S., Khoshbakhti, R., Khalilarya, S., & Jafarmadar, S. (2015). Towards modeling of combined cooling, heating and power system with artificial neural network for exergy destruction and exergy efficiency prognostication of tri-generation components. *Applied Thermal Engineering, 89*, 156–168. https://doi.org/10.1016/j.applthermaleng.2015.06.008

Tsilingiridis, G., & Martinopoulos, G. (2010). Thirty years of domestic solar hot water systems use in Greece-energy and environmental benefits-future perspectives. *Renewable Energy, 35*(2), 490–497. https://doi.org/10.1016/j.renene.2009.05.001

Tsilingiridis, G., Martinopoulos, G., & Kyriakis, N. (2004). Life cycle environmental impact of a thermosyphonic domestic solar hot water system in comparison with electrical and gas water heating. *Renewable Energy, 29*(8), 1277–1288. https://doi.org/10.1016/j.renene.2003.12.007

Tsoutsos, T., Gekas, V., & Frantzeskaki, N. (2005). Environmental impacts from the solar energy technologies. *Energy Policy, 33*, 289–296. https://doi.org/10.1016/S0301-4215(03)00241-6

Wang, Y., Morgan, R. K., & Cashmore, M. (2003). Environmental impact assessment of projects in the People's Republic of China: New law, old problems. *Environmental Impact Assessment Review, 23*(5), 543–579. https://doi.org/10.1016/S0195-9255 (03)00071-4

Zare, V., & Hassanzadeh, M. (2016). Energy and exergy analysis of a closed Brayton cycle based combined cycle for solar power tower plants. *Energy Conversion and Management, 128*, 227–237. https://doi.org/10.1016/j.enconman.2016.09.080

Index

Note: 'Page numbers followed by "f" indicate figures and "t" indicate tables.'

E

F

G

R

S

T

U

V

W

Z

Kolibri 360 GmbH
Ferdinand-Jühlke-Straße 7
99095 Erfurt, Deutschland
produktsicherheit@kolibri360.de